# Suppressing the Mind

For further volumes:
http://www.springer.com/series/7678

Anthony Hudetz · Robert Pearce
Editors

# Suppressing the Mind

Anesthetic Modulation of Memory
and Consciousness

 Humana Press

*Editors*
Anthony Hudetz
Department of Anesthesiology
Medical College of Wisconsin
8701 Watertown Plank Road
Milwaukee WI 53226
USA
ahudetz@mcw.edu

Robert Pearce
Department of Anesthesiology
University of Wisconsin, Madison
B6/319 Clinical Science Center
600 Highland Avenue
Madison WI 53792
USA
rapearce@wisc.edu

ISBN 978-1-60761-463-0        e-ISBN 978-1-60761-462-3
DOI 10.1007/978-1-60761-462-3

Library of Congress Control Number: 2009938710

Printed on acid-free paper

springer.com

# Foreword

Anesthetics produce a reversible state of unconsciousness accompanied by antero-grade amnesia. This remarkable phenomenon brings great relief to surgical patients and wonder to clinicians and scientists. To date, we do not fully understand the mechanisms by which anesthetics ablate conscious sensation and memory. We are, however, making progress.

This book presents original results as well as overviews of the current state of knowledge of the problem. It is authored by investigators who know the field well; their research at a number of levels has contributed substantially to our current understanding of anesthetic modulation of memory and consciousness. Most of the contributors were presenters at two workshops organized by Dr. Pearce and Dr. Hudetz at the 40th Annual Winter Conference on Brain Research, held at Snowmass Village, Colorado, from January 27 through February 2, 2007. One workshop focused on anesthetic modulation of consciousness and another on anesthetic modulation of memory. Seven of the chapters are based on material presented at these symposia – appropriately updated with new relevant findings. This information is supplemented by chapters on anesthesia and sleep, computational analysis of the state of anesthesia, and the clinical phenomenon of "anesthesia awareness," a topic that has recently received much public attention. With these three additional contributions, the book thus includes 10 chapters.

Several excellent books on consciousness and memory have been published in recent years, but none of these has presented a systematic compilation of studies on anesthetic modulation of memory and consciousness – at least in a unified view of the subject matter. Likewise, several texts have been written about fundamental anesthetic mechanisms, focusing on pharmacological, cellular, and molecular changes. However, no volume has bridged molecular, cellular, integrative, and systems-level effects, as we believe will be necessary to address the core issues of anesthetic mechanisms. This book is intended to fill this need. We hope that by building these bridges between bench and clinical research, new ideas and testable hypotheses will emerge, so that future work will ultimately lead to an integrative theory of anesthetic-induced unconsciousness and amnesia.

There is a long history of interest in unraveling the mechanisms of anesthesia. With the recent introduction of several new investigative methodologies, and as new hypotheses have emerged, there has been a surge in interest from the traditional, pharmacological, and clinical neurosciences, as well as newer fields, such as cognitive and computational neurosciences. Considering the incredible significance of understanding the neurobiological basis of consciousness and memory, we expect that this interest will continue to grow. This book should appeal to anesthesiologists, neurologists, psychologists, scientists, and anyone interested in anesthesia, consciousness, or memory. We hope that it serves as a reference for the scientific community and provides a useful perspective for future treatments of the subject. As a summary of the current state of knowledge, it should serve as a useful text for graduate students and researchers who wish to engage in anesthesia research. Although a significant part of the information included here is technical, it is written in a style that we hope makes it accessible to a wider audience than simply scientists who are currently engaged in research in the field.

The editors would like to express their sincere thanks to all contributors for their outstanding work. They appreciate the reviewers' suggestions for the inclusion of additional specific topics. Special thanks are due to Patrick J. Marton and Matthew Giampoala at Springer US for the invitation to prepare this book, and for the editorial assistance of Marnie Filstein.

Milwaukee and Madison, WI                              Anthony Hudetz, DBM, PhD and
                                                                Robert Pearce, MD, PhD

# Contents

# Contributors

**Michael T. Alkire** Department of Anesthesiology and Perioperative Care, University of California, Irvine Medical Center, Orange, CA, USA

**Helen A. Baghdoyan** Department of Anesthesiology, University of Michigan, Ann Arbor, MI, USA

**Matthew I. Banks** Department of Anesthesiology, University of Wisconsin, Madison, WI, USA

**Hugh C. Hemmings, Jr.** Department of Anesthesiology, Weill Cornell Medical College, New York, NY, USA

**Anthony Hudetz** Department of Anesthesiology, Medical College of Wisconsin, Milwaukee, WI, USA

**Ralph Lydic** Department of Anesthesiology, University of Michigan, Ann Arbor, MI, USA

**M. Bruce MacIver** Department of Anesthesia, Stanford University School of Medicine, Stanford, CA, USA

**George A. Mashour** Departments of Anesthesiology and Neurosurgery, University of Michigan Medical School; Ann Arbor, MI, USA

**Robert Pearce** Department of Anesthesiology, University of Wisconsin, Madison, WI, USA

**Misha Perouansky** Department of Anesthesiology, University of Wisconsin, Madison, WI, USA

**Kane O. Pryor** Department of Anesthesiology and Critical Care Medicine, Memorial Sloan-Kettering Cancer Center, New York, NY, USA; Department of Anesthesiology, Weill Cornell Medical College, New York, NY, USA

**Jamie Sleigh** Department of Anaesthesiology, University of Auckland, Hamilton, New Zealand

**Alistair Steyn-Ross** University of Waikato, Hamilton, New Zealand

**Moira Steyn-Ross** University of Waikato, Hamilton, New Zealand

**Robert A. Veselis** Department of Anesthesiology and Critical Care Medicine, Memorial Sloan-Kettering Cancer Center, New York, NY, USA; Department of Anesthesiology, Weill Cornell Medical College, New York, NY, USA

**Logan Voss** Department of Anaesthesiology, University of Auckland, Hamilton, New Zealand

**Christopher J. Watson** Department of Anesthesiology, University of Michigan, Ann Arbor, MI, USA

**Marcus Wilson** University of Waikato, Hamilton, New Zealand

# Chapter 1
# Introduction

Anthony Hudetz and Robert Pearce

Approximately 100,000 patients undergo general anesthesia in the United States every day. It may come as a surprise that, to date, we do not fully understand how general anesthetics work. That is, we do not know how they "put patients to sleep" (as is commonly said) or (as we would say) how they "suppress the mind." We do not really have a much better understanding of how anesthetic agents prevent people or animals from moving, either spontaneously or in response to a painful stimulus, though there is good evidence that the immobilizing effect of anesthetics arises primarily from actions at the level of the spinal cord. Although their ability to prevent movement is evidently of great practical value in permitting surgical procedures to be performed, what patients really desire is that they feel no pain, that they sleep peacefully through their operation, and that they remember nothing afterward. It is quite likely that the hypnotic and amnesic effects of anesthetics, and possibly also their analgesic effects, derive primarily from actions in the brain through their modulation of intricately connected, complex network of hundreds or even thousands of specialized neuron groups. The biochemical and neurophysiological mechanisms of anesthetic action may, therefore, be understandably quite complex. The amnesic (memory erasing) and hypnotic (consciousness erasing) effects of anesthetics remain mysterious for two underlying reasons: we do not understand the neuronal functions that make the brain conscious and allow it to store and recollect events and we also do not know the specific molecular and cellular targets of anesthetics that cause them to interfere with consciousness and memory. Thus, we face two fundamental but interrelated questions: how does the brain work? And how do anesthetics interfere? Understanding the neuronal mechanism of general anesthesia both helps and benefits from research into the neurobiological mechanisms of consciousness and memory.

Why should we be concerned with these questions? Understanding the neurobiological bases of consciousness and memory are arguably two of the greatest challenges for neuroscience. The potential impact of such discoveries for science, medicine, and society is enormous. Knowing what makes people consciously

A. Hudetz (✉)
Department of Anesthesiology, Medical College of Wisconsin, Milwaukee, WI, USA
ahudetz@mcw.edu

A. Hudetz, R. Pearce (eds.), *Suppressing the Mind*, Contemporary Clinical Neuroscience, DOI 10.1007/978-1-60761-462-3_1,
© Humana Press, a part of Springer Science+Business Media, LLC 2010

perceive and behave as they do, and how and why they learn, remember, and forget, not only would revolutionize fields of medicine such as neurology, psychiatry, and anesthesiology but would have far-reaching implications for morality, ethics, law, and education.

We believe that anesthesia research can make major contributions in these research endeavors. Anesthetic agents represent an exclusive class of psychoactive drugs that can be used to modulate the states of consciousness and memory in a safe and reversible manner. A century of experience with anesthetic drugs, and the rapidly expanding knowledge of their molecular, cellular, neurophysiological, and psychological actions, make them unique and useful tools to study the neurobiological bases of consciousness and memory. As anesthesiologists remove and restore human consciousness and memory daily, their methods and experience should arguably be a foundation *par excellence* for a scientific understanding of the nature of states of consciousness and memory.

An obvious and direct benefit of this research for anesthesiology is the potential discovery of novel anesthetic drugs. Although general anesthesia is highly reproducible and reversible, all anesthetic agents also have undesirable side effects. Some can be of significant concern, particularly when they are administered to patients with compromised health. Knowing the key mechanisms leading to loss of consciousness or impairment of memory would open the door to the design of new agents with more specific actions, and therefore safer drugs. Understanding the mechanisms of anesthetic-induced unconsciousness and amnesia would have another important benefit: it would aid in the development of novel anesthetic depth monitors based on sound neurophysiologic principles. Many such monitors, which already are used in spite of their limitations, have attracted significant attention, by both medical specialists and the public, because of their potential benefit in minimizing the occurrence of unwanted intraoperative awareness.

This book represents a compendium of data-driven theories. They are based upon the results obtained by research groups who investigate the mechanisms of general anesthesia from a wide range of perspectives. The experimental approaches range from molecular, cellular, and pharmacological to electrophysiological, behavioral, functional brain imaging, and computer simulation, in humans and animals, in vitro and in vivo. An amalgam that includes this diversity of approaches is not a coincidence; it is consistent with our belief that only such a multilevel and interdisciplinary approach will be successful in revealing the mechanisms of anesthesia, particularly for such complex phenomena as consciousness and memory. Clearly, whole fields of research into anesthesia mechanisms cannot be accounted for in 10 chapters. Nevertheless, we feel that the contributions do represent a reasonable sampling of the range of anesthesia research that exists today.

The past two decades have seen dramatic shifts in our thinking about mechanisms of anesthetic action at multiple levels, including the molecular nature of anesthetic targets, how changes in cellular function lead to altered network activity patterns, and which brain regions are associated with specific behavioral effects. Considerable evidence now supports the notion that anesthetic agents act on multiple specific molecular targets, including various ligand-gated, voltage-gated, and other types of ion channels and proteins, and that their relative abilities to achieve various end points reflect the diversity of these actions. The mechanisms by which volatile agents produce their desired (and undesired) effects are more uncertain than those of intravenous agents; however, significant progress is being made in defining molecular targets as they relate to specific anesthetic end points. Chapter 2 by Hemmings presents an overview of the currently known molecular targets of general anesthetics in the vertebrate nervous system.

A fundamental problem that hinders the bridging of molecular, synaptic, and neurochemical actions of anesthetic drugs to their observable cognitive-behavioral effects on an intact organism is the lack of a rigorous definition of consciousness. What is it exactly that the anesthetics remove? And does the removal of a particular type of brain activity invariably imply the removal of conscious experience? We can observe patient response, movement, and behavior, but not consciousness per se. The definition of memory is also problematic, but at this time memory processes can be more objectively studied. There is a greater difficulty with defining consciousness. Several chapters address the definitional issues for consciousness. To make any progress, a pragmatic approach is necessary. Watson, Baghdoyan, and Lydic define the states of consciousness as "specific traits that include physiological and behavioral measures." Hudetz views consciousness as "subjective experience" but then adopts an objective measure of consciousness in terms of cortical information exchange. Sleigh, in turn, defines the loss of consciousness as a state transition in neuronal population activity.

Clinically, one of the principal assessments of the state of "anesthesia" is the presence or absence of a voluntary response to verbal commands. It is a common supposition that consciousness is lost when the motor response is absent. How secure is this conclusion, and does it apply equally to all anesthetics? The motor response depends not only on sensory perception, but also on the capacity and motivation to respond. Moreover, not all anesthetic agents may affect motor and sensory systems to the same degree, potentially leading to a dissociation of perceptual and volitional consciousness. The most notable "dissociative agents" are the NMDA receptor blockers – ketamine, nitrous oxide, and xenon. However, other agents may share this trait to some extent. Indeed, Banks presents an experimental model in which the effects of isoflurane on motor control and sensory processing are dissociable.

The chapter by Watson et al. presents a comprehensive review of complex subcortical systems that control wakefulness and sleep, including "rapid eye movement" (REM) and non-REM sleep. It emphasizes implications for the mechanism of anesthesia. The authors hypothesize that anesthetics work by altering endogenous neurochemical systems that regulate sleep and wakefulness. There is already clear evidence for this hypothesis, at least for certain receptor-specific agents such as alpha 2 adrenergic agonists. Neuron groups of critical importance are found in various subcortical nuclei, including the thalamus, select regions of the hypothalamus, pontine reticular formation, raphe nuclei, locus ceruleus, nucleus basalis, and other basal forebrain nuclei. Most of these nuclei can exert wake-promoting or sleep-promoting effects, with specific effects on REM and non-REM sleep, depending on the neurotransmitters and receptor subtypes used. It is clear that this is a very complex network and that a great deal more research will be necessary to help assemble the details of current knowledge into a quantitative model of network interactions that regulate sleep–wake transitions. Proper functioning of this system is critical to enable consciousness and memory encoding that takes place in the cortex and the hippocampus. In turn, anesthetics can interfere with consciousness and memory by disrupting this subcortical "enabling system" at multiple points. Which points actually do serve causal roles as anesthetic targets, and how the entire system is impacted by pharmacological activation of specific nuclei, such as the central medial thalamus, through which consciousness, or at least wakeful behavior, can be restored (as described in the chapter by Alkire), are promising avenues for future research.

Several common themes emerge in the contributions by various authors. One of these is the phenomenon of neuronal synchronization. The importance of neuronal synchronization for information processing is now widely recognized. The brain's ability to synchronize the activity of functionally related neuronal populations, evident in local field potentials as well as in single-unit activity, is thought to be critical to the formation of "percepts." Activation of synchronous neuronal inputs also supports neuronal plasticity, and thus learning and memory. It would make sense, then, if synchrony were suppressed under anesthesia, as this might lead to loss of consciousness and impaired memory. A hallmark of both sleep and anesthesia is, however, the appearance of dominant slow, periodic waves in the spontaneous EEG, commonly referred to as "synchronization." Perouansky reports that not only is the hippocampal theta rhythm preserved under amnesic concentrations of inhaled agents, but it is even enhanced by some anesthetics, such as halothane. Banks reports that isoflurane, an anesthetic agent that prolongs postsynaptic inhibitory current, actually increases spike synchrony in cortical neurons. Perhaps the activity of the brain under anesthesia would more appropriately be termed "hypersynchronization"?

Another general finding, one that is addressed by several contributors (and is related to the issue of synchronization), is that under anesthesia at a hypnotic level, the brain preserves its primary reactivity to sensory stimuli. Banks finds that the magnitude of cortical auditory-evoked responses is augmented, and Hudetz finds the same with visual stimulation. MacIver cites similar evidence from other laboratories. In all, this implies that hypnosis (loss of consciousness) should not be equated with thalamic (or other) deafferentation of the cortex. Although anesthetic effects on subcortical relays may alter sensory information en route to the cortex, the emerging view is that changes in the evoked oscillation frequency and unit synchrony may garble the orderly information flow to the point it cannot gain access to consciousness.

Several chapters endorse a "corticocentric" view of anesthetic action. This represents a departure from the more traditional view that emphasizes the role of the ascending activating system, including relevant nuclei of the thalamus, in controlling consciousness. The cerebral cortex as a primary target of volatile anesthetic agents in particular, with loss of connectivity of neuronal populations as a potential mechanism of anesthesia, receives support from multiple quarters. It is supported by studies in functional brain imaging (Alkire), auditory perception (Banks), EEG coherence (Hudetz), and computer simulation (Sleigh). Nevertheless, given the clear role of the thalamus in controlling states of consciousness, and the emerging view that cortical integrative functions may to a large degree depend on thalamo-corticothalamic connections, a possible role for anesthetic modulation of thalamic elements must still be considered. Alkire discusses in detail the possible involvement of the thalamus in anesthesia with respect to unconsciousness. A consistent suppression in thalamic activity is observed with many anesthetic agents, but whether this is due to a direct effect on the thalamus or secondary to a suppression of corticothalamic feedback remains an important unanswered question.

From a number of contributions, a general hypothesis for anesthetic-induced unconsciousness emerges at an integrative level. Several contributors emphasize the role anesthetics may play in disrupting cortical information integration necessary for conscious perception, and possibly consciousness itself. As Alkire and Hudetz point out, there may be certain cortical regions with a particularly critical role in consciousness, for example, association regions of the posterior parietal cortex that are involved in multisensory integration, attention, and resting state activity. Brain imaging studies suggest that the activity of this region may be a primary target of anesthetics. Another emerging integrative model of unconsciousness invokes the role of top-down processes, as argued by both Banks and Hudetz. Top-down or feedback modulation of sensory cortical activity at rest and during sensory input has been postulated to play a critical role in attention and contextual interpretation of sensory information. Both authors report that bidirectional recurrent activity between primary sensory and higher association regions is disrupted by general anesthesia, as observed in EEG coherence, local field potentials, and unit synchronization, in a manner that correlates with anesthetic-induced loss of consciousness.

Computer simulations represent a novel and interesting approach to better understand the effect of anesthetics on neuronal network operations. The complexity and interconnectedness of cortical brain circuits calls for a quantitative modeling to understand the dynamics of activity of this system under normal conditions and under the influence of drugs with known synaptic effects. Sleigh presents a continuum model of the activity of a population of cortical excitatory and inhibitory neurons that predicts a state transition in the form of an abrupt change in the simulated EEG. This is explained by a loss of neuronal connectivity, which is proposed to serve as an indication of the loss of consciousness. Two contributors (Sleigh and Hudetz) elaborate on the significance of network topology as a potential determinant of how and why general anesthetics may disrupt functional systems of the brain. As already recognized, higher cognitive operations, and likely consciousness itself, depend on a high degree of integration that involves large-scale networks of the brain. Network organization appears to follow the "small world" model, characterized by hubs with high local connectivity and tied together by sparse long-range interconnections. This topology is efficient and economical, and it influences how diffuse lesions or drug-induced modulation of neurotransmission may affect connectivity of the system as a whole. It may also highlight vulnerabilities of the system to particular interventions. Relevant studies with anesthetic agents are yet to be performed.

Theories put forward here differ in the way they conceptualize the types of transitions that occur between states of consciousness. An issue that has been previously raised by several authors is whether consciousness is suppressed by anesthetics gradually, i.e., as a "dimmer switch," or abruptly, as an "on/off switch." Anesthetic pharmacology, information theoretic approaches, and related behavioral observations favor a graded transition; the experimental paradigms of cognitive psychology and the neuron population model of Sleigh favor the categorical one. The source of this difference may be a discord between macroscopic and microscopic descriptions or between the criteria adopted for assessing consciousness. Sleigh suggests a resolution of the apparent contradiction by relating the graded state of anesthesia to a gradation of an arousing stimulus.

The clinical significance of the issue of "anesthesia awareness," the case when the mind is not (fully) suppressed during anesthesia, is highlighted by Mashour. He discusses the incidence, risk factors, prevention, and postoperative consequences of anesthesia awareness as a complication feared by patients and anesthetists alike. As institutionally defined, the term anesthesia awareness does not necessarily mean awareness during anesthesia itself, but implies the presence of postoperative memory of some event that occurred during the surgical procedure. However, for scientific purposes, a clear distinction between intraoperative awareness and postoperative memory is necessary. The potential psychological consequences of anesthesia awareness for the patient, its public appreciation, and the socio-legal consequences provide strong impetus for continued research into discovering the neurobiological mechanisms of anesthesia, to develop principled criteria, and to advance the current technology for reliable monitoring of the depth of anesthesia with any agent of choice during surgical operations.

Several chapters focus on anesthetic-induced amnesia. The authors come at this issue from a variety of perspectives and utilize a wide range of approaches – from in vitro to in vivo electrophysiological recordings in animals to brain imaging and behavioral studies in humans. The work is synergistic in illuminating how anesthetics may interfere with learning and memory of various types, including explicit and implicit processes. Veselis and Pryor provide a brief overview of the "taxonomy of memory," describing its separation into "conscious" (explicit, declarative, and episodic) versus "unconscious" (implicit, procedural, and subliminal) processes, which may be familiar to most readers. They also describe an alternative "Serial Parallel Independent" (SPI) taxonomy, which they then use to guide their interpretation of investigations of anesthetic modulation of memory in humans. Recognizing that episodic memories seem to be the most "human" of memory functions, and also the most problematic in relating to experimental work in animals, they consider how underlying processes including synaptic plasticity and neuronal synchrony inform our understanding of memory formation (and decline).

It is now well accepted that the phenomenon of "neuronal plasticity" is the cellular-level substrate of learning and memory. That is, enduring memories are encoded by the patterns of synaptic connections formed as a consequence of experience. The rules that govern these changes, and the molecules, cell types, and brain regions that participate, have been topics of intense interest and experimentation. It is clear that the medial temporal lobe, which importantly includes the hippocampal formation, is critically important for the successful encoding of episodic memories in humans. Its disruption leads to impairment in a variety of memory tasks in experimental animals as well, so it has held the attention of investigators interested in mechanisms of memory for many years. Appropriately, two of the chapters focus on hippocampal physiology. MacIver provides an overview of the molecular and cellular events that lead to "long-term potentiation" (LTP) of synaptic strength in the hippocampus, an experimental paradigm that is perhaps the most intensively studied neurophysiological phenomenon in all of neuroscience. He considers the potential contributions of a wide range of anesthetic targets to anesthetic impairment of LTP and memory. Although neuronal activity in a single brain region does not of course reproduce the complex processes that comprise thought and recollection, the ability of anesthetics to impair LTP in the hippocampus at concentrations that also impair hippocampal-dependent learning and memory provides a promising avenue for relating anesthetic actions on specific molecular targets (as described in the chapter by Hemmings) to the ultimate behavioral consequence of episodic memory impairment.

Although studies of LTP have led to tremendous advances in our understanding of the molecular- and cellular-level changes that underlie synaptic plasticity, the experimental paradigm as it is typically applied (activation of afferent fibers by prolonged high-frequency trains) is quite nonphysiological. Large numbers of stimuli ($\sim$100) are required to elicit persistent changes in synaptic strength. Stimulation patterns that mimic physiological patterns of activity, such as bursts that recur at theta frequency (5 Hz is often used), also lead to changes in synaptic strength, with fewer stimuli ($\sim$20) required to elicit potentiation. If the stimuli are provided to an active circuit that is itself generating theta oscillations, synaptic plasticity can be induced by even a single burst of two to four stimuli. Importantly, whether this input leads to an increase in synaptic strength (LTP) or a decrease (long term depression, or LTD) depends on the phase of the ongoing oscillation at which it is presented. This remarkable finding, which is reminiscent of the phenomenon of "spike timing-dependent plasticity," demonstrates the critical importance of timing in the function of the hippocampus – at least with regard to its ability to undergo synaptic plasticity.

This theme of "critical timing" is developed further, as Perouansky and Pearce present a brief background of the rhythms in the hippocampus and explore the impact of altered oscillations on memory function. Surprisingly, modest changes in the average frequency of ongoing theta oscillations, produced by a variety of experimental manipulations and natural conditions, can impact memory; comparable changes are induced in experimental animals by a range of anesthetics that differentially target receptors that are thought to contribute importantly to the amnestic properties, including $GABA_A$ and NMDA receptors. These observations lead them to hypothesize that anesthetic-induced alterations in theta rhythmicity, brought about by drug actions on a variety of molecular and cellular targets, are casual in impairing hippocampus-dependent memory. Important unanswered questions that will drive future research include the identification of the molecular and cellular targets and brain regions that are modulated by anesthetics to alter timing, the relative importance of actions directly on the hippocampus versus the other structures with which it communicates, and the dynamic processes that occur within the hippocampal and cortical networks that support and require timing-dependent plasticity. Nevertheless, the recognition that "timing matters" provides an organizing principle that should aid in our development of a more complete understanding of both anesthesia and memory.

Although memory and consciousness are often addressed separately – conceptually as well as experimentally – these two cognitive functions are evidently related. The discussions by Mashour and Hudetz raise the issue of mechanistic interrelationships between the two phenomena, and Veselis and Pryor address this relationship explicitly. They present an experimental paradigm that can be used in humans, the "continuous recognition task," to separate the sedative and nonsedative amnestic actions of drugs. Their studies indicate that for some drugs, including ethanol and thiopental, the sedative and amnestic actions go hand in hand. However, for others, including propofol and midazolam (two important agents in widespread use today), amnesia is produced out of proportion to sedation. They present evidence that this amnestic action is more a matter of accelerated forgetting than of successful encoding. Using an electrophysiological measure termed the "event-related potential" (ERPs), which may provide a window into hippocampal processes in humans, they examine the impact of these agents on working memory and recognition memory. What emerges is the proposal that conscious memory is never encoded at higher anesthetic concentrations, or is encoded but not consolidated at lower anesthetic concentrations. This analysis harkens back to previously developed themes of synchrony and timing.

In summary, current evidence from sensory electrophysiology, event-related potentials, and functional brain imaging suggests that the failure of conscious perception at a critical depth of anesthesia may best be interpreted as "information received but not perceived" due to the disintegration of cortical or thalamocortical information processing. Cortico-cortical and thalamocortical information transfer may depend on the synchronization or coherence of gamma oscillations that are likely suppressed by general anesthetic agents; similarly, amnesia may be produced by altering critically important timing provided by the theta rhythm in the hippocampal circuit. There is another attractive symmetry here: explicit memory may require encoding during awareness, and a very short-term memory may be required for consciousness.

The molecular, cellular, and network mechanisms of these effects are beginning to be uncovered. The extent to which different molecular targets contribute and the relative importance of anesthetic effects on cortical and hippocampal targets versus thalamic, septal, and arousal systems need further investigation. It is likely that different anesthetic agents affect different receptors, synapses, and neuronal pathways in their action to produce unconsciousness and amnesia. Nevertheless, a common effect of several volatile anesthetic agents may be the suppression of large-scale cortical connectivity and the feedback loops that link sensory and association regions. Elucidation of the neurofunctional systems involved in anesthetic-induced unconsciousness and amnesia is among the important future challenges for anesthesia research, just as the neuronal basis of consciousness and memory is for neuroscience.

It is always appropriate and desirable to recognize the hard work that so many researchers have previously devoted to this subject. The general problem of memory and awareness in anesthesia has been previously reviewed in a series of books, many of which were based on past scientific meetings on the subject (Rosen and Lunn 1987; Bonke, Fitch, and Millar 1990; Sebel, Bonke, and Winograd 1993; Jordan, Vaughan, and Newton 2000; Ghoneim 2001). These books address a wide variety of definitional issues including the assessment of consciousness, learning and memory, techniques for cerebral function and anesthetic state monitoring, the incidence, risk factors, psychological consequences, and medicolegal aspects of anesthesia awareness. Other notable texts have been devoted to the scientific foundations of anesthesia (Yaksh 1998; Antognini, Carstens, and Raines 2003; Hemmings and Hopkins 2006) with a fairly general scope. *Suppressing the Mind* is more focused, and we hope that the reader will find it a unique multidisciplinary treatment of the current leading theories and state of knowledge of the mechanisms of the effects of general anesthetics on consciousness and memory, from molecular to a systems-integrative level.

# References

Antognini, J. F., E. E. Carstens, and D. E. Raines. 2003. *Neural mechanisms of anesthesia*. Totowa, NJ: Humana Press.

Bonke, B., W. Fitch, and K. Millar. 1990. *Memory and awareness in anesthesia*. Rockland, MA: Sweets & Zeitlinger.

Ghoneim, M. M. 2001. *Awareness during anesthesia*. Oxford; Boston: Butterworth-Heinemann.

Hemmings, H. C., and P. M. Hopkins. 2006. *Foundations of anesthesia: basic sciences for clinical practice*. 2nd ed. Philadelphia: Mosby Elsevier.

Jordan, C., D. J. A. Vaughan, and D. E. F. Newton. 2000. *Memory and awareness in anesthesia IV*. London: Imperial College Press.

Rosen, M., and J. N. Lunn. 1987. *Consciousness, awareness, and pain in general anaesthesia*. London; Boston: Butterworths.

Sebel, P. S., B. Bonke, and E. Winograd. 1993. *Memory and awareness in anesthesia*. Upper Saddle River, NJ: Prentice Hall.

Yaksh, T. L. 1998. *Anesthesia: biologic foundations*. Philadelphia: Lippincott-Raven.

# Chapter 2
# Molecular Targets of General Anesthetics in the Nervous System

Hugh C. Hemmings, Jr.

**Abstract** Current concepts of the molecular and cellular mechanisms that underlie general anesthetic actions are incomplete. This is both surprising, given that leading scientists have approached this problem for more than a century, and unfortunate since this lack of knowledge limits our ability to employ these important drugs with optimal safety and efficacy. Considerable evidence now implicates agent-specific effects on discreet molecular targets and neuronal networks central to specific anesthetic end points. Major progress in understanding the molecular pharmacology of the intravenous anesthetics has been made using modern genetic approaches, but the actions of the inhaled anesthetics have been more difficult to resolve. This chapter provides an overview of the principal molecular targets implicated in mediating the effects of general anesthetics on vertebrate neuronal function.

**Keywords** Mechanisms of anesthesia · pharmacology · general anesthetics · pharmacology · ion channels · synaptic transmission · inhaled anesthetics · intravenous anesthetics

## Criteria for Identifying Anesthetic Targets

Specific criteria have been proposed to evaluate the relevance of the many potential molecular targets proposed for general anesthesia (Franks and Lieb 1994; Hemmings, Akabas et al. 2005). The major requirements are (1) *acutely reversible alteration of target function at relevant concentrations comparable to those achieved* in vivo (Eger et al. 2001) and (2) *expression of the target in appropriate anatomical locations to mediate the specific anesthetic end point.* These criteria vary for specific end points. For example, targets involved in immobility should be sensitive to anesthetics near MAC (minimum alveolar concentration of an inhaled

H.C. Hemmings, Jr. (✉)
Department of Anesthesiology, Weill Cornell Medical College, 1300 York Avenue, New York, NY, 10065, USA
e-mail: hchemmi@med.cornell.edu

A. Hudetz, R. Pearce (eds.), *Suppressing the Mind*, Contemporary Clinical Neuroscience, DOI 10.1007/978-1-60761-462-3_2,
© Humana Press, a part of Springer Science+Business Media, LLC 2010

anesthetic required to prevent movement in response to a noxious stimulus in 50% of subjects, an expression of $ED_{50}$) and expressed in the spinal cord, since immobilization by inhaled agents appears to involve primarily depression of reflex pathways in the spinal cord independent of drug actions in the brain (Antognini and Schwartz 1993; Rampil, Mason, and Singh 1993). In contrast, targets mediating memory impairment should be affected at a fraction of MAC and expressed in forebrain regions involved in memory (Rudolph and Antkowiak 2004). Stereoselectivity and appropriate sensitivity or insensitivity to model anesthetic and nonanesthetic compounds are also useful criteria for selected agents (Koblin et al. 1994; Raines and Miller 1994).

## The Molecular Targets of General Anesthetics

Ion channels have emerged as the most likely molecular targets for general anesthetics. Neurotransmitter-gated ion channels, in particular $GABA_A$, glycine, and NMDA-type glutamate receptors, are leading candidates due to their appropriate CNS distributions, essential physiological roles in inhibitory and excitatory synaptic transmission, and sensitivities to clinically relevant concentrations of anesthetics [for reviews, see (Yamakura et al. 2001; Sonner et al. 2003; Rudolph and Antkowiak 2004; Hemmings, Akabas et al. 2005)]. Other ion channels that are sensitive to anesthetics include the HCN family of channels that gives rise to pacemaker currents (Sirois et al. 2000) and regulate dendritic excitability, two-pore domain ($K_{2P}$) "leak" $K^+$ channels that maintain resting membrane potential in many cells (Patel et al. 1999), and voltage-gated $Na^+$ and $Ca^{2+}$ channels critical to excitability and synaptic transmission (Perouansky et al. 2003).

Broadly speaking, general anesthetic targets vary between the major classes of anesthetics. Inhaled anesthetics can be divided into two classes based on their distinct pharmacological properties: (1) The potent inhaled (volatile) anesthetics, which exhibit positive modulation of $GABA_A$ receptors and also produce significant, anesthesia-compatible effects on a number of other receptors/channels, including enhancement of inhibitory glycine receptors, inhibition of excitatory $N$-methyl-D-aspartate (NMDA) type glutamate and neuronal nicotinic acetylcholine receptors, activation of $K_{2P}$ channels, and inhibition of presynaptic $Na^+$ channels; and (2) The gaseous inhaled anesthetics, including cyclopropane, nitrous oxide, and xenon, which are inactive at $GABA_A$ receptors but block NMDA receptors and activate certain $K_{2P}$ channels at clinical concentrations (Gruss et al. 2004). Intravenous anesthetics like propofol and etomidate represent more potent and specific positive modulators of $GABA_A$ receptors. The intravenous anesthetic ketamine is a potent and specific blocker of NMDA receptors and HCN1 pacemaker channels.

## Ligand-Gated Ion Channels: Targets for All Anesthetics

General anesthetics can act as either positive or negative allosteric modulators of ligand-gated ion channels (LGICs) at clinically effective concentrations [for reviews, see (Krasowski and Harrison 1999; Yamakura et al. 2001)]. Most inhaled anesthetics, including all of the ether anesthetics (e.g., isoflurane, sevoflurane, desflurane, and enflurane) and some of the alkanes (e.g., halothane), enhance GABA$_A$ receptor function by enhancing sensitivity to agonists and by directly opening channels in the absence of agonists at higher anesthetic concentrations. This increases channel opening to enhance inhibition at both synaptic and extrasynaptic receptors. Most intravenous anesthetics, including propofol and etomidate, also modulate GABA$_A$ receptors by enhancing gating of the receptors by agonists. Some intravenous anesthetic agents at higher concentrations also open GABA$_A$ receptors in the absence of GABA. Propofol slows desensitization of GABA$_A$ receptors, an important action during rapid repetitive activation of inhibitory synapses.

In addition to their prominent effects on GABA$_A$ receptors, the potent halogenated inhaled anesthetics depress excitatory synaptic transmission both presynaptically, where their principal action appears to be a reduction in glutamate release (Perouansky et al. 1995; Schlame and Hemmings 1995; MacIver et al. 1996), and/or postsynaptically by blocking ionotropic glutamate receptors (Dickinson et al. 2007). In contrast, the nonhalogenated inhaled anesthetics, such as xenon (Franks et al. 1998), nitrous oxide (Jevtovic-Todorovic et al. 1998), and cyclopropane (Raines et al. 2001), as well as the intravenous anesthetic ketamine (Zeilhofer et al. 1992), have little or no effect on the GABA$_A$ receptors, but depress excitatory glutamatergic synaptic transmission postsynaptically through NMDA glutamate receptor blockade. Many inhaled anesthetics also inhibit neuronal nicotinic acetylcholine receptors, often at subanesthetic concentrations. These receptors could be involved in the nociceptive and/or amnestic effects rather than the immobilizing effects of anesthetics (Flood, Ramirez-Latorre, and Role 1997; Violet et al. 1997).

## Potentiation of Inhibitory GABA$_A$ and Glycine Receptors

The ether anesthetics, the alkane anesthetic halothane, most intravenous anesthetics (including propofol, etomidate, and barbiturates), and the anesthetic neurosteroids enhance GABA$_A$ and glycine receptor (GlyR) function at clinical concentrations (Krasowski and Harrison 1999; Zeller et al. 2008). GABA$_A$ receptors and GlyRs are members of the same Cys-loop ligand-gated ion channel superfamily that also includes the cation-permeable nicotinic acetylcholine and 5-hydroxytristamine (serotonin)-3 (5HT$_3$) receptors. GABA$_A$ receptors are the principal transmitter-gated Cl$^-$ channels in the neocortex and allocortex, while GlyRs fulfill this function in the spinal cord. Activated receptors conduct chloride ions, driving the membrane potential toward the Cl$^-$ equilibrium potential. Both receptors are primarily inhibitory since the Cl$^-$ equilibrium potential is usually more negative than the normal resting membrane potential. This is not always the case, however, since the negative Cl$^-$ equilibrium potential is reversed during early neuronal development during which GABA is excitatory, a time when general anesthetics show their peak neurotoxic potential in animal models (Perouansky and Hemmings In Press).

Most functional GABA$_A$ and GlyRs are heteropentamers, typically consisting of three different GABA$_A$ subunits (e.g., two α two β and one γ or δ) or two different GlyR subunits (three α and two β). The subunit composition of GABA$_A$ receptors determines their physiological and pharmacological properties and varies between and within brain areas as well as between different compartments of individual neurons [see (Zeller et al. 2008)]. Examples include the preferential expression of the α5 subunit in the dendritic field of the hippocampal CA1 area (a region important for memory formation), of the α4 subunit in the thalamus, and of the α6 subunit in the cerebellum. Presence of a γ subunit is required for benzodiazepine modulation of GABA$_A$ receptors and can also influence modulation by inhaled anesthetics.

While the molecular mechanisms of receptor modulation by inhaled anesthetics are not definitively settled, these receptors have been key to our understanding of anesthetic–receptor interactions. Using chimeric receptor constructs between anesthetic-sensitive GABA$_A$ receptor and insensitive GlyR subunits, specific amino acid residues in the second and third transmembrane domains (TM2 and TM3) critical to the action of inhaled anesthetics have been identified (Mihic et al. 1997). This laid the groundwork for the construction of anesthetic-resistant GABA$_A$ receptors and the generation of transgenic mice with altered anesthetic sensitivity (below).

The related cation-permeable 5-hydroxytryptamine (serotonin)-3 (5HT$_3$) receptors are similarly potentiated by inhaled anesthetics (Machu and Harris 1994; Jenkins, Franks, and Lieb 1996). 5HT$_3$ receptors are involved with autonomic reflexes and probably contribute to the emetogenic properties of volatile anesthetics.

## *Inhibition of Excitatory Acetylcholine and Glutamate Receptors*

Neuronal nicotinic acetylcholine receptors (nAChRs), like the other members of the Cys-loop family, are heteropentameric. They are composed of α and β subunits, but functional homomeric receptors can be formed by α7 subunits. In the CNS, nAChRs are localized primarily presynaptically (Role and Berg 1996). Homomeric α7 receptors have high permeability to $Ca^{2+}$ that can exceed that of NMDA receptors. In contrast to GABA$_A$ receptors and GlyRs, nAChRs pass cations when activated and therefore depolarize the membrane potential. Receptors containing α4β2 (but not α7) subunits are very sensitive to block by isoflurane and propofol (Flood, Ramirez-Latorre, and Role 1997; Violet et al. 1997). Relevance of nAChR block to immobilization, sedation, and unconsciousness by inhaled anesthetics is unlikely since nAChRs are also blocked by nonimmobilizers, although it is possible that they do contribute to amnesia (Raines and Miller 1994), possibly by actions in the hippocampus (Westphalen, Gomez, and Hemmings 2009).

NMDA receptors are a major postsynaptic receptor subtype of ionotropic receptors for glutamate, the principal excitatory neurotransmitter in the mammalian CNS (Dingledine et al. 1999). Typical NMDA receptors, defined pharmacologically by their selective activation by the exogenous agonist NMDA, are heteromers consisting of an obligatory NR1 subunit and modulatory NR2 subunits. Channel opening requires glutamate (or another agonist such as NMDA) binding to the NR2 subunit while the co-agonist glycine binds to the NR1 subunit. NMDA receptors also require membrane depolarization to relieve voltage-dependent block by $Mg^{2+}$. Depolarization is typically provided by binding of glutamate to non-NMDA glutamate receptors. Because of this "dual" requirement (transmitter release and postsynaptic depolarization), synaptic NMDA receptors function as coincidence detectors, and this characteristic is thought to be central to their role in synaptic plasticity. NMDA receptors are also involved in the development of chronic pain, perhaps due to similar mechanisms underlying synaptic plasticity, and in ischemia-induced excitotoxicity by virtue of their capacity to allow entry of extracellular $Ca^{2+}$, which is a ubiquitous intracellular signal. The nonhalogenated inhaled anesthetics xenon, nitrous oxide, and cyclopropane, as well as the intravenous anesthetic ketamine, have minimal effects on $GABA_A$ receptors, but depress excitatory glutamatergic synaptic transmission postsynaptically via NMDA glutamate receptor blockade (Franks et al. 1998; Jevtovic-Todorovic et al. 1998; Mennerick et al. 1998). Potent inhaled anesthetics can also inhibit isolated NMDA receptors at higher concentrations (Solt, Eger, and Raines 2006; Dickinson et al. 2007). Along with presynaptic inhibition of glutamate release, this might contribute to their depression of NMDA receptor-mediated excitatory transmission.

A second class of ionotropic glutamate receptors includes the non-NMDA-receptors, which are subdivided into AMPA and kainate receptors based again on their sensitivities to selective exogenous agonists (Dingledine et al. 1999). Inhaled anesthetics have been found to inhibit AMPA receptors weakly in some neurons (Harris et al. 1995), but a recent study found that both NMDA and AMPA receptors were inhibited with comparable potencies by xenon in amygdale neurons (Haseneder et al. 2008). Interestingly, kainate receptors are enhanced by inhaled anesthetics, but this is unlikely to be involved in immobility since MAC is not altered in mice deficient in the GluR6 receptor subunit (Sonner et al. 2005). Most evidence suggests that the principal mechanism for depression of glutamatergic transmission by potent inhaled anesthetics is presynaptic, with variable contributions from postsynaptic receptor blockade [see (Perouansky et al. 2003; Hemmings, Akabas et al. 2005)].

## Synaptic Versus Extrasynaptic GABAergic Actions

$GABA_A$ receptors mediate fast synaptic inhibition by generating transient inhibitory postsynaptic currents. Isoflurane prolongs these inhibitory postsynaptic potentials to enhance net inhibitory current (Jones and Harrison 1993). In addition to this "phasic" inhibition, a "tonic" or persistent inhibitory conductance has been identified (Semyanov et al. 2004) that involves activation of extrasynaptic $GABA_A$ receptors. Extrasynaptic $GABA_A$ receptors have different pharmacological and kinetic properties compared to synaptic receptors due to their distinct subunit compositions. Extrasynaptic $GABA_A$ receptors are tonically exposed to low ambient GABA concentrations, at which the potentiating effect of anesthetics is most significant, have a high affinity for GABA, and show slow desensitization. Increased tonic inhibition mediated by extrasynaptic $GABA_A$ receptors could contribute to the neurodepressive properties of general anesthetics (Bai et al. 2001; Bieda and MacIver 2004; Caraiscos, Elliott et al. 2004).

The effects of anesthetics on tonic inhibition have been characterized in the hippocampus, which is critically involved in learning and memory. Hippocampal neurons generate a robust tonic current via activation of $\alpha5$ subunit-containing $GABA_A$ receptors, which are highly sensitive to propofol, midazolam (Bai et al. 2001; Bieda and MacIver 2004), and isoflurane (Caraiscos, Newell et al. 2004). $GABA_A$ receptors containing the $\alpha5$ subunit are highly sensitive to propofol and isoflurane at low concentrations that produce amnesia but not unconsciousness. The tonic inhibitory conductance mediated by $GABA_A$ receptors is a likely substrate for the amnestic properties of these anesthetics (Cheng et al. 2006). Alternatively, or in addition, slow dendritic inhibitory synapses that incorporate $\alpha5$ subunits (Zarnowska et al. 2009) and that are modulated by low (amnestic) anesthetic concentrations (Dai, Perouansky, and Pearce 2009) might contribute to anesthetic-induced amnesia.


## Voltage-Gated Ion Channels as Anesthetic Targets

### *Na$^+$ Channels*

Voltage-gated Na$^+$ channels are critical to axonal conduction, synaptic integration, and neuronal excitability. Axonal action potentials were initially reported to be relatively resistant to clinical concentrations of volatile anesthetics (Larrabee and Posternak 1952), which was consistent with the relative insensitivity of Na$^+$ currents in squid (Haydon and Urban 1983) and crayfish (Bean, Shrager, and Goldstein 1981) giant axons to volatile anesthetics. This established the widespread notion "that clinical concentrations of general anesthetics almost certainly do not act by blocking Na$^+$ channels" (Bean, Shrager, and Goldstein 1981) or any other voltage-gated ion channel (Franks and Lieb 1994). However, axonal conduction in small ($0.1–0.2 \mu m$) unmyelinated hippocampal axons is significantly depressed by inhaled anesthetics (Berg-Johnsen and Langmoen 1986; Mikulec et al. 1998), and other small diameter structures such as nerve terminals might also be sensitive. Patch clamp recordings of accessible nerve terminals have shown that isoflurane inhibits action potential amplitude (Ouyang and Hemmings 2005) and that reductions in nerve terminal action potential amplitude have significant effects on transmitter release and hence on postsynaptic responses (Wu et al. 2004).

Evidence that mammalian voltage-gated Na$^+$ channels are sensitive to clinically relevant concentrations of general anesthetics has come from careful analysis of anesthetic effects on heterologously expressed channels. The Na$^+$ channel family consists of nine homologous pore-forming $\alpha$ subunits with distinct cellular and subcellular distributions (Yu and Catterall 2004). One neuronal isoform (Na$_v$1.2) is inhibited by multiple potent inhaled anesthetics through a voltage-independent block of peak current and a hyperpolarizing shift in the voltage dependence of steady-state inactivation (Rehberg, Xiao, and Duch 1996). Isoflurane and other inhaled anesthetics inhibit multiple mammalian Na$^+$ channel isoforms (Ouyang and Hemmings 2007), including Na$_v$1.2 (Rehberg, Xiao, and Duch 1996), Na$_v$1.4 and

$Na_V1.6$ (Shiraishi and Harris 2004; Ouyang, Herold, and Hemmings 2009), $Na_V1.5$ (Stadnicka et al. 1999), and $Na_V1.8$ (Herold et al. 2009). Potent inhaled anesthetics also inhibit native $Na^+$ channels in isolated nerve terminals (Ratnakumari and Hemmings 1998; Ouyang, Wang, and Hemmings 2003) and dorsal root ganglion neurons, while the nonimmobilizer F6 is ineffective (Ratnakumari et al. 2000). In contrast, xenon has no detectable effect on $Na^+$, $Ca^{2+}$, or $K^+$ channels in isolated cardiomyocytes (Stowe et al. 2000). The recent demonstration that NaChBac, a prokaryotic homolog of voltage-gated $Na^+$ channels, is also inhibited by volatile anesthetics opens the way for structure-function studies of these channels (Ouyang et al. 2007).

Inhibition of presynaptic $Na^+$ channels has been implicated in the depression of evoked neurotransmitter release by volatile anesthetics from isolated nerve terminals (Schlame and Hemmings 1995; Westphalen and Hemmings 2003) and cultured rat hippocampal neurons (Hemmings, Yan et al. 2005). Isoflurane inhibits nerve terminal $Na^+$ currents and action potential amplitude through $Na^+$ channel blockade in isolated rat neurohypophysial terminals (Ouyang, Wang, and Hemmings 2003; Ouyang and Hemmings 2005). Isoflurane significantly depressed action potential-evoked synaptic vesicle exocytosis and EPSC amplitude with only a small reduction in presynaptic action potential amplitude and no direct effect on $Ca^{2+}$ current in the rat calyx of Held synapse (Wu et al. 2004). Simulated reductions in action potential amplitude reproduced this highly nonlinear relationship between peak $Na^+$ current inhibition and exocytosis. These findings identify presynaptic $Na^+$ channels as important anesthetic targets for potent inhaled anesthetics.

## $Ca^{2+}$ Channels

Multiple cellular functions depend on the tightly controlled concentration of intracellular free $Ca^{2+}$, which is determined by the integrated activity of voltage-gated $Ca^{2+}$ channels, capacitative $Ca^{2+}$ channels, plasma membrane and endo/sarcoplasmic reticulum $Ca^{2+}$-ATPases (pumps), $Na^+/Ca^{2+}$ exchangers, and mitochondrial $Ca^{2+}$ sequestration. Alteration of any of these mechanisms by anesthetics could affect the many cellular processes regulated by the second messenger actions of $Ca^{2+}$, which include synaptic transmission, gene expression, cytotoxicity, and muscle excitation–contraction coupling. Excitable cells translate electrical activity into action by $Ca^{2+}$ fluxes mediated primarily by voltage-gated $Ca^{2+}$ channels in the plasma membrane. Distinct $Ca^{2+}$ channel subtypes are expressed in various cells and tissues and are classified pharmacologically and functionally by the degree of depolarization required to gate the channel, e.g., low voltage-activated (LVA; T-type) and high voltage-activated (HVA; L-, N-, R-, and P/Q-type) channels. More recently, the molecular identity of their pore-forming α-subunits has been used for classification (Catterall 2000). There is convincing evidence that inhaled anesthetics inhibit some $Ca^{2+}$ channel isoforms, but not others [for review, see (Topf et al. 2003)].

Inhibition of presynaptic voltage-gated $Ca^{2+}$ channels coupled to transmitter release has been proposed as a mechanism by which volatile anesthetics might reduce excitatory transmission (Miao, Frazer, and Lynch 1995). Indeed, heterologously expressed N-type ($Ca_v2.2$) and P-type ($Ca_v2.1$) channels, which mediate $Ca^{2+}$ entry coupled to neurotransmitter release, are modestly sensitive to potent inhaled anesthetics (Study 1994; Kameyama, Aono, and Kitamura 1999). But they are not inhibited in all neuron types (Hall, Lieb, and Franks 1994), which suggests the importance of auxiliary subunits, posttranslation modification, or other unknown modulators of anesthetic sensitivity. A modest contribution of R-type $Ca^{2+}$ channels ($Ca_v2.3$) to anesthesia is suggested by their sensitivity to inhaled anesthetics and a small increase in MAC produced by their genetic deletion in mice (Takei et al. 2003). T-type $Ca^{2+}$ channels are particularly sensitive to potent inhaled anesthetics (Joksovic, Bayliss, and Todorovic 2005) and nitrous oxide (Todorovic et al. 2001). However, mutant mice lacking a major neuronal T-type $Ca^{2+}$ channel isoform ($Ca_v3.1$) have normal anesthetic sensitivity, although the onset of anesthesia is delayed (Petrenko et al. 2007). The role that inhibition of these or other $Ca^{2+}$ channels plays in the CNS effects of inhaled anesthetics is still unclear.

## $K^+$ Channels and HCN Channels

Potassium ($K^+$) channels are an extremely diverse ion channel family noted for their varied modes of activation. They regulate electrical excitability, muscle contractility, and neurotransmitter release and are important in determining the input resistance and in driving repolarization, and thus determine excitability and action potential duration. Given the large diversity in $K^+$ channel structure, function, and anesthetic sensitivity, it is not surprising that there is considerable diversity in their sensitivity and response to anesthetics (Yost 1999; Friederich et al. 2001): from relatively insensitive (voltage-gated $K^+$ channels $K_v1.1$, $K_v3$) to sensitive [some members of the two-pore domain $K^+$ channels ($K_{2P}$) family], resulting in either inhibition, activation, or no effect on $K^+$ currents.

Anesthetic activation of certain "leak" $K^+$ channels was first observed for isoflurane in the snail *Lymnaea* (Franks and Lieb 1988), though the molecular identity of the affected ion channels was unknown. Activation of $K_{2P}$ channels by inhaled anesthetics, including xenon, nitrous oxide, and cyclopropane, was subsequently observed in mammals (Franks and Honore 2004). Increased $K^+$ conductance can hyperpolarize neurons, reducing responsiveness to excitatory synaptic input and altering network synchrony. Targeted deletion of the TASK-1, TASK-3, and TREK-1 $K_{2P}$ channels in mice reduces sensitivity to immobilization by potent inhaled anesthetics (Heurteaux et al. 2004; Linden et al. 2006, 2007) and their presynaptic inhibition of transmitter release (Westphalen et al. 2007), implicating these channels as anesthetic targets in vivo. Other members of this large family of $K^+$

channels are also sensitive to xenon and potent inhaled anesthetics (Patel and Honore 2001).

Potent inhaled anesthetics also inhibit HCN channels, reducing the rate of rise of pacemaker potentials and the bursting frequency of certain neurons showing autorhythmicity. They decrease $I_h$ conductance in neurons (Sirois, Lynch, and Bayliss 2002) and modulate recombinant HCN1 and HCN2 channel isoforms at clinically relevant concentrations (Chen et al. 2005). As HCN channels contribute to resting membrane potential, control action potential firing, dendritic integration, neuronal automaticity, and temporal summation, and determine periodicity and synchronization of oscillations in many neuronal networks (Robinson and Siegelbaum 2003), anesthetic modulation of these channels could play an important role in anesthetic effects on neuronal integrative functions. HCN1 channels have also been implicated in the anesthetic actions of propofol and etomidate. Both of these anesthetics inhibit heterologously expressed HCN1 channels and neuronal $I_h$ currents, and their hypnotic potency is greatly reduced in HCN1 knockout mice (Chen, Shu, and Bayliss 2009).

## Intracellular Signaling Mechanisms

Cell signaling mechanisms are critical to all phases of organ function and have been attractive targets for the broad effects of general anesthetics. Anesthetics have poorly understood actions on intracellular cell signaling pathways, which include processes downstream from cell surface receptors and ion channels, such as effects of second messengers, protein phosphorylation pathways, and other regulatory mechanisms (Girault and Hemmings 2006).

A variety of signals, including hormones, neurotransmitters, cytokines, pheromones, odorants, and photons, produce their intracellular actions by interactions with metabotropic receptors that activate heterotrimeric guanine nucleotide (GTP)-binding proteins (G proteins). In contrast to the ionotropic receptors that directly couple to ion-selective channels, G proteins act as indirect molecular switches to relay information from activated plasma membrane receptors to appropriate intracellular targets. Heterotrimeric G proteins consist of a large $\alpha$ subunit and a smaller $\beta\gamma$ subunit dimer, each expressed as multiple isoforms with distinct properties and downstream targets. G proteins regulate a plethora of downstream effectors to control the levels of cytosolic second messengers such as $Ca^{2+}$, cAMP, and inositol trisphosphate ($IP_3$). These, in turn, regulate effector proteins such as ion channels and enzymes, either directly or via second messenger-regulated protein phosphorylation pathways.

Drugs that act through G-protein-coupled receptors (GPCRs), such as agonists for $\mu$ opioid and $\alpha2$ adrenergic receptors, can affect anesthetic sensitivity (reduce MAC). Inhaled anesthetics can also directly affect signaling via GPCRs (Rebecchi and Pentyala 2002). For example, inhaled anesthetics activate multiple rat olfactory

GPCRs in vivo in a receptor- and agent-selective manner (Peterlin et al. 2005). Analogous effects on related GPCRs more relevant to critical anesthetic endpoints are possible, but remain to be demonstrated. The observation that both inhaled anesthetics and nonimmobilizers inhibit mGluR5 glutamate receptors, 5-HT$_{2A}$ serotonin receptors and muscarinic acetylcholine receptors suggests that these GPCR effects do not contribute to anesthetic immobilization (Minami, Minami, and Harris 1997; Minami et al. 1997, 1998).

Phosphorylation of proteins on specific serine, threonine, or tyrosine hydroxyl groups, a post-translational modification involved in the regulation of many anesthetic-sensitive receptors and ion channels, is pivotal to synaptic plasticity (e.g., long-term potentiation). Phosphorylation is controlled by the balance of activity between protein kinases and phosphatases, several of which are plausible anesthetic targets. The protein kinase C (PKC) family of multifunctional protein kinases is activated by the lipid signaling molecule diacylglycerol and is involved in the regulation of many ion channels and receptors. Potent inhaled anesthetics enhance the activity of specific PKC isoforms and stimulate phosphorylation of specific PKC substrates (Hemmings 1998; Gomez, Guatimosim, and Gomez 2003). Structural studies have identified a potential binding site in the diacylglycerol binding domain of PKCδ, consistent with the ability of certain anesthetics to mimic this natural regulator by binding to the activating site (Das et al. 2004). A specific role for a direct pharmacologically relevant effect mediated by anesthetic activation of PKC or of any other kinase has yet to be demonstrated. Intrathecal injection of isoform-specific inhibitors of PKC does not affect sensitivity to halothane in vivo (Shumilla et al. 2004). Knockout mice lacking the PKCγ isoform show normal sensitivity to halothane and desflurane while isoflurane MAC was increased (Sonner et al. 1999), suggesting that PKC is not critical to inhaled anesthetic immobilization.

The effects of anesthetics on the phosphorylation of individual residues in specific substrates can be studied using phosphorylation state-specific antibodies that are able to detect the phosphorylated form of kinase substrates. A comparison of the effects of three mechanistically diverse anesthetics (isoflurane, propofol, and ketamine) on critical intracellular protein phosphorylation signaling pathways that are known to integrate multiple second-messenger systems reveals both shared and agent-specific actions in vivo (Snyder et al. 2007). All three anesthetics reduce phosphorylation of activating sites on NMDA (S897) and AMPA (S831) glutamate receptors and of the downstream extracellular signal-regulated kinase ERK2, which are involved in synaptic plasticity, consistent with depression of normal glutamatergic synaptic transmission in the anesthetized mouse cerebral cortex. This effect is rather selective in that several other PKA substrates examined are not affected, indicating a substrate-specific effect rather than a general inhibition of PKA activity (Hemmings and Adamo 1994). Additional studies will be required to determine which anesthetic effects on kinase pathways represent direct effects, as occurs with PKC, and which are indirect due to anesthetic-induced alterations in signaling molecules known to regulate protein kinase and phosphatase activity such as $Ca^{2+}$ and other second messengers.

# Gene Expression

The ability of general anesthetics to alter gene expression in the brain was first observed for the highly reactive acute early genes *c-fos* and *c-jun* (Marota, Crosby, and Uhl 1992). Anesthetic effects on gene expression have since been observed for multiple anesthetics and organs (Hamaya et al. 2000). In the hippocampus of aged rats, changes in gene expression persisted for up to 2 days in rats exposed to isoflurane and nitrous oxide (Culley et al. 2006) and changes in protein expression have been observed 3 days after exposure to desflurane (Futterer et al. 2004). The significance of these changes in gene and protein expression persisting after recovery from the classical signs of anesthesia remains to be established.

# Identification of Anesthetic Binding Sites

Data from X-ray crystallography, molecular modeling, and structure–function studies indicate that anesthetics bind in hydrophobic cavities formed within proteins (Bertaccini, Trudell, and Franks 2007). The lipophilic (or hydrophobic) nature of these binding sites underlies the Meyer–Overton correlation between anesthetic potency and lipophilicity. An element of amphiphilicity (possessing both polar and nonpolar characteristics) is also required for effective interaction with these cavities, as indicated by improvements in the Meyer–Overton correlation with more polar solvents.

Identifying anesthetic binding sites on plausible target proteins is difficult due to the low affinity interactions of inhaled anesthetics and the paucity of atomic resolution structures for pharmacologically relevant target proteins, most of which are membrane proteins that are difficult to resolve structurally. Consequently, most defined anesthetic binding sites have been identified in model proteins for which three-dimensional atomic resolution structures are available, such as luciferase and albumin (Bertaccini, Trudell, and Franks 2007), rather than in physiologically relevant proteins. These studies indicate that anesthetics bind in pockets with both nonpolar and polar noncovalent interactions. Binding involves weak hydrogen bond interactions with polar amino acid residues and water molecules, nonpolar van der Waals interactions, and a polarizing effect of the amphiphilic binding cavity on the relatively hydrophobic anesthetic molecules. Internal cavities are important for the conformational flexibility involved in ion channel gating and ligand-induced signal transduction of receptor proteins. Occupation of a critical volume within these cavities by anesthetics provides a plausible mechanism for alteration of receptor and ion channel function by selective stabilization of a particular confirmation, e.g., an open or inactivated state of an ion channel. Anesthetics also obtain binding energy from the entropy generated by displacing bound water from these relatively promiscuous binding sites. Studies of glycine, $GABA_A$, and NMDA receptors provide convincing evidence for the existence of anesthetic binding sites in critical neuronal signaling proteins by identifying amino acid residues critical for anesthetic actions (Wick et al. 1998; Koltchine et al. 1999; Jenkins et al. 2001).

Molecular modeling based on structurally homologous proteins has also been used to identify putative anesthetic binding sites in the transmembrane domains of GABA$_A$ and glycine receptors (Trudell and Bertaccini 2004). This model suggests that different drugs might either bind in different orientations within a single amphiphilic cavity or occupy different cavities within the protein, causing similar functional effects. Refinement of these molecular models will continue to provide new insights into the molecular bases for general anesthetic actions that can be experimentally tested.

## Modeling Anesthetic Binding Sites in GABA$_A$ Receptors

The hydrophobic nature and low receptor affinities of most general anesthetics have made identification of pharmacologically relevant binding sites difficult. Evidence supports the existence of anesthetic binding sites between the second and third transmembrane segments (TM2 and TM3) of GABA$_A$ receptor subunits. Construction of chimeric receptors between anesthetic-sensitive and anesthetic-insensitive receptor subunits led to identification of two critical positions corresponding to α1Ser270 in TM2 and α1Ala291 near the extracellular end of TM3 (Mihic et al. 1997). Substitution of larger amino acids at these positions in the α subunit alters potentiation by inhaled anesthetics (Koltchine et al. 1999), whereas mutations in β subunits, particularly at the TM3 position, alter potentiation by propofol (Krasowski et al. 2001). The specific amino acid at the β3 subunit TM2 position (Asn265) has a similar influence on potentiation of GABA$_A$ receptors by etomidate (Belelli et al. 1997). Groundbreaking studies have shown that genetically engineered β3 and β2 knock-in mice bearing the mutation β3Asn265Met or β2Asn265Ser are relatively insensitive (depending on the behavioral endpoint being measured) to etomidate and propofol (Jurd et al. 2003) or etomidate (Reynolds et al. 2003), respectively (Fig. 2.1). A role for this receptor in mediating anesthetic effects on hypothalamic sleep mechanisms has recently been confirmed by showing that propofol effects on tuberomammillary nucleus and perifornical area neurons are greatly attenuated in the β3Asn265Met knock-in mouse (Zecharia et al. 2009). These critical studies have solidified the GABA$_A$ receptor as a principal molecular target relevant to the anesthetic actions of these drugs.

These critical TM2/TM3 residues of GABA$_A$ receptors might contribute to anesthetic binding sites or, alternatively, to the allosteric transduction between anesthetic binding and receptor modulation. Considerable evidence supports these residues as part of a direct anesthetic binding site. For example, the molecular volume of the amino acid substituted for β2Met286 in TM3 correlates inversely with the maximal molecular volume of active propofol derivatives (Krasowski et al. 2001). Similar results were obtained with inhaled anesthetics and alcohols at the aligned position in the α subunit (Wick et al. 1998; Jenkins et al. 2001). Using the substituted cysteine accessibility method (SCAM), propofol protects a cysteinyl residue substituted for the β2Met286 from reaction with a charged sulfhydryl-reactive reagent (Bali and Akabas 2004), which supports a steric block by propofol due to binding in close

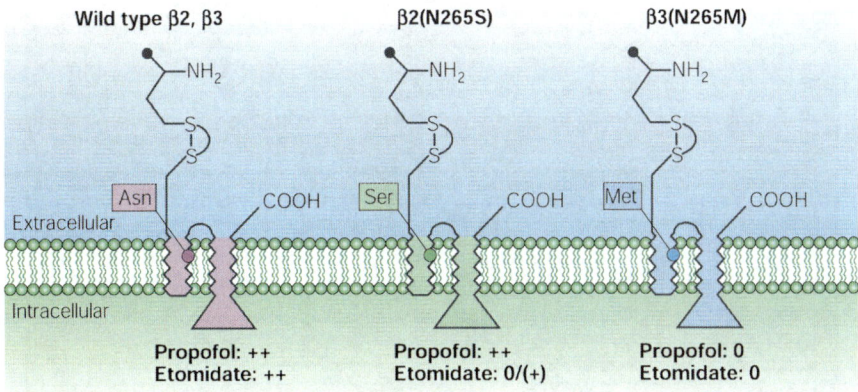

**Fig. 2.1** Single amino acid mutations in GABA$_A$ receptors disrupt anesthetic effects on isolated receptors and in vivo. Wild-type β2 and β3 subunits have an asparagine residue at position 265 (Asn265) in the second transmembrane region. GABA$_A$ receptors that contain these subunits are sensitive to propofol and etomidate. Mutation of Asn265 to serine in the β2 subunit [β2(N265S)] results in receptors that are largely insensitive to etomidate, but are sensitive to propofol. Mice engineered to carry this mutation [β2(N265S) knock-in mice] are also sensitive to propofol, but not to etomidate. Mutation of Ser265 to methionine in the β3 subunit [β3(N265M)] results in receptors that are insensitive to both etomidate and propofol, and knock-in mutations of this receptor result in mice that are insensitive to the hypnotic effects of both anesthetics. Adapted from (Rudolph and Antkowiak 2004).

proximity to β2Met286. In order to promote channel opening, anesthetics must stabilize activated states of the GABA$_A$ receptor. A molecular mechanism of anesthetic action could involve preferential partitioning into the proposed cavity between the TM2–TM3 segments, displacing bound water and thereby stabilizing the activated state of the receptor (Bertaccini, Trudell, and Franks 2007).

Molecular modeling based on structurally homologous proteins has been used to predict anesthetic binding sites in the transmembrane domains of the GABA$_A$ receptor (Fig. 2.2). This model suggests that different drugs might bind in different orientations within a single binding site, e.g., isoflurane (Fig. 2.2A) and propofol (Fig. 2.2B). This provides a potential explanation of how mutation of different amino acid residues could affect potentiation by specific drugs in the different subunits. Alternatively, different drugs might occupy different cavities within the protein, nevertheless causing similar functional effects. Refinement of these molecular models will continue to provide new insights into the molecular basis for general anesthetic action.

## Modeling Anesthetic Binding Sites in NMDA Receptors

Potential sites of interaction of xenon and isoflurane with the NMDA receptor have also been identified using this approach. One site, which might contain up to three

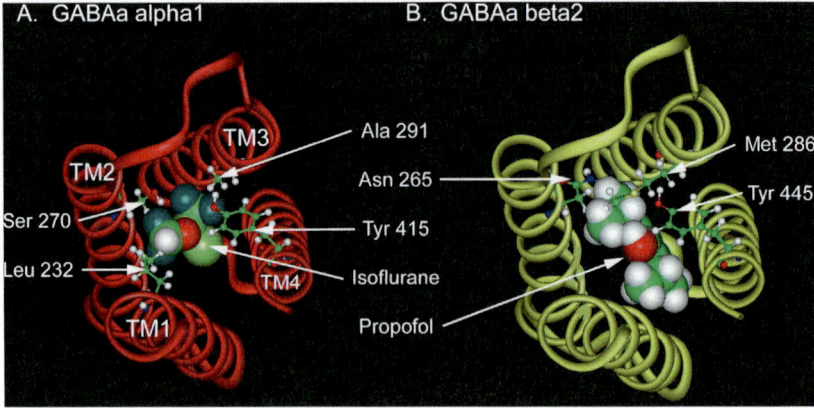

**Fig. 2.2** Putative anesthetic binding sites on GABA$_A$ receptors. A. Model of rat GABA$_A$ α1 sub-unit with Leu232, Ser270, Ala291, and Tyr415 rendered in ball-and-stick. A molecule of isoflurane built at the same scale is positioned in the putative binding site. The transmembrane α helices (TM) are numbered 1–4. B. Corresponding model of GABA$_A$ β2 subunit with Asn265, Met286, and Tyr445 rendered in ball-and-stick. A model of propofol built at the same scale is positioned in the putative binding site. Adapted from (Hemmings, Akabas et al. 2005).

**Fig. 2.3** Crystal structure of the ligand-binding domain of the NMDA receptor showing xenon-binding sites predicted by molecular modeling. The spheres represent xenon atoms at the center of the density clusters that comprise the binding sites. The predicted xenon-binding sites occupy the site normally occupied by glycine. The same site is also large enough to accommodate isoflurane. Modified from (Dickinson et al. 2007).

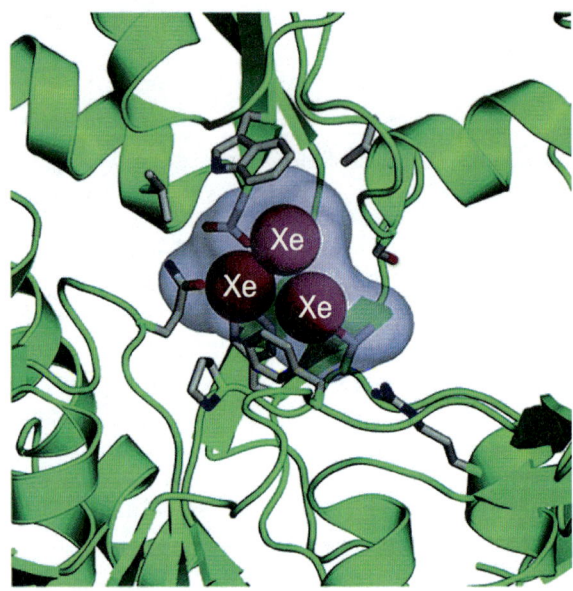

xenon atoms or one molecule of isoflurane, overlaps the known binding site for the co-agonist glycine in the NR1 subunit (Dickinson et al. 2007) (Fig. 2.3). This suggests that two chemically dissimilar inhaled anesthetics inhibit NMDA receptors by direct competitive inhibition of agonist binding at the same site.

# Future Directions

Recent advances in molecular neuroscience continue to provide tremendous opportunities for understanding the molecular actions of general anesthetics on their targets. Targeted mutations of putative anesthetic targets in mice provide an elegant bridge between in vitro observations and whole animal experiments essential for demonstrating anesthetic endpoints. The existence of multiple targets of general anesthetics and genetic redundancy among ion channels make this a more challenging experimental approach for this drug class.

There is now ample evidence that clinical concentrations of most general anesthetics influence the function of ligand-gated ion channels and/or other important ion channels, and we are on the verge of a molecular understanding of the action of these drugs on $GABA_A$ receptors. However, there is still little information, or at least agreement, as to how modulating these channels alters brain function leading to the state of general anesthesia. Advances in systems neurobiology should extend our knowledge of anesthetic effects from the synaptic level to their actions on cognitive and motor functions of the intact organism.

# References

Antognini, J. F., and K. Schwartz. 1993. Exaggerated anesthetic requirements in the preferentially anesthetized brain. *Anesthesiology* 79(6):1244–1249.

Bai, D., G. Zhu, P. Pennefather, M. F. Jackson, J. F. MacDonald, and B. A. Orser. 2001. Distinct functional and pharmacological properties of tonic and quantal inhibitory postsynaptic currents mediated by g-aminobutyric acid A receptors in hippocampal neurons. *Mol Pharmacol* 59: 814–824.

Bali, M., and M. H. Akabas. 2004. Defining the propofol binding site location on the GABAA receptor. *Mol Pharmacol* 65(1):68–76.

Bean, B. P., P. Shrager, and D. A. Goldstein. 1981. Modification of sodium and potassium channel gating kinetics by ether and halothane. *J Gen Physiol* 77:233–253.

Belelli, D., J. J. Lambert, J. A. Peters, K. Wafford, and P. J. Whiting. 1997. The interaction of the general anesthetic etomidate with the g -aminobutyric acid type A receptor is influenced by a single amino acid. *Proc Natl Acad Sci USA* 94(20):11031–11036.

Berg-Johnsen, J., and I. A. Langmoen. 1986. The effect of isoflurane on unmyelinated and myelinated fibres in the rat brain. *Acta Physiol Scand* 127:87–93.

Bertaccini, E. J., J. R. Trudell, and N. P. Franks. 2007. The common chemical motifs within anesthetic binding sites. *Anesth Analg* 104(2):318–324.

Bieda, M. C., and M. B. MacIver. 2004. Major role for tonic GABAA conductances in anesthetic suppression of intrinsic neuronal excitability. *J Neurophysiol* 92(3):1658–1667.

Caraiscos, V. B., E. M. Elliott, Ten You, V. Y. Cheng, D. Belelli, J. G. Newell, M. F. Jackson, J. J. Lambert, T. W. Rosahl, K. A. Wafford, J. F. MacDonald, and B. A. Orser. 2004. Tonic inhibition in mouse hippocampal CA1 pyramidal neurons is mediated by {alpha}5 subunit-containing {gamma}-aminobutyric acid type A receptors. *Proc Natl Acad Sci USA* 101(10): 3662–3667.

Caraiscos, V. B., J. G. Newell, Ten You, E. M. Elliott, T. W. Rosahl, K. A. Wafford, J. F. MacDonald, and B. A. Orser. 2004. Selective enhancement of tonic GABAergic inhibition in murine hippocampal neurons by low concentrations of the volatile anesthetic isoflurane. *J Neurosci* 24(39):8454–8458.

Catterall, W. A. 2000. Structure and regulation of voltage-gated $Ca^{2+}$ channels. *Annu Rev Cell Dev Biol* 16:521–555.

Chen, X., S. Shu, and D. A. Bayliss. 2009. HCN1 channel subunits are a molecular substrate for hypnotic actions of ketamine. *J Neurosci* 29(3):600–609.

Chen, X., J. E. Sirois, Q. Lei, E. M. Talley, C. Lynch, III, and D. A. Bayliss. 2005. HCN subunit-specific and cAMP-modulated effects of anesthetics on neuronal pacemaker currents. *J Neurosci* 25(24):5803–5814.

Cheng, V. Y., L. J. Martin, E. M. Elliott, J. H. Kim, H. T. Mount, F. A. Taverna, J. C. Roder, J. F. MacDonald, A. Bhambri, N. Collinson, K. A. Wafford, and B. A. Orser. 2006. Alpha5GABA$_A$ receptors mediate the amnestic but not sedative-hypnotic effects of the general anesthetic etomidate. *J Neurosci* 26(14):3713–3720.

Culley, D. J., R. Y. Yukhananov, Z. C. Xie, R. R. Gali, R. E. Tanzi, and G. Crosby. 2006. Altered hippocampal gene expression 2 days after general anesthesia in rats. *Eur J Pharmacol* 549(1–3):71–78.

Dai, S., M. Perouansky, and R. A. Pearce. 2009. Amnestic concentrations of etomidate modulate GABA$_{A,slow}$ synaptic inhibition in hippocampus. *Anesthesiology.* Sep7. (Epub a head of print) PMID: 19741493

Das, Joydip, George H. Addona, Warren S. Sandberg, S. Shaukat Husain, Thilo Stehle, and Keith W. Miller. 2004. Identification of a General Anesthetic Binding Site in the Diacylglycerol-binding Domain of Protein Kinase C{δ}. *J Biol Chem* 279(36):37964–37972.

Dickinson, R., B. K. Peterson, P. Banks, C. Simillis, J. C. S. Martin, C. A. Valenzuela, M. Maze, and N. P. Franks. 2007. Competitive inhibition at the glycine site of the n-methyl-d-aspartate receptor by the Anesthetics xenon and Isoflurane. *Anesthesiology* 107(5):756–767.

Dingledine, R., K. Borges, D. Bowie, and S. F. Traynelis. 1999. The glutamate receptor ion channels. *Pharmacol Rev* 51(1):7–61.

Eger, E. I., 2nd, D. M. Fisher, J. P. Dilger, J. M. Sonner, A. R. Evers, N. P. Franks, R. A. Harris, J. J. Kendig, W. R. Lieb, and T. Yamakura. 2001. Relevant concentrations of inhaled anesthetics for in vitro studies of anesthetic mechanisms. *Anesthesiology* 94:915–921.

Flood, P., J. Ramirez-Latorre, and L. Role. 1997. Alpha 4 beta 2 neuronal nicotinic acetylcholine receptors in the central nervous system are inhibited by isoflurane and propofol, but alpha 7-type nicotinic acetylcholine receptors are unaffected [see comments]. *Anesthesiology* 86(4):859–865.

Franks, N. P., R. Dickinson, S. L. de Sousa, A. C. Hall, and W. R. Lieb. 1998. How does xenon produce anaesthesia? *Nature* 396(6709):324.

Franks, N. P., and E. Honore. 2004. The TREK K-2P channels and their role in general anaesthesia and neuroprotection. *Trends Pharmacol Sci* 25(11):601–608.

Franks, N. P., and W. R. Lieb. 1988. Volatile general anaesthetics activate a novel neuronal $K^+$ current. *Nature* 333:662–664.

Franks, N. P. 1994. Molecular and cellular mechanisms of general anaesthesia. *Nature* 367(6464):607–614.

Friederich, P., D. Benzenberg, S. Trellakis, and B. W. Urban. 2001. Interaction of volatile anesthetics with human Kv channels in relation to clinical concentrations. *Anesthesiology* 95(4):954–958.

Futterer, C. D., M. H. Maurer, A. Schmitt, R. E. Feldmann, W. Kuschinsky, and K. F. Waschke. 2004. Alterations in rat brain proteins after desflurane anesthesia. *Anesthesiology* 100(2):302–308.

Girault, JA, and HC Jr Hemmings. 2006. Cell Signaling. In *Foundations of Anesthesia: Basic Sciences for Clinical Practice*, edited by H. C. Hemmings, Jr. and P. M. Hopkins. London: Moby Elsevier.

Gomez, R. S., C. Guatimosim, and M. V. Gomez. 2003. Mechanism of action of volatile anesthetics: Role of protein kinase C. *Cell Mol Neurobiol* 23(6):877–885.

Gruss, M., T. J. Bushell, D. P. Bright, W. R. Lieb, A. Mathie, and N. P. Franks. 2004. Two-pore-domain $K^+$ channels are a novel target for the anesthetic gases xenon, nitrous oxide, and cyclopropane. *Mol Pharmacol* 65(2):443–452.

Hall, A. C., W. R. Lieb, and N. P. Franks. 1994. Insensitivity of P-type calcium channels to inhalational and intravenous general anesthetics. *Anesthesiology* 81(1):117–123.

Hamaya, Y., T. Takeda, S. Dohi, S. Nakashima, and Y. Nozawa. 2000. The effects of pentobarbital, isoflurane, and propofol on immediate-early gene expression in the vital organs of the rat. *Anesth Analg* 90(5):1177–1183.

Harris, R. A., S. J. Mihic, J. E. Dildymayfield, and T. K. Machu. 1995. Actions of anesthetics on ligand-gated ion channels - Role of Receptor Subunit Composition. *FASEB J* 9(14): 1454–1462.

Haseneder, R., S. Kratzer, E. Kochs, V. S. Eckle, W. Zieglgansberger, and G. Rammes. 2008. Xenon reduces N-methyl-D-aspartate and alpha-amino-3-hydroxy-5-methyl-4-isoxazolepropionic acid receptor-mediated synaptic transmission in the amygdala. *Anesthesiology* 109(6): 998–1006.

Haydon, D. A., and B. W. Urban. 1983. The effects of some inhalation anaesthetics on the sodium current of the squid giant axon. *J Physiol (Lond)* 341:429–439.

Hemmings, H. C., Jr. 1998. General anesthetic effects on protein kinase C. *Toxicol Lett* 100–101:89–95.

Hemmings, H. C., Jr., and A. I. Adamo. 1994. Effects of halothane and propofol on purified brain protein kinase C activation. *Anesthesiology* 81(1):147–155.

Hemmings, H. C., Jr., M. H. Akabas, P. A. Goldstein, J. R. Trudell, B. A. Orser, and N. L. Harrison. 2005. Emerging molecular mechanisms of general anesthetic action. *Trends Pharmacol Sci* 26(10):503–510.

Hemmings, H. C., Jr., Perouansky, M. 2009. Neurotoxicity of general anesthetics: cause for concern? *Anesthesiology* (in press).

Hemmings, H. C., Jr., W. Yan, R. I. Westphalen, and T. A. Ryan. 2005. The general anesthetic isoflurane depresses synaptic vesicle exocytosis. *Mol Pharmacol* 67(5):1591–1599.

Herold, K.F., C. Nau, W. Ouyang, and H.C. Jr. Hemmings. 2009. Isoflurane inhibits the tetrodotoxin-resistant voltage-gated sodium channel $Na_V1.8$. Anesthesiology 111:591–599.

Heurteaux, C., N. Guy, C. Laigle, N. Blondeau, F. Duprat, M. Mazzuca, L. Lang-Lazdunski, C. Widmann, M. Zanzouri, G. Romey, and M. Lazdunski. 2004. TREK-1, a K(+) channel involved in neuroprotection and general anesthesia. *EMBO J* 23(13):2684–2695.

Jenkins, A., N. P. Franks, and W. R. Lieb. 1996. Actions of general anaesthetics on 5-HT$_3$ receptors in NIE-115 neuroblastoma cells. *Br J Pharmacol* 117(7):1507–1515.

Jenkins, A., E. P. Greenblatt, H. J. Faulkner, E. Bertaccini, A. Light, A. Lin, A. Andreasen, A. Viner, J. R. Trudell, and N. L. Harrison. 2001. Evidence for a common binding cavity for three general anesthetics within the GABA(A) receptor. *J Neurosci* 21(6):art-RC136.

Jevtovic-Todorovic, V., S. M. Todorovic, S. Mennerick, S. Powell, K. Dikranian, N. Benshoff, C. F. Zorumski, and J. W. Olney. 1998. Nitrous oxide (laughing gas) is an NMDA antagonist, neuroprotectant and neurotoxin. *Nat Med* (4):460–463.

Joksovic, Pavle M., Douglas A. Bayliss, and Slobodan M. Todorovic. 2005. Different kinetic properties of two T-type Ca2+ currents of rat reticular thalamic neurones and their modulation by enflurane. *J Physiol Online* 566(1):125–142.

Jones, M. V., and N. L. Harrison. 1993. Effects of volatile anesthetics on the kinetics of inhibitory postsynaptic currents in cultured rat hippocampal neurons. *J Neurophysiol* 70(4): 1339–1349.

Jurd, R., M. Arras, S. Lambert, B. Drexler, R. Siegwart, F. Crestani, M. Zaugg, K. E. Vogt, B. Ledermann, B. Antkowiak, and U. Rudolph. 2003. General anesthetic actions in vivo strongly attenuated by a point mutation in the GABA(A) receptor beta3 subunit. *FASEB J* 17(2): 250–252.

Kameyama, K., K. Aono, and K. Kitamura. 1999. Isoflurane inhibits neuronal Ca2+ channels through enhancement of current inactivation. *Br J Anaesth* 82(3):402–411.

Koblin, D. D., B. S. Chortkoff, M. J. Laster, E. I. Eger, 2nd, M. J. Halsey, and P. Ionescu. 1994. Polyhalogenated and perfluorinated compounds that disobey the meyer-overton hypothesis. *Anesth Analg* 79(6):1043–1048.

Koltchine, V. V., S. E. Finn, A. Jenkins, N. Nikolaeva, A. Lin, and N. L. Harrison. 1999. Agonist gating and isoflurane potentiation in the human gamma-aminobutyric acid type A receptor determined by the volume of a second transmembrane domain residue. *Mol Pharmacol.* 56(5):1087–1093.

Krasowski, M. D., and N. L. Harrison. 1999. General anaesthetic actions on ligand-gated ion channels. *Cell Mol Life Sci* 55(10):1278–1303.

Krasowski, M. D., K. Nishikawa, N. Nikolaeva, A. Lin, and N. L. Harrison. 2001. Methionine 286 in transmembrane domain 3 of the GABAA receptor beta subunit controls a binding cavity for propofol and other alkylphenol general anesthetics. *Neuropharmacology* 41(8):952–964.

Larrabee, M. G., and J. M. Posternak. 1952. Selective action of anesthetics on synapses and axons in mammalian sympathetic ganglia. *J Neurophysiol* 15:91–114.

Linden, A. M., M. I. Aller, E. Leppa, O. Vekovischeva, T. itta-Aho, E. L. Veale, A. Mathie, P. Rosenberg, W. Wisden, and E. R. Korpi. 2006. The in vivo contributions of TASK-1-containing channels to the actions of inhalation anesthetics, the alpha(2) adrenergic sedative dexmedetomidine, and cannabinoid agonists. *J Pharmacol Exp Ther* 317(2):615–626.

Linden, A. M., C. Sandu, M. I. Aller, O. Y. Vekovischeva, P. H. Rosenberg, W. Wisden, and E. R. Korpi. 2007. TASK-3 knockout mice exhibit exaggerated nocturnal activity, impairments in cognitive functions, and reduced sensitivity to inhalation anesthetics. *J Pharmacol Exp Ther* 323(3):924–934.

Machu, T. K., and R. A. Harris. 1994. Alcohols and anesthetics enhance the function of 5-hydroxytryptamine(3) receptors expressed in Xenopus-laevis oocytes. *J Pharmacol Exp Ther* 271(2):898–905.

MacIver, M. B., A. A. Mikulec, S. M. Amagasu, and F. A. Monroe. 1996. Volatile anesthetics depress glutamate transmission via presynaptic actions. *Anesthesiology* 85:823–834.

Marota, J. J., G. Crosby, and G. R. Uhl. 1992. Selective effects of pentobarbital and halothane on c-fos and jun-B gene expression in rat brain. *Anesthesiology* 77(2):365–371.

Mennerick, S., V. Jevtovic-Todorovic, S. M. Todorovic, W. Shen, J. W. Olney, and C. F. Zorumski. 1998. Effect of nitrous oxide on excitatory and inhibitory synaptic transmission in hippocampal cultures. *J Neurosci* 26(23):9716–9726.

Miao, N., M. J. Frazer, and C. Lynch, 3rd. 1995. Volatile anesthetics depress $ca^{2+}$ transients and glutamate release in isolated cerebral synaptosomes. *Anesthesiology* 83(3):593–603.

Mihic, S. J., Q. Ye, M. J. Wick, V. V. Koltchine, M. D. Krasowski, S. E. Finn, M. P. Mascia, C. F. Valenzuela, K. K. Hanson, E. P. Greenblatt, R. A. Harris, and N. L. Harrison. 1997. Sites of alcohol and volatile anaesthetic action on GABA(A) and glycine receptors. *Nature* 389(6649):385–389.

Mikulec, A. A., S. Pittson, S. M. Amagasu, F. A. Monroe, and M. B. MacIver. 1998. Halothane depresses action potential conduction in hippocampal axons. *Brain Res* 796(1–2): 231–238.

Minami, K., R. W. Gereau, M. Minami, S. F. Heinemann, and R. A. Harris. 1998. Effects of ethanol and anesthetics on type 1 and 5 metabotropic glutamate receptors expressed in Xenopus laevis oocytes. *Mol Pharmacol.* 53(1):148–156.

Minami, K., M. Minami, and R. A. Harris. 1997. Inhibition of 5-hydroxytryptamine type 2A receptor-induced currents by n-alcohols and anesthetics. *J Pharmacol Exp Ther.* 281(3): 1136–1143.

Minami, K., T. W. Vanderah, M. Minami, and R. A. Harris. 1997. Inhibitory effects of anesthetics and ethanol on muscarinic receptors expressed in Xenopus oocytes. *Eur J Pharmacol.* 339(2–3):237–244.

Ouyang, W., and H. C. Hemmings, Jr. 2005. Depression by isoflurane of the action potential and underlying voltage-gated ion currents in isolated rat neurohypophysial nerve terminals. *J Pharmacol Exp Ther* 312(2):801–808.

Ouyang, W. 2007. Isoform-selective effects of isoflurane on voltage-gated $Na^+$ channels. *Anesthesiology* 107(1):91–98.

Ouyang, W., K. F. Herold, and H. C. Hemmings, Jr. 2009. Comparative effects of halogenated inhaled anesthetics on voltage-gated $Na^+$ channel function. *Anesthesiology* 110(3): 582–590.

Ouyang, W., T. Y. Jih, T. T. Zhang, A. M. Correa, and H. C. Hemmings, Jr. 2007. Isoflurane inhibits NaChBac, a prokaryotic voltage-gated sodium channel. *J Pharmacol Exp Ther* 322(3): 1076–1083.

Ouyang, W., G. Wang, and H. C. Hemmings. 2003. Isoflurane and propofol inhibit voltage-gated sodium channels in isolated rat neurohypophysial nerve terminals. *Mol Pharmacol* 64(2): 373–381.

Patel, A. J., and E. Honore. 2001. Anesthetic-sensitive 2P domain $K^+$ channels. *Anesthesiology* 95(4):1013–1021.

Patel, A. J., E. Honore, F. Lesage, M. Fink, G. Romey, and M. Lazdunski. 1999. Inhalational anesthetics activate two-pore-domain background $K^+$ channels. *Nat Neurosci* 2(5):422–426.

Perouansky, M., and H. C. Hemmings, Jr. 2009. Neurotoxicity of general anesthetics: Cause for concern? *Anesthesiology*, in press.

Perouansky, M., D. Baranov, M. Salman, and Y. Yaari. 1995. Effects of halothane on glutamate receptor-mediated excitatory postsynaptic currents. a patch-clamp study in adult mouse hippocampal slices. *Anesthesiology* 83(1):109–119.

Perouansky, M., H. C. Hemmings. 2003. Presynaptic actions of general anesthetics. In *Neural Mechanisms of Anesthesia*, edited by J. F. Antognini, E. E. Carstens, and D. E. Raines. Totowa, NJ: Humana Press.

Peterlin, Zita, Yumiko Ishizawa, Ricardo Araneda, Roderic Eckenhoff, and Stuart Firestein. 2005. Selective activation of G-protein coupled receptors by volatile anesthetics. *Mol Cell Neurosci* 30(4):506–512.

Petrenko, A. B., M. Tsujita, T. Kohno, K. Sakimura, and H. Baba. 2007. Mutation of alpha(1G) T-type calcium channels in mice does not change anesthetic requirements for loss of the righting reflex and minimum alveolar concentration but delays the onset of anesthetic induction. *Anesthesiology* 106(6):1177–1185.

Raines, D. E., R. J. Claycomb, M. Scheller, and S. A. Forman. 2001. Nonhalogenated alkane anesthetics fail to potentiate agonist actions on two ligand-gated ion channels. *Anesthesiology* 95(2):470–477.

Raines, D. E., and K. W. Miller. 1994. On the importance of volatile agents devoid of anesthetic action. *Anesth Analg* 79(6):1031–1033.

Rampil, I. J., P. Mason, and H. Singh. 1993. Anesthetic potency (MAC) is independent of forebrain structures in the rat. *Anesthesiology* 78(4):707–712.

Ratnakumari, L., and H. C. Hemmings. 1998. Inhibition of presynaptic sodium channels by halothane. *Anesthesiology* 88(4):1043–1054.

Ratnakumari, L., T. N. Vysotskaya, D. S. Duch, and H. C. Hemmings. 2000. Differential effects of anesthetic and nonanesthetic cyclobutanes on neuronal voltage-gated sodium channels. *Anesthesiology* 92(2):529–541.

Rebecchi, M. J., and S. N. Pentyala. 2002. Anaesthetic actions on other targets:protein kinase C and guanine nucleotide-binding proteins. *Br J Anaesth* 89(1):62–78.

Rehberg, B., Y. H. Xiao, and D. S. Duch. 1996. Central nervous system sodium channels are significantly suppressed at clinical concentrations of volatile anesthetics. *Anesthesiology* 84(5):1223–1233.

Reynolds, D. S., T. W. Rosahl, J. Cirone, G. F. O'Meara, A. Haythornthwaite, R. J. Newman, J. Myers, C. Sur, O. Howell, A. R. Rutter, J. Atack, A. J. Macaulay, K. L. Hadingham, P. H. Hutson, D. Belelli, J. J. Lambert, G. R. Dawson, R. McKernan, P. J. Whiting, and K. A. Wafford. 2003. Sedation and anesthesia mediated by distinct GABA(A) receptor isoforms. *J Neurosci* 23(24):8608–8617.

Robinson, R. B., and S. A. Siegelbaum. 2003. Hyperpolarization-activated cation currents: from molecules to physiological function. *Annu Rev Physiol* 65:453–480.

Role, L. W., and D. K. Berg. 1996. Nicotinic receptors in the development and modulation of CNS synapses. *Neuron* 16(6):1077–1085.

Rudolph, U., and B. Antkowiak. 2004. Molecular and neuronal substrates for general anaesthetics. *Nat Rev Neurosci* 5(9):709–720.

Schlame, M., and H. C. Hemmings, Jr. 1995. Inhibition by volatile anesthetics of endogenous glutamate release from synaptosomes by a presynaptic mechanism. *Anesthesiology* 82(6):1406–1416.

Semyanov, A., M. C. Walker, D. M. Kullmann, and R. A. Silver. 2004. Tonically active GABA(A) receptors: modulating gain and maintaining the tone. *Trends Neurosci* 27(5):262–269.

Shiraishi, M., and R. A. Harris. 2004. Effects of alcohols and anesthetics on recombinant voltage-gated Na+ channels. *J Pharmacol Exp Ther* 309(3):987–994.

Shumilla, Jennifer A., Sarah M. Sweitzer, Edmond I. Eger, II, Michael J. Laster, and Joan J. Kendig. 2004. Inhibition of spinal protein kinase C-{epsilon} or -{gamma} isozymes does not affect halothane minimum alveolar anesthetic concentration in rats. *Anesth Analg* 99(1):82–84.

Sirois, J. E., Q. Lei, E. M. Talley, C. Lynch, III, and D. A. Bayliss. 2000. The TASK-1 two-pore domain K+ channel is a molecular substrate for neuronal effects of inhalation anesthetics. *J Neurosci* 20(17):6347–6354.

Sirois, J. E., C. Lynch, III, and D. A. Bayliss. 2002. Convergent and reciprocal modulation of a leak K+ current and I(h) by an inhalational anaesthetic and neurotransmitters in rat brainstem motoneurones. *J Physiol* 541(Pt 3):717–729.

Snyder, G. L., S. Galdi, J. P. Hendrick, and H. C. Hemmings, Jr. 2007. General anesthetics selectively modulate glutamatergic and dopaminergic signaling via site-specific phosphorylation in vivo. *Neuropharmacology* 53(5):619–630.

Solt, K., E. I. Eger, and D. E. Raines. 2006. Differential modulation of human N-Methyl-Δ-aspartate receptors by structurally diverse general anesthetics. *Anesth Analg* 102(5):1407–1411.

Sonner, J. M., J. F. Antognini, R. C. Dutton, P. Flood, A. T. Gray, R. A. Harris, G. E. Homanics, J. Kendig, B. Orser, D. E. Raines, I. J. Rampil, J. Trudell, B. Vissel, and E. I. Eger. 2003. Inhaled anesthetics and immobility: mechanisms, mysteries, and minimum alveolar anesthetic concentration. *Anesth Analg* 97(3):718–740.

Sonner, J. M., D. Gong, J. Li, E. I. Eger, and M. J. Laster. 1999. Mouse strain modestly influences minimum alveolar anesthetic concentration and convulsivity of inhaled compounds. *Anesth Analg* 89(4):1030–1034.

Sonner, J. M., B. Vissel, G. Royle, A. Maurer, D. Gong, N. V. Baron, N. Harrison, M. Fanselow, and E. I. Eger. 2005. The effect of three inhaled anesthetics in mice harboring mutations in the GluR6 (kainate) receptor gene. *Anesth Analg* 101(1):143–148, table.

Stadnicka, A., W. M. Kwok, H. A. Hartmann, and Z. J. Bosnjak. 1999. Effects of halothane and isoflurane on fast and slow inactivation of human heart hH1a sodium channels. *Anesthesiology* 90(6):1671–1683.

Stowe, D. F., G. C. Rehmert, W. M. Kwok, H. U. Weigt, M. Georgieff, and Z. J. Bosnjak. 2000. Xenon does not alter cardiac function or major cation currents in isolated guinea pig hearts or myocytes. *Anesthesiology* 92(2):516–522.

Study, R. E. 1994. Isoflurane inhibits multiple voltage gated calcium currents in hippocampal pyramidal neurons. *Anesthesiology* 81(1):104–116.

Takei, T., H. Saegusa, S. Zong, T. Murakoshi, K. Makita, and T. Tanabe. 2003. Increased sensitivity to halothane but decreased sensitivity to propofol in mice lacking the N-type $Ca^{2+}$ channel. *Neurosci Lett* 350(1):41–45.

Todorovic, S. M., V. Jevtovic-Todorovic, S. Mennerick, E. Perez-Reyes, and C. F. Zorumski. 2001. Ca(v)3.2 channel is a molecular substrate for inhibition of T-type calcium currents in rat sensory neurons by nitrous oxide. *Mol Pharmacol* 60(3):603–610.

Topf, N., E. Recio-Pinto, T. J. Blanck, and H. C. Hemmings. 2003. Actions of general anesthetics on voltage-gated ion channels. In *Neural Mechanisms of Anesthesia*, edited by J. F. Antognini, E. E. Carstens, and D.E. Raines. Totowa, NJ: Humana Press.

Trudell, J. R., and E. Bertaccini. 2004. Comparative modeling of a GABA$_A$ alpha1 receptor using three crystal structures as templates. *J Mol Graph Model* 23(1):39–49.

Violet, J. M., D. L. Downie, R. C. Nakisa, W. R. Lieb, and N. P. Franks. 1997. Differential sensitivities of mammalian neuronal and muscle nicotinic acetylcholine receptors to general anesthetics. *Anesthesiology* 86(4):866–874.

Westphalen, R. I., R. S. Gomez, and H. C. Hemmings, Jr. 2009. Nicotinic receptor-evoked hippocampal norepinephrine release is highly sensitive to inhibition by isoflurane. *Br J Anaesth* 102(3):355–360.

Westphalen, R. I., and H. C. Hemmings. 2003. Selective depression by general anesthetics of glutamate versus GABA release from isolated cortical nerve terminals. *J Pharmacol Exp Ther* 304(3):1188–1196.

Westphalen, R. I., M. Krivitski, A. Amarosa, N. Guy, and H. C. Hemmings, Jr. 2007. Reduced inhibition of cortical glutamate and GABA release by halothane in mice lacking the K(+) channel, TREK-1. *Br J Pharmacol* 152(6):939–945.

Wick, M. J., S. J. Mihic, S. Ueno, M. P. Mascia, J. R. Trudell, S. J. Brozowski, Q. Ye, N. L. Harrison, and R. A. Harris. 1998. Mutations of gamma-aminobutyric acid and glycine receptors change alcohol cutoff: evidence for an alcohol receptor? *Proc Natl Acad Sci USA* 95(11): 6504–6509.

Wu, X. S., J. Y. Sun, A. S. Evers, M. Crowder, and L. G. Wu. 2004. Isoflurane inhibits transmitter release and the presynaptic action potential. *Anesthesiology* 100(3):663–670.

Yamakura, T., E. Bertaccini, J. R. Trudell, and R. A. Harris. 2001. Anesthetics and ion channels: molecular models and sites of action. *Ann Rev Pharmacol Toxicol* 41:23–51.

Yost, C. S. 1999. Potassium channels: basic aspects, functional roles, and medical significance. *Anesthesiology* 90(4):1186–1203.

Yu, F. H., and W. A. Catterall. 2004. The VGL-chanome: a protein superfamily specialized for electrical signaling and ionic homeostasis. *Sci STKE* 2004(253):re15.

Zarnowska, E. D., R. Keist, U. Rudolph, and R. A. Pearce. 2009. GABA$_A$ receptor alpha5 subunits contribute to GABA$_A$, slow synaptic inhibition in mouse hippocampus. *J Neurophysiol* 101(3):1179–1191.

Zecharia, A. Y., L. E. Nelson, T. C. Gent, M. Schumacher, R. Jurd, U. Rudolph, S. G. Brickley, M. Maze, and N. P. Franks. 2009. The involvement of hypothalamic sleep pathways in general anesthesia: testing the hypothesis using the GABA$_A$ receptor {beta}3N265M knock-in mouse. *J Neurosci* 29(7):2177–2187.

Zeilhofer, H. U., D. Swandulla, G. Geisslinger, and K. Brune. 1992. Differential effects of ketamine enantiomers on NMDA receptor currents in cultured neurons. *Eur J Pharmacol* 213(1): 155–158.

Zeller, A., R. Jurd, S. Lambert, M. Arras, B. Drexler, C. Grashoff, B. Antkowiak, and U. Rudolph. 2008. Inhibitory ligand-gated ion channels as substrates for general anesthetic actions. *Handb Exp Pharmacol* 182:31–51.

# Chapter 3
# A Neurochemical Perspective on States of Consciousness

**Christopher J. Watson, Helen A. Baghdoyan, and Ralph Lydic**

**Abstract** The foundation of anesthesia rests upon discoveries made by chemists. This is illustrated by Joseph Preistly's and Humphrey Davy's research on nitrous oxide; Paracelsus' work with ether; the studies of chemists Eugène Soubeiran, Justus von Liebig, Samual Gutherie, and Jean-Baptiste Dumas with chloroform; and more recently Paul Janssen's development of fentanyl. Anesthetics are synthetic compounds that exert effects on one or more endogenous neurochemical systems to produce a behavioral state characterized by traits that include unconsciousness, amnesia, and analgesia. The mechanisms by which anesthetics cause the loss of consciousness (hypnosis) are not known, but there is compelling evidence that anesthetics alter the endogenous neurochemical systems that regulate sleep and wakefulness (Keifer et al. 1994; Lydic 1996; Vanini et al. 2008). Advances in analytical chemistry now make it possible to begin a systematic characterization of the endogenous molecules that regulate states of consciousness such as sleep and anesthesia. This chapter provides a brief overview of sleep neurobiology and its unique relevance for efforts to understand the neurochemical mechanisms of anesthesia. Readers are referred elsewhere for detailed reviews on sleep (Lydic and Biebuyck 1994; España and Scammell 2004; Steriade and McCarley 2005; Datta and MacLean 2007; McCarley 2007; Monti, Pandi-Perumal, and Sinton 2008).

**Keywords** Sleep · anesthesia · neurochemistry · acetylcholine · GABA · adenosine · neuropeptide

## Defining States of Consciousness

The state of any system at any moment is defined by the numerical values of its set of variables at that time point (Hobson 1978). If all the variables for a system are known, it should be possible to produce an exact mathematical model of a

C.J. Watson (✉)
Department of Anesthesiology, University of Michigan, Ann Arbor, MI, USA
e-mail: watsoncj@umich.edu

A. Hudetz, R. Pearce (eds.), *Suppressing the Mind*, Contemporary Clinical
Neuroscience, DOI 10.1007/978-1-60761-462 3_3,
© Humana Press, a part of Springer Science+Business Media, LLC 2010

system, including its state at any moment. Mathematical models are an ultimate scientific goal because they have predictive power. For states of consciousness, the system can be modeled only in part because the neurochemical variables modulating states of sleep and wakefulness are incompletely understood. States of consciousness are defined by specific traits that include physiological and behavioral measures.

Polysomnography combines recordings of the electroencephalogram (EEG), electromyogram (EMG), and electrooculogram (EOG) to monitor activity of the cerebral cortex, skeletal muscle tone, and eye movements, respectively. A typical EEG contains waves at multiple frequencies that depend on the level of consciousness. EEG waves are grouped in frequency bands and defined as delta (1–4 Hz), theta (5–8 Hz), alpha (9–12 Hz), beta (13–26 Hz), and gamma (27–70 Hz) waves. Beta waves may be further classified as low beta (13–15 Hz), beta (16–18 Hz), or high beta (19–26 Hz) waves. Figure 3.1 shows how measures of EEG, EMG, and EOG are combined to objectively identify states of sleep and wakefulness.

Rechtschaffen and Kales used polysomnography to pioneer the systematic classification of sleep–wake states in 1968 (Rechtschaffen and Kales 1968), and their manual is considered the gold standard for human sleep classification. In 2007, the American Academy of Sleep Medicine updated the sleep classification system (Iber et al. 2007; Silber et al. 2007). Criteria for wakefulness (Fig. 3.1A) include alpha waves present in more than 50% of an epoch recorded from the occipital electrodes ($O_1$–$M_2$ and $O_2$–$M_1$; eighth and ninth trace from top on panels **A–E**). If alpha waves are not observed, the presence of 0.5–2 Hz eye blinks ($E_1$–$M_2$ and $E_2$–$M_2$; top two traces on panels **A–E**), eye movements associated with reading, or irregular eye movements combined with normal to high muscle tone (Chin1–Chin2; third trace from the top on panels **A–E**) are used to identify wakefulness. Classification of Stage 1 non-rapid eye movement (NREM) sleep (Fig. 3.1B) requires the reduction of alpha waves and a simultaneous increase in mixed-frequency (4–7 Hz), low-amplitude waves. Other indicators of Stage 1 NREM sleep include a greater than 1 Hz slowing of background frequencies as compared to wakefulness, V waves (vertex sharp waves with a duration less than 0.5 s), and slow, sinusoidal eye movements. K-complexes (unassociated with an arousal event) and trains of sleep spindles ($C_3$–$M_2$ or $C_4$–$M_1$; sixth and seventh traces from top in panels **A–E**) are hallmarks of Stage 2 NREM sleep (Fig. 3.1C). Stage 3 NREM sleep (Fig. 3.1D) is identified by the presence of delta waves in more than 20% of an epoch ($C_3$–$M_2$ or $C_4$–$M_1$). Requirements for scoring an epoch as rapid eye movement (REM) sleep (Fig. 3.1E) include the presence of a mixed-frequency, low-amplitude EEG that includes sawtooth waves (2–6 Hz, maximum amplitude measured from $C_3$–$M_2$ and $C_4$–$M_1$). Classification as REM sleep also requires low muscle tone (Chin1–Chin2) and rapid eye movements ($E_1$–$M_2$ and $E_2$–$M_1$).

The diseases for which biomedical research has made the most progress are those for which there are animal models. Sleep research has greatly benefited from the use of animal models because the traits that comprise states of animal sleep are similar to the traits defining the sleep of humans. For instance, during wakefulness (Fig. 3.2A) the EEG is characterized by a mixed-frequency, low-amplitude signal. The EMG

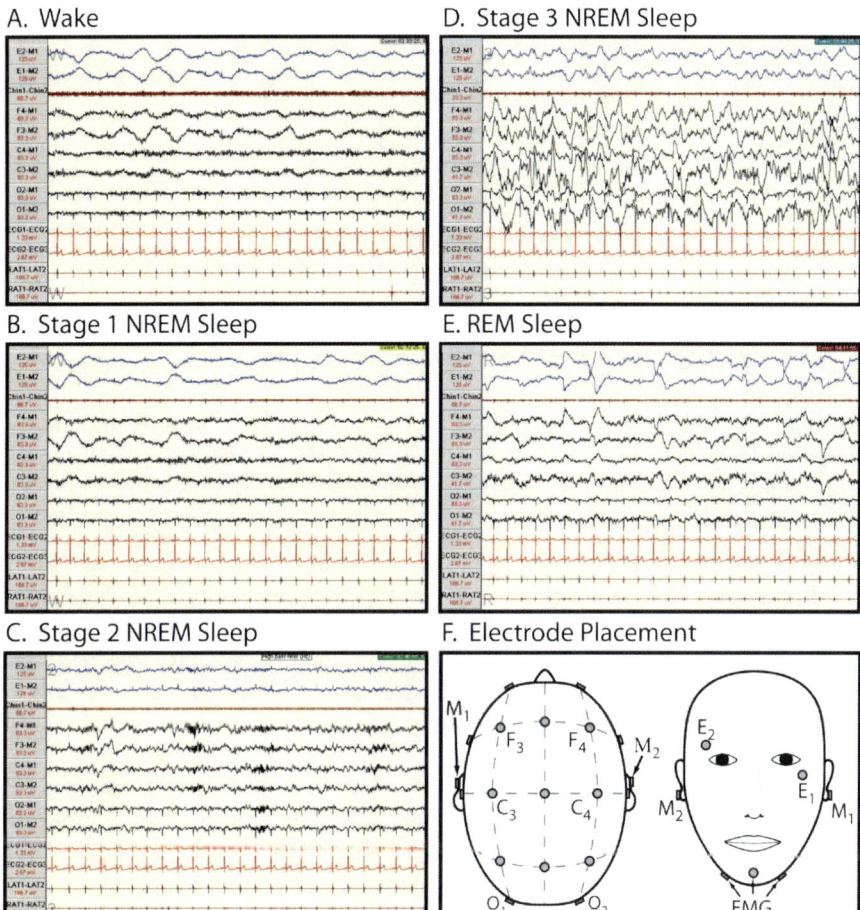

**Fig. 3.1** Polygraphic recording of human sleep. All polygraphic recordings are from a 25-year-old male in good health. Each panel (**A–E**) represents a 30 s recording. In each of these panels, the top two traces show EOG, the third trace shows EMG, and traces 4–9 show EEG. States of wakefulness (**A**), stage 1 NREM sleep (**B**), stage 2 NREM sleep (**C**), stage 3 NREM sleep (**D**), and REM sleep (**E**) were scored in 30 s epochs according to guidelines established by the American Academy of Sleep Medicine (Iber et al. 2007) (**F**). Diagrams adapted from the AASM Sleep Scoring Manual (Iber et al. 2007) show the placement of electrodes for recording arousal state. Polygraphic recordings (**A–E**) were graciously provided by Dr. Naricha Chirakalwasan of the University of Michigan Department of Neurology

trace displays high amplitude and spiking, which correspond to high muscle tone and movement. The EOG (data not shown) shows the presence of conjugate eye movements. NREM sleep (Fig. 3.2B) is characterized by high-amplitude waves in the delta frequency range. No eye movements are observed, and breathing rate is slow and regular. EMG amplitude during NREM sleep is lower than during wakefulness. During the REM phase of sleep (Fig. 3.2C), there is an activated cortical

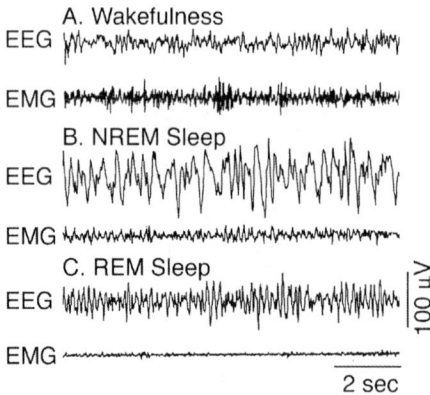

**Fig. 3.2** Polygraphic recording of rodent sleep. EEG and EMG traces recorded from a mouse during states of wakefulness (**A**), NREM sleep (**B**), and REM sleep (**C**). Each representative sample is 10 s in length. See text for details. Recordings provided kindness of Professor Subimal Datta of the Boston University Department of Psychiatry and Behavioral Neuroscience

EEG characterized by low amplitude and a dominant frequency in the theta range. REM sleep is also referred to as "active" or "paradoxical" sleep. Muscle atonia, a classic trait of REM sleep, appears as a flat line on the EMG trace, and eye movements are present. Studies using animals have demonstrated that wakefulness, NREM sleep, and REM sleep are actively generated by anatomically distributed neuronal networks that are neurochemically diverse.

Modified EEG recordings are now used as a method of monitoring and predicting anesthetic depth. Perhaps the most clinically tested monitor is the bispectral index (BIS) monitor. BIS® monitoring uses an algorithm that rejects identified artifacts in the EEG and then incorporates analyses of numerous variables contained within the EEG recording (including measures of burst suppression, power spectrum analysis, bispectrum analysis of frequencies in the delta and theta ranges, the synchronicity of fast and slow waves, and a ratio of frequencies in the beta range) to generate a dimensionless number between 0 and 100 (Johansen 2006; Mashour 2006). The EEG signal may also be processed by the Narcotrend monitor®. The Narcotrend stage (A–F) or index (100–0) is calculated by artifact rejection followed by a unique algorithm that analyzes variables that include entropy measures as well as autoregressive and spectral parameters (Kreuer 2006). As with the BIS monitor, the Narcotrend monitor is approved by the FDA. Another index for monitoring anesthetic depth is the SNAP monitor®. The following steps are used to generate a SNAP index value: (1) the EEG wave is subjected to fast Fourier transformation, (2) data in the high- (80–420 Hz) and low- (0.1–40 Hz) frequency range are weighted (0–1 and 0–100, respectively), (3) the resulting values for each frequency range are multiplied together, and (4) the resulting number is subtracted from 100 (Bischoff 2006). The Patient State Index or SedLine® monitor analyzes the EEG using a proprietary algorithm that incorporates analysis of the power of various frequency bands, alterations in synchronization across brain regions, and frontal cortex inhibition or activation (Drover 2006). Each of the above monitors uses complex algorithms to rapidly process a raw EEG signal into a value that is simple to understand and correlates to the anesthetic state of the patient. Currently, these monitors provide information about the past state of the patient. Ideally, anesthetic monitors will evolve to a point where they have predictive power.

In spite of advances in anesthetic pharmacology, the commonly accepted traits used to define states of anesthesia have not changed much in 50 years (Woodbridge 1957; Orser 2007). Anesthesia is defined by unconsciousness (also called hypnosis), analgesia, immobility, blunted autonomic responses, and amnesia. Amnesia is a key trait because of the need to suppress intraoperative awareness. Current use of amnestic drugs does not always eliminate the possibility of psychological damage incurred from intraoperative awareness (Hudetz 2008).

Anesthetics were initially thought to produce a sleep-like state, and anesthesiologists tell their patients they will "go to sleep" because this is a comforting metaphor (Shafer 1995). To the casual observer, for limited intervals of time, states of sleep and anesthesia may be indistinguishable. Unconsciousness resulting from sleep, anesthesia, vegetative states, or coma displays similar decreases in brain metabolism (Baars, Ramsoy, and Laureys 2003), indicating similar underlying mechanisms. Although sleep and anesthesia display some similar traits, they are, in fact, very different states. Figure 3.3 describes some similarities and differences between sleep and anesthesia.

## Relevance of Sleep for Anesthesia

The present volume and recent reviews (Hobson 2002; Lydic and Baghdoyan 2005; Mashour 2008; Laureys and Tononi 2009) document the now established view that sleep is relevant for anesthesia. Anesthetics induce or suppress immediate early gene expression in brain regions known to regulate pain, sleep, and wakefulness (Lu et al. 2008) and act at receptors for neurotransmitters that regulate sleep and wakefulness (Campagna, Miller, and Forman 2003; Alkire, Hudetz, and Tononi 2008; Franks 2008). Obstructive sleep apnea (OSA) occurs when the pharyngeal airway repetitively narrows or collapses during sleep, causing recurrent awakenings throughout the night (Eckert and Malhotra 2008; Punjabi 2008). OSA is classified as a sleep disorder because these patients breathe normally when they are awake. Anesthesiologists are now well aware that OSA patients have a greater risk of perioperative respiratory complications [for review, see (Chung, Yuan, and Chung 2008)]. Risk factors associated with OSA include male gender, obesity, advanced age, craniofacial anatomy, genetic predisposition, menopause, race, smoking, and alcohol ingestion [reviewed in (Eckert and Malhotra 2008; Punjabi 2008)].

Advances in sleep neurobiology and sleep disorders medicine are also directly relevant for efforts to elucidate anesthetic mechanisms. The sleep disorder narcolepsy, which presents with excessive daytime sleepiness and disrupted nighttime sleep, results from defective hypocretinergic neurotransmission (Nishino and

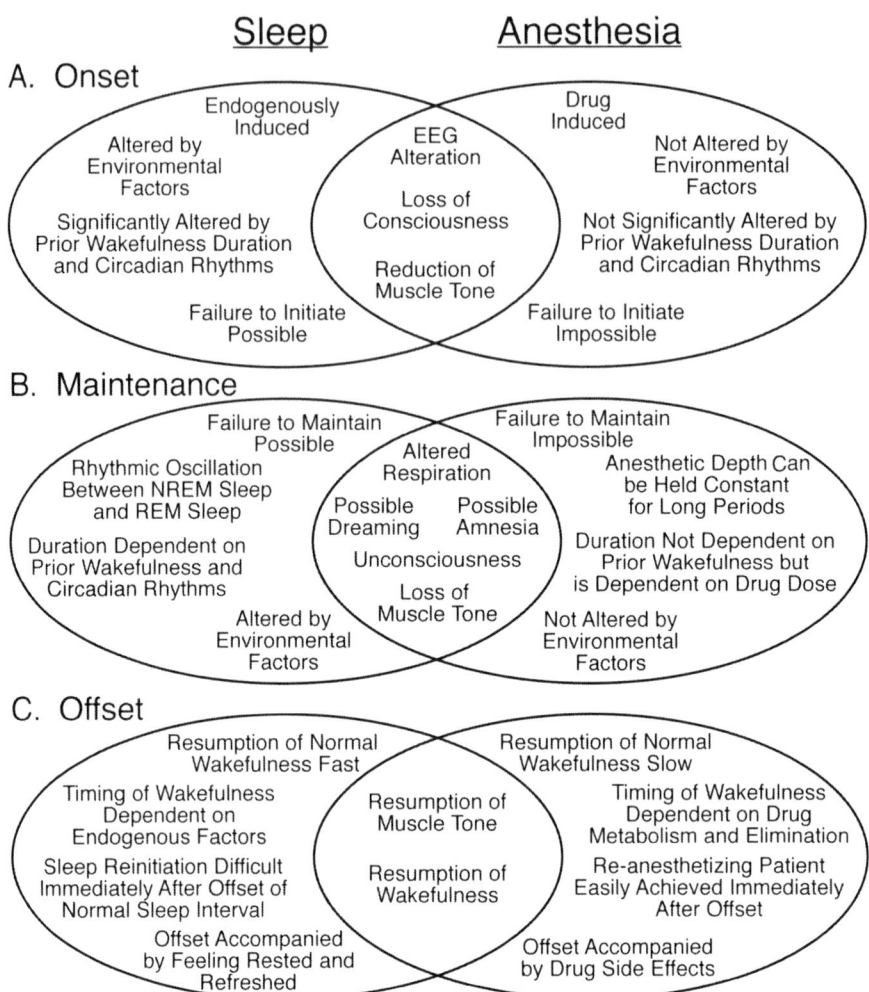

## Sleep                    Anesthesia

### A. Onset

Endogenously Induced

Altered by Environmental Factors

Significantly Altered by Prior Wakefulness Duration and Circadian Rhythms

Failure to Initiate Possible

EEG Alteration

Loss of Consciousness

Reduction of Muscle Tone

Drug Induced

Not Altered by Environmental Factors

Not Significantly Altered by Prior Wakefulness Duration and Circadian Rhythms

Failure to Initiate Impossible

### B. Maintenance

Failure to Maintain Possible

Rhythmic Oscillation Between NREM Sleep and REM Sleep

Duration Dependent on Prior Wakefulness and Circadian Rhythms

Altered by Environmental Factors

Altered Respiration

Possible Dreaming    Possible Amnesia

Unconsciousness

Loss of Muscle Tone

Failure to Maintain Impossible

Anesthetic Depth Can be Held Constant for Long Periods

Duration Not Dependent on Prior Wakefulness but is Dependent on Drug Dose

Not Altered by Environmental Factors

### C. Offset

Resumption of Normal Wakefulness Fast

Timing of Wakefulness Dependent on Endogenous Factors

Sleep Reinitiation Difficult Immediately After Offset of Normal Sleep Interval

Offset Accompanied by Feeling Rested and Refreshed

Resumption of Muscle Tone

Resumption of Wakefulness

Resumption of Normal Wakefulness Slow

Timing of Wakefulness Dependent on Drug Metabolism and Elimination

Re-anesthetizing Patient Easily Achieved Immediately After Offset

Offset Accompanied by Drug Side Effects

**Fig. 3.3** Differences and similarities between sleep and anesthesia. Venn diagrams illustrate the similarities (intersections of sleep and anesthesia ellipses) occurring during the onset (**A**), maintenance (**B**), and offset (**C**) of both sleep and anesthesia. Traits lying outside of the intersection are unique to sleep or anesthesia

Sakurai 2005). Evidence from preclinical studies shows that hypocretin decreases recovery time from anesthesia, but does not alter anesthesia induction time (Kelz et al. 2008). These results support the interpretation that the mechanisms underlying anesthetic induction and emergence are not the same. From a patient's perspective, the goal of anesthesia is to prevent the perception of pain. Many reviews have been published regarding the link between pain and sleep (Menefee et al. 2000; Moldofsky 2001; Smith and Haythornthwaite 2004; Roehrs and Roth 2005; Lydic and Baghdoyan 2007), and there is evidence that hypocretin, which is thought to promote wakefulness, has analgesic effects on acute and chronic pain models (Bingham et al. 2001; Yamamoto, Nozaki-Taguchi, and Chiba 2002; Mobarakeh et al. 2005; Watanabe et al. 2005) and antiallodynic effects on a postoperative pain model (Cheng et al. 2003).

## A Brief History of Sleep Neurobiology

Polysomnography began with the first recordings of brain electrical activity in the late 1800s (Caton 1875; Walker 1998). In 1875, Caton observed fluctuations in the potentials measured from electrodes placed on the gray matter of rabbit and monkey brains. He determined that these fluctuations originated from brain tissue (Walker 1998) and were not due to cardiac or respiratory rhythms. Fifty years ensued before systematic reports on human brain electrical activity appeared. In 1929, Hans Berger described changes in brain electrical activity (Berger 1929) that could be recorded when a patient transitioned from wakefulness (which appeared as a low-voltage, high-frequency recording) to sleep (which appeared as a high-voltage, low-frequency recording). Berger was the first to use the term encephalography for recording cortical brain activity. In the mid-to-late 1930s, a series of systematic studies further explored information contained in an EEG (Loomis, Harvey, and Hobart 1935, 1935, 1937, 1938). Loomis and colleagues distinguished periods of wakefulness, sleep, and dreaming based on EEG characteristics. These studies identified five types or stages of sleep (categorized as A–E) and coined the terms "K-complex" and "spindles." Although EEG analysis was refined by these studies, it was not until the mid-1950s that the REM stage of sleep was described in humans (Aserinsky and Kleitman 1953). Subsequent studies associated the psychological experience of dreaming with the physiological characteristics of REM sleep (Dement and Kleitman 1957). Shortly thereafter, Dement (1958) identified a state of activated sleep in cats characterized by muscle atonia and a low-voltage, high-frequency EEG. Jouvet and coworkers then described a similar sleep state in cats (Jouvet and Michel 1959; Jouvet, Michel, and Courjon 1959, 1959). Jouvet subsequently named this state paradoxical sleep, referring to the paradox that although the EEG displayed the low-voltage, fast frequency pattern of wakefulness, the animal was asleep (Jouvet 1965). Thus, the terms REM sleep, activated sleep, and paradoxical sleep are all names for the same state.

## *Sleep–Wake Regulating Brain Regions Identified by Lesion and Stimulation Studies*

Von Economo provided one of the first clinical reports linking specific brain regions to sleep regulation (von Economo 1930). In postmortem brains of encephalitis lethargica patients who suffered from insomnia, von Economo discovered that lesions were localized to the preoptic area and the anterior hypothalamus. These findings suggested that one or both of these areas play an important role in generating and/or maintaining sleep. In patients presenting with hypersomnia, postmortem exams revealed lesions of the posterior hypothalamus, suggesting that this area contributes to generating or maintaining wakefulness. These findings were later verified by preclinical studies demonstrating that lesioning the anterior hypothalamus (Nauta 1946) or the basal forebrain and preoptic area (McGinty and Sterman 1968; Sallanon et al. 1989) caused insomnia. Preclinical studies also showed that lesioning the posterior hypothalamus produced hypersomnia (Ranson 1939; Nauta 1946; Sweet and Hobson 1968; McGinty 1969). More recently, it has been shown that damaging the lateral hypothalamus causes an increase in both NREM sleep and REM sleep and a decrease in wakefulness (Gerashchenko et al. 2003; Gerashchenko and Shiromani 2004). These data indicate that the lateral hypothalamus promotes wakefulness.

Early brain transection and stimulation studies revealed that nuclei within the pontine brainstem contribute to the generation of sleep and wakefulness. Through a series of studies, Moruzzi and Magoun demonstrated that electrical stimulation of specific brainstem regions produced an "activated" EEG (Moruzzi and Magoun 1949; Moruzzi 1964). Jouvet showed specific alterations in sleep–wake states were caused by transections at different levels of the brainstem (Jouvet 1962). These studies suggested that REM sleep-generating neurons were localized to the pons (Jouvet 1962). This finding has since been supported by a number of human case studies linking pontine injuries to REM sleep behavior disorder (Kimura et al. 2000) or the absence of REM sleep (Feldman 1971; Markland and Dyken 1976; Lavie et al. 1984; Autret et al. 1988; Valldeoriola et al. 1993; Gironell et al. 1995). With refinements in technique, Jouvet demonstrated that extensive lesions of the raphé "complex" caused long-lasting insomnia (Jouvet 1969). Jouvet interpreted these results to suggest that some component of the raphé "complex" is a sleep-promoting brain region. However, recording of dorsal raphé neuron discharge overturned this interpretation and revealed that dorsal raphé neurons discharge during wakefulness and cease firing during REM sleep, which suggests that they are wakefulness promoting (McGinty and Harper 1976; Trulson and Jacobs 1979; Lydic, McCarley, and Hobson 1983).

Selective, localized lesions of smaller regions of the pons have revealed specific REM sleep-regulating nuclei. For instance, two studies suggested that the laterodorsal tegmental (LDT) and pedunculopontine tegmental (PPT) nuclei of the pons play a role in REM sleep generation and/or maintenance. Lesioning of cholinergic LDT/PPT neurons decreases REM sleep proportional to the extent of cholinergic cell loss (Webster and Jones 1988), and electrical stimulation of the LDT increases REM sleep (Thakkar, Portas, and McCarley 1996). Yet, another study found that a unilateral lesion to the locus coeruleus produces a 50% increase in REM sleep (Caballero and de Andrés 1986). Together, these early studies demonstrated that many brain regions contribute to the regulation of sleep and wakefulness.

## *Single Cell Recordings Reveal Brain Region Activity During Wakefulness and Sleep*

By using electrodes to record the discharge of individual neurons in specific brain regions, it became possible to derive a cellular understanding of sleep and wakefulness. If a neuron plays a role in controlling a particular state, one would anticipate that the discharge of that neuron will vary across the sleep–wake cycle (Hobson 1978). Electrophysiological studies have identified multiple types of neuronal discharge activity in relation to states of sleep and wakefulness. One pattern is called Wake-On/REM-Off. These neurons display maximal firing during active wakefulness, a decreased firing during NREM sleep, and an absence of firing during REM sleep. Another group of neurons shows a REM-On firing pattern. REM-On neurons do not fire during wakefulness, show some firing activity during NREM sleep, and show the highest rate of firing during REM sleep. A third type of neuronal firing pattern is Wake-On/REM-On. An additional firing pattern is NREM-On, which is characterized by minimal activity during wakefulness and REM sleep and maximal firing rates during NREM sleep. Table 3.1 summarizes the firing patterns of major sleep- and wakefulness-regulating brain regions.

Locus coeruleus neurons predominantly use norepinephrine for signaling, and multiple studies have shown that locus coeruleus neurons exhibit a Wake-On/REM-Off discharge profile (Chu and Bloom 1973; Hobson, McCarley and Wyzinski 1975; McCarley, and Hobson 1975; Foote, Aston-Jones, and Bloom 1980; Jacobs 1986). Recordings across the sleep–wake cycle show that locus coeruleus neurons also increase discharge rate just prior to resumption of wakefulness (Aston-Jones and Bloom 1981). Together, these data suggest that neurons in the locus coeruleus inhibit REM sleep and also promote wakefulness. Neurons in the tuberomammillary nucleus of the posterior hypothalamus also display a Wake On discharge profile (Vanni-Mercier, Sakai, and Jouvet 1984; Ko et al. 2003; Takahashi, Lin, and Sakai 2006). Coupled with the studies which showed that lesions of the posterior hypothalamus result in hypersomnolence (von Economo 1930; Ranson 1939; Nauta

**Table 3.1**  State-dependent neuronal discharge patterns

| Discharge profile | Cell location |
| --- | --- |
| Wake-On/Rem-Off | basal forebrain; dorsal raphé nucleus; lateral hypothalamus; locus coeruleus; posterior hypothalamus |
| Wake-On/Rem-On | LDT/PPT; basal forebrain; dorsal raphé nucleus; midbrain reticular formation; suprachiasmatic nucleus |
| REM-On | hypothalamic preoptic area; LDT/PPT |
| NREM-On | anterior hypothalamus; basal forebrain; cortex; hypothalamic area |
| State Independent | ventral tegmental area; substantia nigra |

This table provides a summary of neuronal discharge patterns measured in brain regions known to regulate states of sleep and wakefulness. See text for details.

1946; Sweet and Hobson 1968; McGinty 1969), these data support a wakefulness-promoting role for the posterior hypothalamus. Other cell groups that have the highest discharge activity during wakefulness are located in the lateral hypothalamus (Steininger et al. 1999; Alam et al. 2002; Koyama et al. 2002; Lee, Hassani, and Jones 2005; Takahashi, Lin, and Sakai 2008), dorsal raphé nucleus (McGinty and Harper 1976; Trulson and Jacobs 1979; Lydic, McCarley, and Hobson 1983), and basal forebrain (Alam et al. 1999).

One population of basal forebrain neurons exhibits its highest firing frequency during wakefulness and REM sleep (Szymusiak and McGinty 1986). Cell groups that demonstrate a distinct Wake-On/REM-On discharge profile include neurons of the LDT/PPT (Thakkar, Strecker, and McCarley 1998), the midbrain reticular formation (Manohar, Noda, and Adey 1972; Steriade, Oakson, and Ropert 1982), a subpopulation of dorsal raphé neurons (Urbain, Creamer, and Debonnel 2006), and the suprachiasmatic nucleus (Deboer et al. 2003).

The neuronal discharge rate of most brain regions has not been studied across states of consciousness. Recordings to date note that cell groups firing with a NREM-On discharge pattern are not as numerous as the cell groups displaying a Wake-On firing pattern. One group of neurons known to have a NREM-On discharge pattern has been localized to the preoptic area of the anterior hypothalamus of rats (Lincoln 1969; Koyama and Hayaishi 1994; Szymusiak et al. 1998; Suntsova et al. 2002), cats (Kaitin 1984), and rabbits (Findlay and Hayward 1969). This firing pattern fits well with studies demonstrating that lesions of the preoptic area and anterior hypothalamus induce insomnia (von Economo 1930; Nauta 1946; McGinty and Sterman 1968; Sallanon et al. 1989) and supports the interpretation that these brain regions play a role in sleep generation and maintenance. The basal forebrain of rats (Alam et al. 1999) and cats (Szymusiak and McGinty 1986) also contains a small population of neurons that fire with a NREM-On discharge profile, indicating that some basal forebrain neurons are sleep promoting. Another technique, immediate early gene FOS expression, has identified a sleep-active subpopulation of neurons in the cortex of three separate species that is correlated to slow-wave activity and NREM sleep delta energy (Gerashchenko et al. 2008). This suggests that a subpopulation of neurons within the cortex may also be NREM-On neurons.

## Neurochemical and Neuropharmacological Techniques Provide Unique Insights into States of Consciousness

Advances in chemical neuroanatomy and neurochemistry have made it possible to identify the functional roles of neurotransmitters in brain regions regulating sleep and wakefulness. General anesthetics produce their effects, in part, by binding to receptors for GABA, glutamate, and acetylcholine (Campagna, Miller, and Forman 2003; Alkire, Hudetz, and Tononi 2008; Franks 2008; Van Dort, Baghdoyan, and Lydic 2008). This section discusses the techniques used to sample or manipulate these and other endogenous ligands and reviews some of the neurochemicals that participate in generating sleep and wakefulness.

## Sampling and Delivery Techniques

Microdialysis [reviewed in (Watson, Venton, and Kennedy 2006)] is a sampling technique that allows the simultaneous collection of multiple analytes from the same brain region using intact, behaving animals. Samples collected with microdialysis can be analyzed with a variety of techniques such as high performance liquid chromatography (HPLC), capillary electrophoresis (CE), and enzyme assays. This means that virtually any molecule collected using microdialysis has the potential to be quantified with a high degree of selectivity and sensitivity. A typical microdialysis probe has an inlet port, outlet port, probe shaft, and sampling tip for analyte collection. The sampling tip (Fig. 3.4A) is encompassed by a semi-permeable membrane that allows molecules (illustrated by dark circles) within a given kDalton range (ions, neurotransmitters, and neuropeptides) to diffuse across the membrane while blocking the diffusion of larger molecules (proteins). For sample collection, a vehicle solution with an ionic composition similar to cerebral spinal fluid is perfused through the sampling tip (arrows moving out of probe). As the solution passes through the sampling tip, neurochemicals in the vicinity of the probe diffuse through the membrane down a concentration gradient and are then swept out of the sampling tip to the outlet of the microdialysis probe for analysis. Drugs of interest can also be delivered into a given brain region by adding the drug to the vehicle dialysis solution. In this instance, when the vehicle solution containing a drug passes through the sampling tip, the drug diffuses out of the probe into the surrounding brain tissue. Currently, a typical microdialysis probe has a sampling area that is 1–4 mm long and 0.24 mm in diameter. The size of the sampling area can be a limitation in two ways. First, some brain regions involved in controlling states of consciousness are smaller than this sampling area. Second, larger brain areas may have heterogeneous functionality that would be masked by the spatial resolution of a microdialysis probe. Methods of addressing both of these issues work by improving the spatial resolution of the sampling probe.

Direct sampling of neurotransmitters (not shown) is an exciting direction for neurochemistry (Kennedy, Thompson, and Vickroy 2002). For this method, an experimenter inserts a single fused silica capillary tube into the brain region of interest. A small vacuum is applied to the tube so that fluid is removed at a flow rate of 1 nL/min. The filled capillary is connected to a capillary electrophoresis instrument for neurochemical analysis. Direct sampling offers at least a 500-fold improvement in spatial resolution as compared to microdialysis (Kennedy, Thompson, and Vickroy 2002). Another method for measuring endogenous neurotransmitters utilizes low-flow, push–pull perfusion offline (Kottegota, Shaik, and Shippy 2002) or online (Cellar et al. 2005) analysis of amino acid neurotransmitters. Offline push–pull perfusion probes (Fig. 3.4B) use a syringe pump to infuse (outer tube) brain tissue with Ringer's solution and a vacuum pump to remove fluid (inner tube) from the same region. Using this technology, flow rates of 10–50 nL/min have been used to successfully sample endogenous brain chemicals (Kottegota, Shaik, and Shippy 2002). As with microdialysis, fluid collected via push–pull perfusion can be analyzed using a variety of techniques. Online analysis allows sampling at

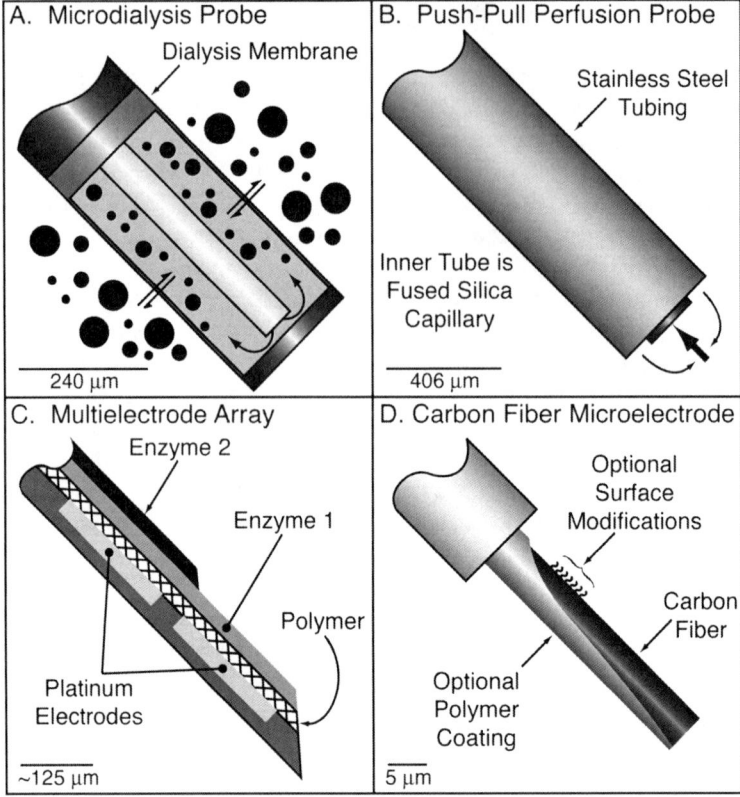

**Fig. 3.4** Techniques for monitoring neurochemicals in vivo. Understanding the neurochemical regulation of states of consciousness requires quantification of state-dependent changes in the release of neurotransmitters and neuromodulators. Each of the four panels illustrates a technique used to measure neurotransmitters. Scale bars in the lower left corner of each box represent the overall diameter and/or thickness of the corresponding sampling device and do not correspond to the length of the sampling device or the thickness of individual layers. The overall thickness for each device was taken from Burmeister, Moxon, and Gerhardt (2000); Kottegota, Shaik, and Shippy (2002); Watson, Lydic, and Baghdoyan (2007); and Hermans et al. (2008). A detailed description of each sampling technique is provided in the text

50 nL/min and offers a 65-fold improvement in spatial resolution as compared to a microdialysis probe with a 4 mm long sampling area (Cellar et al. 2005).

Improvements in temporal resolution also will provide new insights into regulating states of consciousness. Typical HPLC separations require more than 10 μL of sample solution to be injected for analysis. At a flow rate of 2 μL/min, this means that the temporal resolution is greater than 5 min. Because neurochemical signaling is known to occur on a much shorter timescale, valuable information is likely not obtained using in vivo microdialysis and HPLC. Two approaches have been developed to address this issue. Coupling a microdialysis probe online to

a capillary electrophoresis instrument allows separation and detection of 6 analytes every 15 s (Smith et al. 2004) or over 60 analytes every minute (Shou et al.
2006). These techniques have a temporal resolution, or the ability to respond to step
changes in concentration, on the order of 25 to 160 s (depending on the flow rate
through the microdialysis probe). Using segmented flow, the temporal resolution
of microdialysis-based systems has been shortened to 15 s (Wang et al. 2008). If
such measurements can be coupled to the direct or peristaltic push–pull sampling
systems, it will be possible for the first time to simultaneously monitor state-
dependent changes in multiple transmitters localized in arousal-regulating brain
regions. These emerging developments in analytical chemistry will significantly
advance the mechanistic understanding of sleep and anesthesia.

Biosensors (see Fig. 3.4C,D) provide another method of monitoring neurochemicals in vivo [for a review of electrochemical sensors, see (Phillips and Wightman
2003; Robinson et al. 2008)]. A biosensor utilizes electrochemistry to detect the
analyte of interest directly, as with nitric oxide (Barbosa et al. 2008) and dopamine
(Hermans et al. 2008; Johnson et al. 2008), or utilizes enzymes to break down the
analyte of interest and then detects electroactive by-products, as is the case with
acetylcholine (Bruno et al. 2006; Burmeister et al. 2008) or glutamate (Hascup et al.
2008). Biochemical sensors typically have a polymer exclusion layer composed of
nafion (to repel anionic compounds), $m$-polyphenylene, or $o$-phenylenediamine, all
of which act as size exclusion filters (Bruno et al. 2006; Barbosa et al. 2008; Hascup
et al. 2008). Depending on the analyte being monitored, the use of these coatings
greatly enhances the selectivity of the sensor. Figure 3.4C schematizes microelectrode arrays (MEAs) that are stereotaxically placed within a brain region of interest.
MEAs use a self-referencing technique and a polymer layer to achieve analyte selectivity and enzyme layers for analyte detection and quantification. Two common
MEAs measure glutamate (Burmeister et al. 2002; Hascup et al. 2008) and acetylcholine (Bruno et al. 2006; Burmeister et al. 2008). Glutamate MEAs (Burmeister
et al. 2002; Hascup et al. 2008) contain pairs of electrodes. One electrode is
coated with glutamate oxidase (immobilized in a matrix) to convert glutamate to α-
ketoglutarate, ammonia, and hydrogen peroxide ($H_2O_2$). The $H_2O_2$ diffuses through
the polymer layer to the platinum electrode, which is held at constant voltage. The
current changes due to oxidation of $H_2O_2$ are proportional to the amount of glutamate present. The second electrode (sentinel electrode) is coated with the matrix
material only. Current originating from this electrode is subtracted from the current
signal generated from the glutamate electrode to give a current resulting only from
the presence of glutamate. Acetylcholine MEAs (Burmeister et al. 2002; Hascup
et al. 2008) also use pairs of electrodes coated with a polymer layer for enhanced
selectivity. The sentinel electrode is covered with choline oxidase (enzyme layer 1),
which converts choline to betaine and $H_2O_2$. The acetylcholine electrode is coated
first with choline oxidase and then with acetylcholine esterase (enzyme layer 2).
Acetylcholinesterase converts acetylcholine into choline and acetate. To obtain the
acetylcholine-specific current, the current at the sentinel electrode is subtracted from
the current at the acetylcholine electrode.

Carbon fiber electrodes (Fig. 3.4D) are used to detect electroactive species such as dopamine (Hermans et al. 2008). Dopamine detection is typically performed using fast scan cyclic voltammetry. With this technique, the potential applied to the carbon fiber is swept between a specific range of potentials in a triangular waveform while the resulting current is monitored. Dopamine oxidizes to dopamine-$o$-quinone during the anodic scan, and dopamine-$o$-quinone is reduced back to dopamine during the cathodic scan. The resulting cyclic voltammogram is characteristic of the compound being analyzed. The selectivity of the electrode can be enhanced with polymer coatings or chemical modifications to the surface of the electrode.

Biochemical-based sensors and multiple electrode arrays have a much better temporal resolution (around ms to 1 s) than microdialysis sampling (Burmeister et al. 2002; Burmeister et al. 2008) and provide an improvement in spatial resolution. For example, a biosensor uses four platinum electrodes, each of which is 50 $\mu$m wide by 150 $\mu$m long for glutamate (Burmeister et al. 2002) and 15 $\mu$m wide by 333 $\mu$m long for acetylcholine (Burmeister et al. 2008). The main drawbacks of biosensors are their ability to monitor only one compound at a time and the fact that they are available for only a few bioactive compounds (Watson, Venton, and Kennedy 2006). Even with these limitations, biosensors provide an exciting frontier for sleep and anesthesia research. The subsequent paragraphs highlight neurotransmitters and neuromodulators that contribute to the regulation of states of behavioral arousal. Figure 3.5 summarizes neurochemicals used by and measured in brain regions known to regulate sleep and wakefulness.

## *Acetylcholine*

Cholinergic signaling originating from the LDT/PPT (Woolf and Butcher 1986, 1989; Jones and Beaudet 1987; Lydic and Baghdoyan 1993) and the basal forebrain (Woolf, Eckenstein, and Butcher 1984) modulates arousal (Hernandez-Peon et al. 1963; George, Haslett, and Jenden 1964; Jasper and Tessier 1971; Baghdoyan, Monaco et al. 1984; Baghdoyan, Rodrigo-Angulo et al. 1984; Szymusiak and McGinty 1986; Baghdoyan et al. 1987; Day, Damsma, and Fibiger 1991; Kurosawa et al. 1993; Bourgin et al. 1995; Marrosu et al. 1995; Thakkar, Portas, and McCarley 1996; Giovannini et al. 1998; Kshatri, Baghdoyan, and Lydic 1998; Thakkar, Strecker, and McCarley 1998; Alam et al. 1999; Baghdoyan and Lydic 1999; Materi, Rasmusson, and Semba 2000; Sarter and Bruno 2000; Vazquez and Baghdoyan 2001; Lydic, Douglas, and Baghdoyan 2002; Coleman, Lydic, and Baghdoyan 2004). Cholinergic neurotransmission occurs via muscarinic and nicotinic receptors. Neuronal nicotinic acetylcholine receptors are ligand-gated cation channels comprised of five transmembrane protein subunits selected from a pool of 12 possible subunits, designated $\alpha 2$–$\alpha 10$ and $\beta 2$–$\beta 4$ (Hogg, Raggenbass, and Bertrand 2003). Although nicotinic receptors play a role in arousal state control, a majority of research investigating cholinergic regulation of arousal state focuses on muscarinic receptors. Muscarinic receptors are G-protein coupled and five subtypes

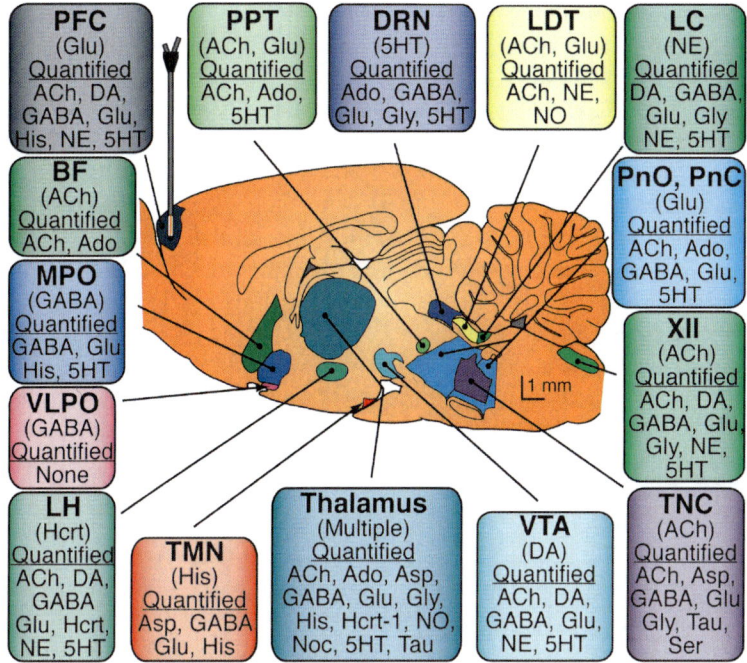

**Fig. 3.5** Sagittal schematic of the rat brain [modified from (Paxinos and Watson 2007)] indicating the location, shape, and size of multiple sleep-related brain regions (collapsed to 0.9 mm lateral to the midline). Bold print denotes the name of the brain region, parentheses denote main neurotransmitter used for signaling to other brain regions, and "quantified" denotes arousal state modulating neurochemical analytes that have been measured in that brain region. The microdialysis probe is drawn to scale and is shown sampling from the prefrontal cortex. Note that at a 1 mm active length (as shown), the microdialysis probe is larger than many arousal-regulating brain regions and indicates a need for improvements in spatial resolution. Abbreviations: XII – hypoglossal nucleus; BF – basal forebrain; DRN – dorsal raphé nucleus; LC – locus coeruleus; LDT – laterodorsal tegmental nucleus; LH – lateral hypothalamus; MPO – medial preoptic area; PFC – prefrontal cortex; PPT – pedunculopontine tegmental nucleus; PnC – pontine reticular formation, caudal part; PnO – pontine reticular formation, oral part; TMN – tuberomamillary nucleus; TNC – trigeminal nucleus complex; VLPO – ventrolateral preoptic area; VTA – ventral tegmental area; 5HT – serotonin; ACh – acetylcholine; Ado –adenosine; Asp – aspartate; DA – dopamine; GABA – γ-aminobutyric acid; Glu – glutamate; Gly – glycine; His – histamine; Hcrt – hypocretin; NE – norepinephrine; NO – nitric oxide; Noc – nociceptin; Ser – serine; 5HT – serotonin; Tau – taurine

($M_1$–$M_5$) of these receptors have been identified (Ishii and Kurachi 2006). $M_1$, $M_3$, and $M_5$ receptors, which couple to excitatory $G_{q/11}$ proteins, activate phospholipase C, resulting in mobilization of intracellular $Ca^{2+}$ (via inositol 1,4,5-triphosphate) and initiation of protein kinase C signaling via diacylglycerol (Ishii and Kurachi 2006). $M_2$ and $M_4$ receptors couple to inhibitory $G_{i/o}$ proteins and cause inhibition of adenylyl cyclase and a decrease in cyclic adenosine monophosphate (Ishii and Kurachi 2006). Activation of $M_2$ and $M_4$ receptors can hyperpolarize cellular membranes when associated with G-protein-gated potassium channels (Ishii and Kurachi 2006).

Intravenous (IV) administration of physostigmine, an acetylcholinesterase inhibitor, reverses unconsciousness induced by the general anesthetics propofol (Meuret et al. 2000) and sevoflurane (Plourde et al. 2003). Using cross-approximate entropy of the EEG to monitor arousal state, it has been shown that intracerebroventricular injection of neostigmine, an acetylcholinesterase inhibitor, reverses isoflurane-induced anesthesia (Hudetz, Wood, and Kampine 2003). These data indicate that general anesthetics work, at least in part, by suppressing cholinergic neurotransmission. Microinjection of nicotine into rat thalamus reverses the loss of righting reflex induced by the anesthetic sevoflurane (Alkire et al. 2007). This finding is consistent with the interpretation that one method by which sevoflurane causes unconsciousness may be inhibition of the thalamic cholinergic system.

Increasing extracellular acetylcholine levels with an IV injection of physostigmine also induces REM sleep if the injection is given during NREM sleep (Sitaram et al. 1976). Systemic administration of arecoline, a muscarinic and nicotinic receptor agonist, decreases REM latency (Sitaram, Moore, and Gillin 1978). When given systemically, the muscarinic antagonist scopolamine increases REM sleep latency (Sagales, Erill, and Domino 1969; Sitaram, Moore, and Gillin 1978). Together, these data demonstrate that one role of systemic acetylcholine is promoting REM sleep. But does cholinergic signaling specifically from the basal forebrain or LDT/PPT cause REM sleep?

LDT/PPT neurons discharge with either a Wake-On/REM-Off or a Wake-On/REM-On profile (Thakkar, Strecker, and McCarley 1998). LDT/PPT neurons project to many wakefulness-promoting brain regions, including the thalamus, dorsal raphé nucleus, locus coeruleus, and pontine reticular formation (Woolf and Butcher 1986, 1989; Mitani et al. 1988). The pontine reticular formation, a brain region known to promote REM sleep, contains muscarinic receptors (Baghdoyan, Carlson, and Roth 1994; Baghdoyan 1997; Capece, Baghdoyan, and Lydic 1998; DeMarco, Baghdoyan, and Lydic 2003; Brischoux, Mainville, and Jones 2008) and receives cholinergic inputs mainly from the LDT/PPT (Mitani et al. 1988; Shiromani, Armstrong, and Gillin 1988; Reinoso-Suarez et al. 2001). Using acetylcholine and cholinergic agonists, it has been well established that cholinergic transmission in the pontine reticular formation induces REM sleep (Hernandez-Peon et al. 1963; George, Haslett, and Jenden 1964; Baghdoyan, Monaco et al. 1984; Baghdoyan, Rodrigo-Angulo et al. 1984; Baghdoyan et al. 1987; Baghdoyan et al. 1989; Bourgin et al. 1995; Kshatri, Baghdoyan, and Lydic 1998; Baghdoyan and Lydic 1999; Lydic, Douglas, and Baghdoyan 2002; Coleman, Lydic, and Baghdoyan 2004). Electrical stimulation of the PPT increases pontine reticular formation acetylcholine release (Lydic and Baghdoyan 1993), and electrical stimulation of the LDT increases REM sleep (Thakkar, Portas, and McCarley 1996). Furthermore, pontine reticular formation acetylcholine release is higher during REM sleep than during wakefulness or NREM sleep (Kodama, Takahashi, and Honda 1990; Leonard and Lydic 1995; Leonard and Lydic 1997). Taken together, these data demonstrate that cholinergic projections from the LDT/PPT to the pontine reticular formation promote REM sleep.

Although cholinergic signaling from the LDT/PPT promotes REM sleep, cholinergic signaling from the basal forebrain promotes wakefulness. Basal forebrain neurons have the highest firing frequency during wakefulness and REM sleep (Szymusiak and McGinty 1986; Alam et al. 1999). Basal forebrain cholinergic neurons project throughout the entire cerebral cortex (Woolf and Butcher 1989; Materi, Rasmusson, and Semba 2000; Sarter and Bruno 2000; Cooper, Bloom, and Roth 2002). Acetylcholine release in the basal forebrain is highest during REM sleep, lower during quiet wakefulness, and lowest during NREM sleep (Vazquez and Baghdoyan 2001). Acetylcholine release in the cortex, a measure of basal forebrain cholinergic signaling, is higher during wakefulness (Jasper and Tessier 1971; Day, Damsma, and Fibiger 1991; Kurosawa et al. 1993; Marrosu et al. 1995; Giovannini et al. 1998; Materi, Rasmusson, and Semba 2000; Sarter and Bruno 2000) and REM sleep (Jasper and Tessier 1971; Marrosu et al. 1995) than during NREM sleep. These data support the view that basal forebrain acetylcholine promotes cortex activation during wakefulness and REM sleep.

## Adenosine

Adenosine is a product of metabolism of adenosine triphosphate and is located throughout the brain. To date, four adenosine receptors have been identified: $A_1$, $A_{2A}$, $A_{2B}$, and $A_3$. Each receptor is composed of seven transmembrane proteins and is coupled to G proteins. Indirect evidence that adenosine acts as a sleep-promoting compound is the ubiquitous use of coffee and tea to maintain wakefulness and alertness. The active wakefulness-promoting molecule in these beverages is caffeine, which is an adenosine receptor antagonist. In humans, caffeine taken prior to bedtime has been shown to increase sleep latency and reduce sleep efficiency (Landolt, Dijk et al. 1995). Furthermore, caffeine intake in the morning can decrease sleep efficiency and overall sleep during the subsequent night (Landolt, Werth et al. 1995). Similar to human, systemic administration of caffeine to rat results in an increase in sleep latency (Schwierin, Borbely, and Tobler 1996).

Prolonged wakefulness causes an increase in basal forebrain adenosine levels (Porkka-Heiskanen et al. 1997; Porkka-Heiskanen, Strecker, and McCarley 2000), and this adenosine increase is proportional to the length of prolonged wakefulness (Porkka-Heiskanen et al. 1997). Prolonged wakefulness also increases the density of adenosine $A_1$ receptors in rat parietal and motor cortices (Elmenhorst et al. 2009). A causal relationship between adenosine levels and arousal state can be demonstrated by manipulating adenosine levels while monitoring states of sleep and wakefulness. Increasing endogenous basal forebrain adenosine levels with the adenosine transport inhibitor S-(4-nitrobenzyl)-6-thioinosine decreases wakefulness and increases NREM sleep and REM sleep (Porkka-Heiskanen et al. 1997). Intracerebroventricular administration of adenosine increases NREM sleep and decreases wakefulness (Virus et al. 1983), and basal forebrain administration of adenosine also decreases wakefulness and increases sleep (Strecker et al. 2000). Taken together, these data support a sleep-promoting role for adenosine. To determine if the sleep-promoting effects are receptor mediated, selective agonists and antagonists for adenosine receptors have been used (Van Dort, Baghdoyan, and Lydic 2009).

Adenosine inhibits hypocretinergic neurons in the lateral hypothalamus via $A_1$ receptors (Liu and Gao 2007), and bilateral microinjection of the $A_1$ receptor antagonist 1,3-dipropyl-8-phenylxanthine to the perifornical lateral hypothalamus of rat increases wakefulness and decreases sleep (Thakkar et al. 2008). These data suggest that adenosine in the lateral hypothalamus is sleep promoting. Bilateral microinjection of adenosine, $N^6$-cyclopentyladenosine (CPA, an $A_1$ receptor agonist), or coformycin (an adenosine deaminase inhibitor) into the tuberomammillary nucleus (TMN) of rat increases NREM sleep (Oishi et al. 2008), suggesting that adenosinergic signaling in the TMN is sleep promoting. Adenosine, N-ethyl-carboxamidoadenosine (an $A_1/A_2$ receptor agonist), or CPA bilaterally administered to the preoptic area of rats increase NREM sleep, whereas the selective $A_2$ agonist 2-phenylaminoadenosine had no effect on sleep (Ticho and Radulovacki 1991). Reduction of $A_1$ receptors in the basal forebrain of rats by bilateral administration of antisense oligonucleotides decreases EEG delta frequency and NREM sleep time measured during a postsleep deprivation recovery period (Thakkar, Winston, and McCarley 2003), suggesting that $A_1$ receptors play a role in the homeostatic regulation of sleep. Immunohistochemical studies have shown that the basal forebrain (specifically the horizontal band of Broca, the substantia innominata, and the magnocellular preoptic area) contains $A_1$ receptors, but not $A_{2A}$ receptors (Basheer et al. 2001), indicating that adenosine regulation of sleep in the cholinergic portion of the basal forebrain occurs via the $A_1$ receptor. Dialysis administration of $N^6$-$p$-sulfophenyladenosine, an adenosine $A_1$ receptor agonist, to the prefrontal cortex of isoflurane-anesthetized mice causes a concentration-dependent decrease in prefrontal cortex and pontine reticular formation acetylcholine release, a concentration-dependent increase in resumption of righting after isoflurane anesthesia, and an increase in EEG delta power (Van Dort, Baghdoyan, and Lydic 2009). These results support the interpretation that adenosine $A_1$ receptor activation in the prefrontal cortex promotes sleep. This interpretation is supported by data showing that dialysis administration of the adenosine $A_1$ receptor antagonist 8-cyclopentyl-1,3-dipropylxanthine (DPCPX) to the prefrontal cortex of isoflurane-anesthetized mice causes a concentration-dependent increase in prefrontal cortex and pontine reticular formation acetylcholine release, reduces the time needed for resumption of righting after anesthesia, and activates the cortical EEG (Van Dort, Baghdoyan, and Lydic 2009). Similar to DPCPX, caffeine dialyzed into the prefrontal cortex of isoflurane-anesthetized mice increased acetylcholine release in the pontine reticular formation (Van Dort, Baghdoyan, and Lydic 2009). Finally, prefrontal cortex microinjection of DPCPX or the nonspecific adenosine receptor antagonist caffeine in awake mice caused an increase in wakefulness and a decrease in NREM sleep (Van Dort, Baghdoyan, and Lydic 2009), another indication that adenosinergic signaling via the adenosine $A_1$ receptor in the prefrontal cortex promotes sleep.

Adenosine $A_{2A}$ receptors have also been shown to regulate sleep and wakefulness. Dialysis administration of 2-$p$-(2-carboxyethyl)phenethylamino-5'-$N$-ethylcarboxamidoadenosine hydrochloride (CGS; an $A_{2A}$ receptor agonist) to the pontine reticular formation of mice caused an increase in acetylcholine release, NREM sleep, and REM sleep and a decrease in wakefulness (Coleman, Baghdoyan, and Lydic 2006). This suggests that adenosine in the pontine reticular formation may promote sleep via a cholinergic mechanism. Dialysis administration of CGS to the prefrontal cortex of isoflurane-anesthetized mice caused a concentration-dependent increase of prefrontal cortex and pontine reticular formation acetylcholine release, a concentration-dependent decrease in resumption of righting reflex, and an activation of the cortical EEG, all positive measures of wakefulness (Van Dort, Baghdoyan, and Lydic 2009). These data suggest that adenosine signaling via $A_{2A}$ receptors in the prefrontal cortex may promote wakefulness through activation of cholinergic neurotransmission.

## Biogenic Amines

A considerable amount of evidence supports the conclusion that serotonergic neurons localized to the dorsal raphé nucleus promote wakefulness. Extracellular recordings from dorsal raphé neurons revealed that these cells discharge at their highest rates during wakefulness and actually cease discharging during REM sleep (McGinty and Harper 1976; Trulson and Jacobs 1979; Lydic, McCarley, and Hobson 1983). Serotonin (5-hydroxytyrptamine; 5HT) release in the preoptic area of rats is highest during wakefulness, decreases during NREM sleep, and is lowest during REM sleep (Python et al. 2001). Serotonin may also play a permissive role in REM sleep generation. Serotonin receptors are G-protein-coupled receptors, and 14 serotonin receptors, divided into seven families ($5HT_1$–$5HT_7$), have been identified [for review see (Fink and Gothert 2007)]. Serotonin release in the dorsal raphé nucleus is highest during wakefulness (Portas et al. 1998; Fiske et al. 2006). Furthermore, electrical stimulation of the dorsal raphé nucleus increases wakefulness (Houdouin, Cespuglio, and Jouvet 1991), and decreasing serotonin levels in the raphé nucleus increases REM sleep (Portas et al. 1996). Microinjection of 8-hydroxy-2-(di-$n$-propylamino) tetralin, a $5HT_{1A}$ receptor agonist, into the dorsal raphé nucleus of rat increases wakefulness (Monti and Jantos 1992). In addition, systemic administration of $5HT_1$-like receptor agonists increases wakefulness and decreases sleep in rats (Dzoljic, Ukponmwan, and Saxena 1992). Another study has shown that intraperitoneal (IP) injection of the $5HT_{2A}$ receptor antagonist ($\pm$) 2,3 dimethoxyphenyl-1-[2-(4-piperidine)-methanol] or the $5HT_6$ receptor antagonist 3-benzenesulfonyl-7-(4-methyl-piperazin-1-yl)-1H-indole to rats increases NREM sleep and decreases wakefulness without affecting REM sleep (Moriarty et al. 2008). These data are consistent with the view that serotonin is wakefulness promoting. A recent study has shown that the time at which the serotonergic system is activated may contribute significantly to serotonin's arousal-promoting effect. When serotonin is given IP to mice during the light phase, wakefulness increases and NREM sleep and REM sleep decrease (Morrow et al. 2008). However, if the same injection is given at the start of the dark phase, sleep is increased and wakefulness is decreased (Morrow et al. 2008). Genetically modified mice have also been used to explore the role of serotonin in sleep and wakefulness. Mice lacking the genes for the $5HT_{1B}$ receptor showed an increase in REM sleep and a decrease in

NREM sleep, and agonists of the $5HT_{1B}$ receptor reduced the amount of REM sleep (Boutrel et al. 1999). These data indicate that serotonin acting at the $5HT_{1B}$ receptor plays a role in the suppression of REM sleep. Mice lacking the $5HT_{1A}$ receptor gene also displayed an increase in REM sleep, and REM sleep was reduced in nonknockout animals or humans that were given $5HT_{1A}$ receptor agonists (Boutrel et al. 2002; Wilson et al. 2005). Subcutaneous (Monti and Jantos 2006b) or dorsal raphé microinjection (Monti and Jantos 2006a) of the $5HT_{2A/2C}$ receptor agonist [1-(2,5-demethoxy-4-iodophenyl)-2-aminopropane)] into rats also suppresses REM sleep, with the systemic injection also inducing an increase in wakefulness and light sleep and a decrease in slow-wave sleep (Monti and Jantos 2006b). These data illustrate that serotonergic signaling via multiple 5HT receptors modulates sleep and wakefulness.

Noradrenergic cells that regulate wakefulness have been localized to the locus coeruleus (Cooper, Bloom, and Roth 2002) and project to a variety of brain regions including the hypothalamus, thalamus, basal forebrain, and cortex (Berridge and Waterhouse 2003). Noradrenergic signaling occurs via $\alpha_1$-, $\alpha_2$- , and $\beta$-adrenergic receptors, each of which is a G-protein coupled receptor [reviewed in (Hein 2006)]. There is good agreement between results from lesion (Caballero and de Andrés 1986) and single-unit recording (Chu and Bloom 1973; Hobson, McCarley, and Wyzinski 1975; McCarley and Hobson 1975; Foote, Aston-Jones, and Bloom 1980; Jacobs 1986) studies, showing that cells in the locus coeruleus inhibit REM sleep and promote wakefulness. Local administration of noradrenaline or $\alpha$- and $\beta$- receptor agonists to the medial septal area (Berridge and Foote 1996; Berridge, Isaac, and España 2003) or the medial preoptic area (Kumar et al. 1986; Sood et al. 1997) increases wakefulness. Stimulation of locus coeruleus neurons increases noradrenaline in the prefrontal cortex of anesthetized rats (Florin-Lechner et al. 1996; Berridge and Abercrombie 1999) and contributes to cortical activation. These data are consistent with the view that noradrenaline promotes wakefulness. However, bilateral microinjection of an $\alpha_1$-antagonist (prazosin), an $\alpha_2$-agonist (clonidine), or a $\beta$-antagonist (propranolol) into the PPT increases REM sleep with little to no effect on NREM sleep or wakefulness (Pal and Mallick 2006).

Wakefulness-promoting stimulants like amphetamine, cocaine, and methylphenidate act by increasing endogenous dopamine levels [reviewed in (Boutrel and Koob 2004)], and recent studies have shown that in humans sustained-release methamphetamine pills increase self-reports of sleep loss in cocaine-dependent individuals (Mooney et al. 2009). These data support the view that dopamine is a wakefulness-promoting neurotransmitter. The cell bodies of dopaminergic neurons that play a role in sleep–wake regulation are localized to the ventral tegmental area and the substantia nigra pars compacta (Cooper, Bloom, and Roth 2002). These neurons project to the dorsal raphé nucleus (Sakai et al. 1977), basal forebrain (Fallon and Moore 1978; Jones and Cuello 1989; Zaborszky, Cullinan, and Braun 1991), locus coeruleus (Simon et al. 1979), thalamus (Freeman et al. 2001), and laterodorsal tegmental nucleus (Cornwall, Cooper, and Phillipson 1990). Dopamine activates G-protein coupled receptors comprised of seven transmembrane proteins. Five dopaminergic receptors have been cloned (D1–D5) and are separated into two groups based on facilitation (D1 and D5) or inhibition (D2–D4) of adenylyl cyclase activity (Missale et al. 1998). Dopaminergic D2–D4 receptors also hyperpolarize cells by increasing outward potassium currents and inhibiting inward calcium currents [reviewed in (Missale et al. 1998)]. Even though substantia nigra (Miller et al. 1983; Steinfels et al. 1983) and ventral tegmental area (Miller et al. 1983) dopaminergic neurons in rats and cats do not change firing rates as animals transition between states of quiet wakefulness, NREM sleep, and REM sleep, dopamine is known to promote wakefulness. Mice lacking the dopamine transporter show an increase in wakefulness and a decrease in NREM sleep (Wisor et al. 2001), and single-unit recordings in cat show that substantia nigra dopamine neurons have the greatest firing frequency during active wakefulness (Steinfels et al. 1983). Systemic (Monti, Fernandez, and Jantos 1990) or intracerebroventricular administration (Isaac and Berridge 2003) of D1 receptor agonists increases wakefulness in rats, and wakefulness is decreased by systemic administration of D1 receptor antagonists in rats (Monti, Fernandez, and Jantos 1990). Intracerebroventricular administration of D2 agonists also increases wakefulness in rats (Isaac and Berridge 2003), but systemic administration of a D2 receptor agonist causes biphasic effects, with low doses decreasing wakefulness and high doses increasing wakefulness (Monti, Hawkins et al. 1988; Monti, Jantos, and Fernandez 1989). Systemic administration of D-amphetamine to rats increases wakefulness and decreases NREM sleep and REM sleep when given acutely and chronically (Anderson et al. 2009). Taken together, the preceding data support the interpretation that dopamine is wakefulness promoting.

Histaminergic cell bodies in the brain are localized to the tuberomammillary nucleus of the posterior hypothalamus (Panula, Yang, and Costa 1984; Watanabe et al 1984) and have diffuse projections throughout the brain (Inagaki et al. 1988; Panula et al. 1989). Lesion studies (von Economo 1930; Ranson 1939; Nauta 1946; Sweet and Hobson 1968; McGinty 1969) and single-unit recordings (Vanni-Mercier, Sakai, and Jouvet 1984; Ko et al. 2003; Takahashi, Lin, and Sakai 2006) have shown the posterior hypothalamus to be a wakefulness-promoting brain region. Three histaminergic receptors, denoted $H_1$, $H_2$, and $H_3$, have been found in the brain (Bouthenet et al. 1988; Martinez-Mir et al. 1990; for review, see Haas and Panula 2003) [for review, see (Haas and Panula 2003)]. All three receptor subtypes are G-protein coupled, with $H_1$ and $H_2$ receptors having excitatory effects and $H_3$ receptors providing autoinhibition to histaminergic neurons. $H_1$ receptors activate $G_{q/11}$ proteins, which act through phospholipase C and inositol-1,4,5-triphosphate to increase intracellular $Ca^{2+}$. $H_2$ receptors activate G proteins, which act through adenylyl cyclase and cyclic adenosine monophosphate to increase protein kinase C. $H_3$ receptors activate $G_{i/o}$ proteins, which inhibit adenylyl cyclase and high-voltage-activated $Ca^{2+}$ currents. $H_1$ receptor antagonists, especially first-generation antagonists such as the commonly used drug diphenhydramine, cause drowsiness (sedation) and impaired performance in humans (Nicholson and Stine 1986) and rats (Saitou et al. 1999; Kaneko et al. 2000). The potent $H_1$ receptor antagonist doxepin improved subjective and objective sleep measures in patients with chronic primary insomnia without causing next-day sedation or psychomotor impairments

(Roth et al. 2007). Decreasing histamine by IP injection of α-fluoromethylhistidine, an inhibitor of histidine decarboxylase, causes a significant decrease in waking, an increase in NREM sleep, and has no effect on REM sleep in rats (Kiyono et al. 1985; Monti, D'Angelo et al. 1988) and cats (Lin, Sakai, and Jouvet 1988). The sleep effects paralleled the decrease in brain histamine levels induced by α-fluoromethylhistidine. IP injection of the $H_1$ receptor antagonist mepyramine caused a significant decrease in wakefulness and REM sleep and a significant increase in NREM sleep (Lin, Sakai, and Jouvet 1988). Cyproheptadine, an $H_1$ receptor antagonist, given orally to sleep-disturbed rats decreased sleep latency and wakefulness and increased sleep (Tokunaga et al. 2007). These data suggest that histaminergic signaling via the $H_1$ receptor is wakefulness promoting. New therapies for sleep disorders and maintaining vigilance include $H_3$ receptor antagonists and inverse agonists, which have been shown to promote wakefulness (Barbier et al. 2004; Ligneau et al. 2007; Parmentier et al. 2007; Le et al. 2008).

## γ-Aminobutyric Acid (GABA)

Most investigations into the arousal-regulating roles of GABA have focused on signaling at the $GABA_A$ receptor. These receptors are ligand-gated anionic channels with a heteropentameric structure assembled from seven possible subunits, with each subunit comprising several proteins: $\alpha 1$–$\alpha 6$, $\beta 1$–$\beta 3$, $\gamma 1$–$\gamma 3$, $\epsilon 1$–$\epsilon 3$, $\delta$, $\theta$, and $\pi$ (Jacob, Moss, and Jurd 2008). Typically, two $\alpha$ subunits, two $\beta$ subunits, and one $\gamma$ or $\delta$ combine to form a $GABA_A$ receptor, and the variety of subunits comprising a $GABA_A$ receptor determines the selectivity and functionality of that receptor (Jacob, Moss, and Jurd 2008). The use of GABAergic drugs as sleep medications is well documented. For example, gaboxadol, an agonist specific for extrasynaptic $GABA_A$ receptors containing a $\delta$ subunit (Winsky-Sommerer et al. 2007), increases sleep and decreases wakefulness when systemically administered to humans (Faulhaber, Steiger, and Lancel 1997; Walsh et al. 2007; Walsh et al. 2008) and rats (Lancel 1997). Zolpidem and eszopiclone, two nonbenzodiazepine hypnotics that act at $GABA_A$ receptors, increase NREM sleep and decrease wakefulness when given systemically to guinea pig (Xi and Chase 2008). IP injection of zolpidem and E-6199, a putative hypnotic compound, increases NREM sleep and decreases wakefulness in mice (Alexandre et al. 2008). Oral administration of nitrazepam, a benzodiazepine hypnotic, increases NREM sleep and decreases wakefulness in sleep-disturbed rats (Tokunaga et al. 2007). Tiagabine, a selective GABA uptake inhibitor, dose-dependently increases NREM sleep in patients with primary insomnia (Walsh et al. 2006). The sodium salt of γ-hydroxybutyrate (sodium oxybate) is FDA approved for the treatment of daytime sleepiness. Tiagabine has also been shown to provide pain relief and improve function in patients suffering from fibromyalgia (Russell et al. 2009), suggesting another link between sleep and pain.

Microinjecting muscimol, a $GABA_A$ receptor agonist, into the pontine reticular formation increases wakefulness in cats (Xi, Morales, and Chase 1999) and rats (Camacho-Arroyo et al. 1991). Inhibition of GABAergic signaling within the pontine reticular formation via microinjection of bicuculline, a GABA$_A$ receptor antagonist, increases REM sleep and decreases wakefulness in cats (Xi, Morales, and Chase 1999) and rats (Sanford et al. 2003; Marks, Sachs, and Birabil 2008). Furthermore, local administration of nipecotic acid, a GABA uptake inhibitor, to rat pontine reticular formation increases wakefulness (Watson, Lydic, and Baghdoyan 2008) and increases pontine reticular formation GABA levels (Watson, Lydic, and Baghdoyan 2007). Pontine reticular formation administration of 3-mercaptopropionic acid, a glutamic acid decarboxylase inhibitor known to decrease extracellular GABA levels (Kehr and Ungerstedt 1988), decreases wakefulness and increases sleep in rats (Watson, Lydic, and Baghdoyan 2008). These findings demonstrate a wakefulness-promoting role for pontine reticular formation GABAergic transmission. Decreasing endogenous pontine reticular formation GABA levels increases sleep. Evidence suggests that isoflurane eliminates waking consciousness by decreasing GABA levels in the pontine reticular formation. Compared to wakefulness, isoflurane anesthesia significantly decreases GABA levels in cat pontine reticular formation (Vanini et al. 2008). Isoflurane induction time is significantly increased or decreased by increasing or decreasing, respectively, endogenous pontine reticular formation GABA levels (Vanini et al. 2008). Together these data suggest that isoflurane may exert its anesthetic effects by decreasing pontine reticular formation GABA levels.

Although GABAergic transmission in the pontine reticular formation promotes wakefulness, GABAergic transmission in other brain regions promotes sleep. For instance, dialysis delivery of bicuculline to the dorsal raphé nucleus of rats increases serotonin release, increases wakefulness, and decreases NREM sleep and REM sleep (Fiske et al. 2006). GABAergic neurons localized to the ventrolateral preoptic area project to the tuberomammillary nucleus (Sherin et al. 1996; Sherin et al. 1998), and microinjection of muscimol into the tuberomammillary nucleus increases NREM sleep and suppresses wakefulness (Lin et al. 1989; Sakai et al. 1990). Ventrolateral preoptic area GABAergic neurons also project to the locus coeruleus and raphé nucleus (Steininger et al. 2001), which, as noted previously, are wakefulness-promoting brain regions. In the medial preoptic area, microinjections of ethanol (Mendelson 2001), triazolam (Mendelson et al. 1989), and propofol (Tung, Bluhm, and Mendelson 2001) increase NREM sleep. In the posterior hypothalamus, microinjecting muscimol increases NREM sleep with a short latency (Lin et al. 1989). Microdialysis studies reveal increased GABA levels in the posterior hypothalamus during NREM sleep as compared to wakefulness and REM sleep (Nitz and Siegel 1996). Bilateral microinjection of muscimol into the mesencephalic tegmentum (ventrolateral periaqueductal gray and dorsal mesencephalic reticular formation) causes an increase in NREM sleep and REM sleep, whereas microinjection of bicuculline causes an increase in wakefulness (Vanini et al. 2007). Dialysis delivery of gaboxadol into the perifornical hypothalamus increases NREM sleep and decreases wakefulness (Thakkar, Winston, and McCarley 2008). These data show that the wakefulness- or sleep-promoting effects of GABAergic transmission are brain region specific.

## Glutamate

Glutamate is the main excitatory neurotransmitter in the brain, and arousal-regulating brain regions utilize glutamate for signaling (Franks 2008). It is surprising, therefore, that relatively little is known about alterations in glutamate release across states of consciousness. Recently, glutamate levels in the tuberomammillary nucleus have been shown to be higher during active wakefulness and REM sleep than during quiet waking and NREM sleep (John, Ramanathan, and Siegel 2008). There is evidence suggesting that within the LDT/PPT, the pontine reticular formation, and the medullary brainstem, glutamate contributes to arousal state control (Lai and Siegel 1988, 1991; Stevens, McCarley, and Greene 1992; Onoe and Sakai 1995; Datta, Patterson, and Spoley 2001; Datta, Spoley, and Patterson 2001; Datta et al. 2002). Glutamatergic neurons are present in rat pontine reticular formation (Kaneko et al. 1989), and a vast majority of neurons within this brain region are capable of synthesizing glutamate for use as a neurotransmitter (Jones 2003). Glutamate elicits excitatory responses from pontine reticular formation neurons in rats (Núñez, Buño, and Reinoso-Suárez 1998) and cats (Greene and Carpenter 1985), and glutamatergic and cholinergic transmission in the pontine reticular formation of rats interacts synergistically to potentiate catalepsy (Elazar and Berchanski 2001). Glutamate acts on three types of ionotropic receptors: α-amino-3-hydroxy-5-methyl-4-isoxazole propionic acid (AMPA), kainate, and N-methyl-d-aspartate (NMDA). Given individually, agonists for each of these receptors elicit excitatory responses from pontine reticular formation neurons (Stevens, McCarley, and Greene 1992). Dialysis delivery of the NMDA receptor antagonists ketamine or (5R,10S)-(+)-5-Methyl-10,11-dihydro-5H-dibenzo[a,d]cyclohepten-5,10-imine (MK-801) to cat pontine reticular formation decreases acetylcholine release in the pontine reticular formation and disrupts breathing (Lydic and Baghdoyan 2002). Microinjection of kainic acid into the midbrain reticular formation of cats causes cortical activation and behavioral arousal (Kitsikis and Steriade 1981).

## Hypocretin-1 and -2

In 1998, two groups independently discovered two excitatory peptides produced solely in the dorsolateral hypothalamus and named the peptides hypocretin-1 and -2 (de Lecea et al. 1998) and orexin A and B (Sakurai et al. 1998). In the following year, canine narcolepsy was linked to a mutation in the hypocretin receptor-2 gene (Lin et al. 1999), suggesting that hypocretinergic signaling is an important wakefulness-promoting neurotransmitter. Disruptions of hypocretinergic signaling caused by knocking out the peptide (Chemelli et al. 1999; Willie et al. 2003) and induced by selective lesioning of hypocretinergic neurons with hypocretin/ataxin-3 (Hara et al. 2001; Beuckmann et al. 2004) or hypocretin/saporin (Gerashchenko et al. 2003; Gerashchenko and Shiromani 2004; Murillo-Rodriguez et al. 2008) injections produce a narcoleptic phenotype. The cerebrospinal fluid of human narcoleptic patients with cataplexy has extremely low levels of hypocretin-1 (Nishino et al. 2000; Nishino and Kanbayashi 2005), and narcoleptic patients show greatly diminished or absent levels of hypocretin mRNA and peptide immunoreactivities (Peyron et al. 2000; Thannickal et al. 2000). The preceding findings further suggest that hypocretin is wakefulness promoting. By what mechanisms might hypocretin enhance wakefulness?

Hypocretinergic neurons fire with a Wake-On discharge pattern (Lee, Hassani, and Jones 2005; Mileykovskiy, Kiyashchenko, and Siegel 2005), and immuno-histochemical studies show that hypocretinergic neurons project to many arousal-regulating brain regions (Peyron et al. 1998; Nambu et al. 1999; Taheri et al. 1999; Zhang, Sampogna et al. 2004; Núñez et al. 2006). Hypocretin-1 and -2 receptors have been localized to sleep–wake regulating brain regions including the LDT/PPT nuclei, pontine reticular formation, dorsal raphé nucleus, and locus coeruleus (Greco and Shiromani 2001; Hervieu et al. 2001; Marcus et al. 2001; Brischoux, Mainville, and Jones 2008). In vitro studies demonstrate that hypocretin-1 and/or hypocretin-2 excite neurons in the dorsal raphé nucleus (Brown et al. 2001; Brown et al. 2002; Liu, van den Pol, and Aghajanian 2002; Soffin et al. 2004), LDT/PPT nuclei (Burlet, Tyler, and Leonard 2002; Takahashi et al. 2002; Kim et al. 2009), tuberomammillary nucleus (Bayer et al. 2001; Eriksson et al. 2001), locus coeruleus (Hagan et al. 1999; Horvath et al. 1999; Bourgin et al. 2000), and cholinergic neurons of the basal forebrain (Eggerman et al. 2001). Similarly, in vivo recordings have demonstrated that hypocretin-1 excites pontine reticular formation neurons (Xi et al. 2002). These data support the interpretation that hypocretinergic regulation of sleep and wakefulness is receptor mediated.

Hypocretin-1 increases wakefulness when microinjected into the lateral pre-optic area (Methippara et al. 2000), the LDT nucleus (Xi, Morales, and Chase 2001), the dorsal oral pontine tegmentum (Moreno Balandran et al. 2008), and the pontine reticular formation (Watson, Lydic, and Baghdoyan 2008). Microinjection of hypocretin-1 into the pontine reticular formation of cat during NREM sleep increases REM sleep (Xi et al. 2002), but the same microinjection into cat during wakefulness suppresses REM sleep with no effect on wakefulness (Moreno-Balandran et al. 2008). The wakefulness-promoting effect of hypocretin in the pontine reticular formation is further supported by evidence that delivery of antisense oligonucleotides against the hypocretin-2 receptor to the pontine reticular formation of rats enhanced REM sleep and induced behavioral cataplexy (Thakkar et al. 1999). Finally, intracerebroventricular administration of hypocretin-1 increases arousal in rats (Hagan et al. 1999; Piper et al. 2000), and local administration of hypocretin-1 and -2 to the basal forebrain causes a concentration-dependent increase in wakefulness (España et al. 2001; Thakkar et al. 2001).

Microdialysis studies provide insight into how hypocretin-1 may function as a wakefulness-promoting neuropeptide. In cats, hypothalamic hypocretin-1 levels are higher during wakefulness and REM sleep than during NREM sleep, and hypocretin-1 levels in the basal forebrain are higher during REM sleep than during NREM sleep (Kiyashchenko et al. 2002). Using microdialysis to monitor cortical acetylcholine has shown that basal forebrain administration of hypocretin-1 increases cortical acetylcholine release (Fadel, Pasumarthi, and Reznikov 2005; Dong et al. 2006). Intracerebroventricular injection of hypocretin-1 increases histamine release in rodent frontal cortex (Hong et al. 2005) and anterior hypothalamus (Ishizuka, Yamamoto, and Yamatodani 2002). Intracerebroventricular or ventral tegmental area injection of hypocretin-1 increases both wakefulness and prefrontal cortex dopamine release in rats (Vittoz and Berridge 2006). Microdialysis delivery of hypocretin-1 to the dorsal raphé nucleus of rat increases dorsal raphé serotonin release (Tao et al. 2006), and microdialysis delivery of hypocretin-1 to rat pontine reticular formation increases pontine reticular formation acetylcholine release (Bernard, Lydic, and Baghdoyan 2003, 2006) and GABA levels (Watson, Lydic, and Baghdoyan 2008). These data support the classification of hypocretin-1 as a wakefulness-promoting neuropeptide. An alternative hypothesis is that a primary function of hypocretin is to enhance activity in motor systems, and the increase in wakefulness is secondary. This hypothesis is supported by data showing that hypocretin-1 concentrations in cerebrospinal fluid are significantly greater during active wakefulness with movement than during quiet wakefulness with no movement (Kiyashchenko et al. 2002). However, when given orally, ACT-078573, a hypocretin-1 and -2 receptor antagonist, increases NREM sleep and REM sleep in rats, dogs, and humans (Brisbare-Roch et al. 2007).

Hypocretinergic signaling also plays a role in anesthesia. Intracerebroventricular administration of hypocretin-1 and/or -2 decreases total barbiturate (Kushikata et al. 2003) and ketamine (Tose et al. 2009) anesthesia time. Also, a recent study disrupted hypocretinergic signaling by either genetic ablation of hypocretin neurons or administration of the hypocretin-1 receptor antagonist SB-334867-A and investigated the effects on isoflurane or sevoflurane anesthesia induction and emergence times (Kelz et al. 2008). Mice with compromised hypocretinergic signaling showed no differences in induction time, but did take significantly longer to emerge from anesthesia. Two results from this paper are of particular relevance to anesthesia. First, this report suggests that the mechanisms of anesthetic induction and emergence are distinct from one another. Second, these data indicate hypocretin plays a role in emergence from anesthesia, but not in anesthetic induction.

## Other Neuropeptides

In addition to hypocretin, many other peptides have been shown to play a role in sleep–wake regulation. Investigations into orphan G-protein-coupled receptors have led to the discovery of new arousal-regulating peptides. One such peptide is neuropeptide S. Neuropeptide S is expressed primarily from cells in the dorsolateral pons located between the locus coeruleus and Barrington's nucleus, and receptors for this peptide are expressed in the cortex, hypothalamus, midbrain, pons, and medulla (Xu et al. 2004). Intracerebroventricular administration of neuropeptide S increases wakefulness and decreases NREM sleep and REM sleep (Xu et al. 2004), demonstrating a wakefulness-promoting role for this peptide.

Neuropeptide Y also plays a role in regulating sleep and wakefulness. Neuropeptide Y signals via six inhibitory G-protein-coupled receptors ($Y_1$–$Y_6$)

to inhibit adenylyl cyclase activity (Silva, Cavadas, and Grouzmann 2002). Microinjection of neuropeptide Y into the basal forebrain causes a significant increase in wakefulness and a decrease in NREM sleep (Tóth et al. 2007). When injected at light onset, intracerebroventricular or lateral hypothalamus administration of neuropeptide Y causes a decrease in both NREM sleep and REM sleep (Szentirmai and Krueger 2006). However, IP injection of peptide tyrosine–tyrosine, a neuropeptide Y $Y_2$ receptor specific agonist, increases NREM sleep and decreases wakefulness (Akanmu et al. 2006), indicating that the sleep–wake effects of neuropeptide Y may be receptor subtype dependent.

The neuropeptide cortistatin-14 is expressed throughout the cerebral cortex, has a peptide sequence similar to somatostatin, and binds to the somatostatin receptors (de Lecea et al. 1996). Intracerebroventricular administration of this peptide causes a significant increase in NREM sleep and a decrease in REM sleep (de Lecea et al. 1996). These data indicate that cortistatin-14 is a sleep-promoting neuropeptide. Urotensin II is another neuropeptide that regulates sleep and wakefulness (Huitron-Resendiz et al. 2005). Receptor mRNA for this peptide is expressed in the LDT/PPT and is colocalized with choline acetyltransferase (Clark et al. 2001), possibly indicating that urotensin II regulates sleep and wakefulness via modulating cholinergic signaling. In fact, whole-cell recordings have shown that urotensin II excites cholinergic neurons in the PPT (Huitron-Resendiz et al. 2005). Microinjections of urotensin II into the PPT of rats significantly increased the time spent in REM sleep, with minimal effects on wakefulness and NREM sleep (Huitron-Resendiz et al. 2005). However, intracerebroventricular administration of urotensin II increased the amount of time spent in wakefulness and REM sleep and significantly decreased the amount of time spent in the deeper stages of NREM sleep (Huitron-Resendiz et al. 2005). These data show that urotensin II signaling in the PPT promotes REM sleep, and when administered intracerebroventricularly, urotensin II promotes wakefulness and REM sleep.

A number of peptides that regulate pain perception also play a role in regulating states of consciousness. Nociceptin (orphanin FQ) is the endogenous ligand for the G-protein-coupled opioid receptor-like 1 receptor (nociceptin receptor). The protein sequences of this receptor are closely matched to those of classic opioid receptors ($\mu$, $\delta$, and $\kappa$), but ligands for $\mu$, $\delta$, and $\kappa$ opioid receptors have low affinity for the nociceptin receptor (Byford et al. 2007). Intravenous administration of nociceptin receptor agonists in rats caused a dose-dependent loss of righting response (a measure of hypnosis), an increase in time of resumption to righting response, burst suppression in the EEG (similar to inhaled anesthetics), and showed antinociceptive effects during a formalin paw test (Byford et al. 2007). These results suggest the need for experiments designed to determine whether nociceptin is a sleep-promoting peptide. Although nociceptin shows some effects similar to inhaled anesthetics, a study using nociceptin receptor knockout mice revealed that the absence of this receptor did not alter the minimum alveolar concentration (MAC) values of halothane, isoflurane, or sevoflurane (Himukashi et al. 2005), suggesting that nociceptin signaling does not play a role in volatile anesthetic potency.

Substance P, another pain-regulating peptide, signals via the neurokinin 1 receptor (Regoli et al. 1989). Strong evidence exists for the ability of substance P to increase pain sensitivity, decrease pain thresholds, and possibly play a role in fibromyalgia [reviewed in (Anderson et al. 2006)]. Intracerebroventricular administration of substance P to mice at a dose that does not decrease pain response threshold causes a decrease in NREM sleep and an increase in the amount of wakefulness, number of wakefulness episodes, NREM sleep latency, and REM sleep latency (Anderson et al. 2006). Pain disrupts sleep, and the sleep effects of substance P may be due to direct actions on sleep–wake systems or direct actions on pain pathways, which then cause a disruption in sleep. Bilateral administration of substance P to the ventrolateral preoptic area, a sleep-promoting brain region, decreases wakefulness and increases NREM sleep without affecting REM sleep (Zhang, Wang et al. 2004). These sleep effects were blocked by 3-mercaptopropionic acid, a glutamic acid decarboxylase inhibitor known to decrease GABA levels, suggesting that in the ventrolateral preoptic area substance P may alter arousal via a GABAergic mechanism. In vitro studies using pontine reticular formation slices have shown that substance P is colocalized within cholinergic neurons, causes depolarization when applied to pontine reticular formation neurons (similar to acetylcholine), and that the effects of coadministering substance P and acetylcholine, even at maximum concentrations, are additive (Kohlmeier et al. 2002). To date there are no data showing that substance P applied to the pontine reticular formation causes REM sleep.

## Opioids

Opioids play a major role in the clinical management of pain and, as reviewed elsewhere, systemically administered opioids inhibit the REM phase of sleep and disrupt sleep architecture (Lydic and Baghdoyan 2007). Opioid receptors are G-protein coupled and are categorized as $\mu$, $\delta$, or $\kappa$ (Trescot et al. 2008). Endogenous opioids include the endorphin, enkephalin, and dynorphin classes of peptides (Trescot et al. 2008). Systemic administration of morphine to rats causes an increase in muscle tone, abolishes normal sleep, increases delta wave power in the cortical EEG, and produces an immobilized behavioral state (Osman, Baghdoyan, and Lydic 2005; Watson, Lydic, and Baghdoyan 2007). Whole-cell recordings show that morphine and met-enkephalin inhibit hypocretin neurons, suggesting one mechanism by which opioids may produce mental lethargy and sedation (Li and van den Pol 2008). Microinjection of morphine into cat (Keifer, Baghdoyan, and Lydic 1992) or rat (Watson, Lydic, and Baghdoyan 2007) pontine reticular formation increases wakefulness and decreases REM sleep. Because morphine has activity at $\mu$, $\delta$, and $\kappa$ opioid receptors, these studies did not reveal which receptor subtype caused sleep disruption. The $\mu$-opioid receptor agonist [D-Ala$_2$, N-Met-Phe$_4$, Gly$_5$] enkephalin (DAMGO) activates G proteins in the pontine reticular formation (Tanase et al. 2001). Using subtype selective opioid receptor agonists demonstrates that the sleep-disrupting effects of pontine reticular formation morphine administration are due to the $\mu$ opioid receptor (Cronin et al. 1995). In the ventrolateral preoptic area, a sleep-promoting brain region, over 85% of neurons express mRNA for $\mu$ and κ receptors (Greco et al. 2008). Microinjection of the κ-opioid receptor endogenous agonist dynorphin (during subjective night) causes a dose-dependent increase in NREM sleep, whereas microinjecting DAMGO during subjective day increased wakefulness and decreased NREM sleep and REM sleep (Greco et al. 2008). It is possible that morphine, which acts at both μ and κ receptors, disrupts sleep by simultaneously activating sleep- and wakefulness-promoting pathways. Opioid administration also disrupts sleep in humans. A single IV infusion of morphine to healthy volunteers decreases stages 3 and 4 NREM sleep and REM sleep and increases stage 2 NREM sleep (Shaw et al. 2005), and a nighttime dose of morphine or methadone also decreases stages 3 and 4 NREM sleep while increasing stage 2 NREM sleep (Dimsdale et al. 2007). Constant infusion of analgesic doses of remifentanil overnight decreases REM sleep in healthy volunteers (Bonafide et al. 2008).

Brain regions regulating sleep and wakefulness overlap with appetite-regulating brain regions and certain peptides regulate both arousal state and appetite (Lydic and Baghdoyan 2005). For instance, decreased levels of leptin (an appetite-suppressing hormone) and increased levels of ghrelin (an appetite-stimulating hormone) are linked to a decrease in sleep duration (Taheri et al. 2004; Spiegel et al. 2004). Participants in the reduced sleep group also showed increases in hunger, appetite, and body mass index, likely due to both decreased leptin and increased ghrelin levels. Compared to wild-type controls, obese (*ob/ob*) mice (which have reduced leptin levels) show significant disruptions in sleep architecture including increased arousals, shorter duration sleep bouts (Laposky et al. 2006), and impaired responses to cholinergic enhancement of REM sleep (Douglas et al. 2005). Similarly, when compared to wild-type controls, *db/db* mice (which are leptin resistant) exhibit increases in NREM sleep and REM sleep during the dark phase; increases in stage shifts, arousals, and number of arousal bouts; and a decrease in the duration of wakefulness and NREM sleep bouts (Laposky et al. 2008). Leptin may also play a role in pain modulation as shown by a recent report demonstrating that leptin replacement to leptin deficient *ob/ob* mice rescues supraspinal cholinergic antinociception (Wang, Baghdoyan, and Lydic 2009). Whole-cell recordings demonstrate that ghrelin depolarizes PPT neurons, suggesting an arousal modulating role for this hormone (Kim et al. 2009). Ghrelin microinjected into rat lateral hypothalamus, medial preoptic area, or paraventricular nucleus increases wakefulness, decreases NREM sleep, and increases food intake (Szentirmai, Kapás, and Krueger 2007). These data indicate that leptin and ghrelin are wakefulness promoting.

## Conclusions and Future Directions

Many brain regions and molecules have not yet been studied in relation to states of consciousness. Future studies should include exploration of these brain regions to determine whether or not they play a functional role in the regulation of wakefulness, sleep, and anesthesia. In addition, neurochemicals that have been shown to contribute to the regulation of sleep and wakefulness have not been studied in relation to anesthesia-induced alterations in states of consciousness. Systematic characterization of the similarities in amount and time course of analyte changes during anesthesia compared to similar measures during sleep will further characterize similarities and differences between sleep and anesthesia as altered states of consciousness. Improvements in spatial and temporal resolution of neurotransmitter sampling in the brain will offer a new and exciting frontier for both anesthesia and sleep research. Identification and testing of the specific protein subunit composition of a given neurotransmitter receptor may offer the ability to design consciousness-regulating compounds (anesthetics and sedative hypnotic medications for sleep disorders) that maximize the beneficial effects of the drug while minimizing the unwanted side effects. For instance, selectively targeting components of extrasynaptic $GABA_A$ receptors, which are known to be especially sensitive to numerous sedative hypnotics (Orser 2006), may lead to such consciousness-regulating compounds. Finally, there is recent support for the view that glial cells are involved in regulating states of consciousness (Halassa et al. 2009). Consideration of glia as a functional component of the tripartite synapse (Halassa, Fellin, and Haydon 2007), comprised of an astrocyte and a pre- and postsynaptic terminal, has the potential to revolutionize views on the neurochemistry of consciousness. With this in mind, the fields of anesthesia, sleep, and chemistry will not only share an illustrious past and present, but an extremely bright future for years to come.

**Disclosure Statement** This work supported by National Institutes of Health grants: HL40881, MH45361, HL57120, HL65272, and the Department of Anesthesiology. We thank Mary A. Norat, Giancarlo Vanini, and Sarah L. Watson for critical comments on this chapter. This work was not an industry-supported study, and the authors have no financial conflicts of interest.

# References

Akanmu M. A., O. E. Ukponmwan, Y. Katayama, and K. Honda. 2006. Neuropeptide-Y $Y_2$-receptor agonist, $PYY_{3-36}$ promotes non-rapid eye movement sleep in rat. *Neurosci Res* 54:165–170.

Alam, M. N., H. Gong, T. Alam, R. Jaganath, D. J. McGinty, and R. Szymusiak. 2002. Sleep-waking discharge patterns of neurons recorded in the rat perifornical lateral hypothalamic area. *J Physiol* 538:619–631.

Alam, M. N., R. Szymusiak, H. Gong, J. King, and D. J. McGinty. 1999. Adenosinergic modulation of rat basal forebrain neurons during sleep and waking: neuronal recording with microdialysis. *J Physiol* 521.3:679–690.

Alexandre, C., A. Dordal, R. Aixendri, A. Guzman, M. Hoamon, and J. Adrien. 2008. Sleep-stabilizing effects of E-6199, compared to zopiclone, zolpidem, and THIP in mice. *Sleep* 31:259–270.

Alkire, M. T., A. G. Hudetz, and G. Tononi. 2008. Consciousness and anesthesia. *Science* 322: 876–880.

Alkire, M. T., J. R. McReynolds, E. L. Hahn, and A. N. Trivedi. 2007. Thalamic microinjection of nicotine reverses sevoflurane-induced loss of righting reflex in the rat. *Anesthesiology* 107: 264–272.

Anderson, M. L., R. Margis, B. N. Frey, L. M. F. Giglio, F. Kapczinski, and S. Tufik. 2009. Electrophysiological correlates of sleep disturbance induced by acute and chronic administration of D-amphetamine. *Brain Res* 1249:162–172.

Anderson, M. L., D. C. Nascimento, R. B. Machado, S. Roizenblatt, H. Moldofsky, and S. Tufik. 2006. Sleep disturbance induced by substance P in mice. *Behav Brain Res* 167:212–218.

Aserinsky, E., and N. Kleitman. 1953. Regularly occurring periods of eye motility, and concomitant phenomena, during sleep. *Science* 118:273–274.

Aston-Jones, G., and F. E. Bloom. 1981. Activity of norepinephrine-containing locus coeruleus neurons in behaving rats anticipates fluctuations in the sleep-waking cycle. *J Neurosci* 1: 876–886.

Autret, A., F. Laffont, B. de Toffol, and H. P. Cathala. 1988. A syndrome of REM and non-REM sleep reduction and lateral gaze paresis after medial tegmental pontine stroke. Computed tomographic scans and anatomical correlations in four patients. *Arch Neurol* 45: 1236–1242.

Baars, B. J., T. Z. Ramsoy, and S. Laureys. 2003. Brain, conscious experience and the observing self. *Trends Neurosci* 26:671–675.

Baghdoyan, H. A. 1997. Location and quantification of muscarinic receptor subtypes in rat pons: implications for REM sleep generation. *Am J Physiol* 273:R896–R904.

Baghdoyan, H. A., B. X. Carlson, and M. T. Roth. 1994. Pharmacological characterization of muscarinic cholinergic receptors in cat pons and cortex. Preliminary study. *Pharmacology* 48:77–85.

Baghdoyan, H. A., and R. Lydic. 1999. M2 muscarinic receptor subtype in the feline medial pontine reticular formation modulates the amount of rapid eye movement sleep. *Sleep* 22:835–847.

Baghdoyan, H. A., R. Lydic, C. W. Callaway, and J. A. Hobson. 1989. The carbachol-induced enhancement of desynchronized sleep signs is dose dependent and antagonized by centrally administered atropine. *Neuropsychopharmacol* 2:67–79.

Baghdoyan, H. A., A. Monaco, M. L. Rodrigo-Angulo, F. Assens, R. W. McCarley, and J. A. Hobson. 1984. Microinjection of neostigmine into the pontine reticular formation of cats enhances desynchronized sleep signs. *J Pharmacol Exp Ther* 231:173–180.

Baghdoyan, H. A., M. L. Rodrigo-Angulo, R. W. McCarley, and J. A. Hobson. 1984. Site-specific enhancement and suppression of desynchronized sleep signs following cholinergic stimulation of three brain stem regions. *Brain Res* 306:39–52.

Baghdoyan, H. A., M. L. Rodrigo-Angulo, R. W. McCarley, and J. A. Hobson. 1987. A neuroanatomical gradient in the pontine tegmentum for the cholinoceptive induction of desynchronized sleep signs. *Brain Res* 414:245–261.

Barbier, A. J., C. Berridge, C. Dugovic, A. D. Laposky, S. J. Wilson, J. Boggs, L. Aluisio, B. Lord, C. Mazur, C. M. Pudiak, X. Langlois, W. Xiao, R. Apodaca, and N. I. Carruthers. 2004. Acute wake-promoting actions of JNJ-5207852, a novel, diamine-based H3 antagonist. *Br J Pharmacol* 143:649–661.

Barbosa, R. M., C. F. Lourenço, R. M. Santos, F. Pomerleau, P. Huettl, G. A. Gerhardt, and J. Laranjinha. 2008. In vivo real-time measurement of nitric oxide in anesthetized rat brain. *Methods Enzymol* 441:351–367.

Basheer, R., L. Halldner, L. Alanko, R. W. McCarley, B. B. Fredholm, and T. Porkka-Heiskanen. 2001. Opposite changes in adenosine $A_1$ and $A_{2A}$ receptor mRNA in the rat following sleep deprivation. *Neuroreport* 12:1577–1580.

Bayer, L., E. Eggerman, M. Serafin, B. Saint-Mleux, D. Machard, B. E. Jones, and M. Muhlethaler. 2001. Orexins (hypocretins) directly excite tuberomammillary neurons. *Eur J Neurosci* 14:1571–1575.

Berger, H. 1929. Uber das elektroenkephalogramm des menchen. *Arch Psychiatr Nervenkr* 87:527–570.

Bernard, R., R. Lydic, and H. A. Baghdoyan. 2003. Hypocretin-1 causes G protein activation and increases ACh release in rat pons. *Eur J Neurosci* 18:1775–1785.

Bernard, R, R. Lydic, and H. A. Baghdoyan. 2006. Hypocretin (orexin) receptor subtypes differentially enhance acetylcholine release and activate G protein subtypes in rat pontine reticular formation. *J Pharmacol Exp Ther* 317:163–171.

Berridge, C. W., and E. D. Abercrombie. 1999. Relationship between locus coeruleus discharge rates and rates of norepinephrine release within the neocortex as assessed by in vivo microdialysis. *Neuroscience* 93:1263–1270.

Berridge, C. W., and S. L. Foote. 1996. Enhancement of behavioral and electroencephalographic indices of waking following stimulation of noradrenergic beta-receptors within the medial septal region of the basal forebrain. *J Neurosci* 16:6999–7009.

Berridge, C. W., S. O. Isaac, and R. A. España. 2003. Additive wake-promoting actions of medial basal forebrain noradrenergic α1- and β-receptor stimulation. *Behav Neurosci* 117: 350–359.

Berridge, C. W., and B. D. Waterhouse. 2003. The locus coeruleus-noradrenergic system: modulation of behavioral state and state-dependent cognitive processes. *Brain Res Rev* 42: 33–84.

Beuckmann, C. T., C. M. Sinton, S. C. Williams, J. A. Richardson, R. E. Hammer, T. Sakurai, and M. Yanagisawa. 2004. Expression of a poly-glutamine-ataxin-3 transgene in orexin neurons induces narcolepsy-cataplexy in the rat. *J Neurosci* 24:4469–4477.

Bingham, S., P. T. Davey, A. J. Babbs, E. A. Irving, M. J. Sammons, M. Wyles, P. Jeffrey, L. Cutler, I. Riba, A. Johns, R. A. Porter, N. Upton, A. J. Hunter, and A. A. Parsons. 2001. Orexin-A, an hypothalamic peptide with analgesic properties. *Pain* 92:81–90.

Bischoff, P. 2006. Monitoring methods: SNAP. *Best Pract Res Clin Anaesthesiol* 20:141–146.

Bonafide, C. P., N. Aucutt-Walter, N. Divittore, T. King, E. O. Bixler, and A. J. Cronin. 2008. Remifentanil inhibits rapid eye movement sleep but not the nocturnal melatonin surge in humans. *Anesthesiology* 108:627–633.

Bourgin, P., P. Escourrou, C. Gaultier, and J. Adrien. 1995. Induction of rapid eye movement sleep by carbachol infusion into the pontine reticular formation in the rat. *Neuroreport* 6:532–536.

Bourgin, P., S. Huitron-Resendiz, A. D. Spier, V. Fabre, B. Morte, J. R. Criado, J. G. Sutcliffe, S. J. Henriksen, and L. de Lecea. 2000. Hypocretin-1 modulates rapid eye movement sleep through activation of locus coeruleus neurons. *J Neurosci* 20:7760–7765.

Bouthenet, M. L., M. Ruat, N. Sales, and M. Garbarg. 1988. A detailed mapping of histamine $H_1$-receptors in guinea-pig central nervous system established by autoradiography with [$^{125}$]iodobopyramine. *Neuroscience* 26:553–600.

Boutrel, B., B. Franc, R. Hen, M. Hamon, and J. Adrien. 1999. Key role of 5-HT1B receptors in the regulation of paradoxical sleep as evidenced in 5-HT1B knock-out mice. *J Neurosci* 19:3204–3212.

Boutrel, B., and G. F. Koob. 2004. What keeps us awake: the neuropharmacology of stimulants and wakefulness-promoting medications *Sleep* 27:1181–1194.

Boutrel, B., C. Monaca, R. Hen, M. Hamon, and J. Adrien. 2002. Involvement of 5-HT1A receptors in homeostatic and stress-induced adaptive regulations of paradoxical sleep: studies in 5-HT1A knock-out mice. *J Neurosci* 22:4686–4692.

Brisbare-Roch, C., J. Dingemanse, R. Koberstein, P. Hoever, H. Aissaoui, S. Flores, C. Mueller, O. Nayler, J. van Gervan, S. L. de Haas, P. Hess, C. Qiu, S. Buchmann, M. Scherz, T. Weller, W. Fischli, M. Clozel, and F. Jenck. 2007. Promotion of sleep by targeting the orexin system in rats, dogs, and humans. *Nat Med* 13:150–155.

Brischoux, F., L. Mainville, and B. E. Jones. 2008. Muscarinic-2 and orexin-2 receptors on GABAergic and other neurons in the rat mesopontine tegmentum and their potential role in sleep-wake control. *J Comp Neurol* 510:607–630.

Brown, R. E., O. Sergeeva, K. S. Eriksson, and H. L. Haas. 2001. Orexin A excites serotonergic neurons in the dorsal raphé nucleus of the rat. *Neuropharmacology* 40:457–459.

Brown, R. E., O. Sergeeva, K. S. Eriksson, and H. L. Haas. 2002. Convergent excitation of dorsal raphé serotonin neurons by multiple arousal systems (orexin/hypocretin, histamine and noradrenaline). *J Neurosci* 22:8850–8859.

Bruno, J. P., C. Gash, B. Martin, A. Zmarowski, F. Pomerleau, J. Burmeister, P. Huettl, and G. A. Gerhardt. 2006. Second-by-second measurement of acetylcholine release in prefrontal cortex. *Eur J Neurosci* 24:2749–2757.

Burlet, S., C. J. Tyler, and C. S. Leonard. 2002. Direct and indirect excitation of laterodorsal tegmental neurons by hypocretin/orexin peptides: implications for wakefulness and narcolepsy. *J Neurosci* 22:2862–2872.

Burmeister, J. J., K. Moxon, and G. A. Gerhardt. 2000. Ceramic-based multisite microelectrodes for electrochemical recordings. *Anal Chem* 72:187–192.

Burmeister, J. J., F. Pomerleau, P. Huettl, C. R. Gash, C. E. Werner, J. P. Bruno, and G. A. Gerhardt. 2008. Ceramic-based multisite microelectrode arrays for simultaneous measures of choline and acetylcholine in CNS. *Biosens Bioelectron* 23:1382–1389.

Burmeister, J. J., F. Pomerleau, M. Palmer, B. K. Day, P. Huettl, and G. A. Gerhardt. 2002. Improved ceramic-based multisite microelectrode for rapid measurements of L-glutamate in the CNS. *J Neurosci Methods* 119:163–171.

Byford, A. J., A. Anderson, P. S. Jones, R. Palin, and A. K. Houghton. 2007. The hypnotic, electroencephalographic, and antinociceptive properties of nonpeptide ORL1 receptor agonists after intravenous injection in rodents. *Anesth Analg* 104:174–179.

Caballero, A., and I. de Andrés. 1986. Unilateral lesions in locus coeruleus area enhance paradoxical sleep. *Electroenceph Clin Neurophysiol* 64:339–346.

Camacho-Arroyo, I., R. Alvarado, J. Manjarrez, and R. Tapia. 1991. Microinjections of muscimol and bicuculline into the pontine reticular formation modify the sleep-waking cycle in the rat. *Neurosci Lett* 129:95–97.

Campagna, J. A., K. W. Miller, and S. A. Forman. 2003. Mechanisms of actions of inhaled anesthetics. *N Engl J Med* 348:2110–2124.

Capece, M. L., H. A. Baghdoyan, and R. Lydic. 1998. Carbachol stimulates [$^{35}$S]guanylyl 5'-($\gamma$-thio)-triphosphate binding in rapid eye movement sleep-related brainstem nuclei of rat. *J Comp Neurol* 18:3779–3785.

Caton, R. 1875. The electric currents of the brain. *Br Med J* 2:278.

Cellar, N. A., S. T. Burns, J. C. Meiners, H. Chen, and R. T. Kennedy. 2005. Microfluidic chip for low-flow push–pull perfusion sampling in-vivo with on-line analysis of amino acids. *Anal Chem* 77:7067–7073.

Chemelli, R. M., J. T. Willie, C. M. Sinton, J. K. Elmquist, T. E. Scammell, C. Lee, J. A. Richardson, S. C. Williams, Y. Xiong, Y. Kisanuki, T. E. Fitch, M. Nakazato, R. E. Hammer, C. B. Saper, and M. Yanagisawa. 1999. Narcolepsy in orexin knockout mice: molecular genetics of sleep regulation. *Cell* 98:437–451.

Cheng, J.-K., R. C.-C. Chou, L.-L. Hwang, and L.-C. Chiou. 2003. Antiallodynic effects of intrathecal orexins in a rat model of postoperative pain. *J Pharmacol Exp Ther* 307: 1065–1071.

Chu, N., and F. E. Bloom. 1973. Norepinephrine-containing neurons: changes in spontaneous discharge patterns during sleeping and waking. *Science* 179:908–910.

Chung, S. A., H. Yuan, and F. Chung. 2008. A systematic review of obstructive sleep apnea and its implications for anesthesiologists. *Anesth Analg* 107:1643–1663.

Clark, S. D., H-P. Nothacker, Z. Wang, Y. Saito, F. M. Leslie, and O. Civelli. 2001. The urotensin II receptor is expressed in the cholinergic mesopontine tegmentum of the rat. *Brain Res* 923: 120–127.

Coleman, C. G., H. A. Baghdoyan, and R. Lydic. 2006. Dialysis delivery of an adenosine $A_{2A}$ agonist into the pontine reticular formation of C57BL/6J mouse increases pontine acetylcholine release and sleep. *J Neurochem* 96:1750–1759.

Coleman, C. G., R. Lydic, and H. A. Baghdoyan. 2004. M2 muscarinic receptors in pontine reticular formation of C57BL/6J mouse contribute to rapid eye movement sleep generation. *Neuroscience* 126:821–830.

Cooper, J. R., F. E. Bloom, and R. H. Roth. 2002. *Biochemical Basis of Neuropharmacology*. 8th ed. Oxford: Oxford University Press.

Cornwall, J., J. D. Cooper, and O. T. Phillipson. 1990. Afferent and efferent connections of the laterodorsal tegmental nucleus in the rat. *Brain Res Bull* 25:271–284.

Cronin, A., J. C. Keifer, H. A. Baghdoyan, and R. Lydic. 1995. Opioid inhibition of rapid eye movement sleep by a specific mu receptor agonist. *Br J Anaesth* 74:188–192.

Datta, S., and R. R. MacLean. 2007. Neurobiological mechanisms for the regulation of mammalian sleep-wake behavior: reinterpretation of historical evidence and inclusion of contemporary cellular and molecular evidence. *Neurosci Biobehav Rev* 31:775–824.

Datta, S., E. H. Patterson, and E. E. Spoley. 2001. Excitation of the pedunculopontine tegmental NMDA receptors induces wakefulness and cortical activation in the rat. *J Neurosci Res* 66(1):109–116.

Datta, S., E. E. Spoley, V. K. Mavanji, and E. H. Patterson. 2002. A novel role of pedunculopontine tegmental kainate receptors: a mechanism of rapid eye movement sleep generation in the rat. *Neuroscience* 114:157–164.

Datta, S., E. E. Spoley, and E. H. Patterson. 2001. Microinjection of glutamate into the pedunculopontine tegmentum induces REM sleep and wakefulness in the rat. *Am J Physiol Regul Integr Comp Physiol* 280:R752–R759.

Day, J., G. Damsma, and H. C. Fibiger. 1991. Cholinergic activity in the rat hippocampus, cortex and striatum correlates to locomotor activity: an in vivo microdialysis study. *Pharmacol Biochem Behav* 38:723–729.

de Lecea, L., J. R. Criado, Ó. Prospero-Garcia, K. M. Gautvik, P. Schweitzer, P. E. Danielson, C. L. M. Dunlop, G. R. Siggins, S. J. Henriksen, and J. G. Sutcliffe. 1996. A cortical neuropeptide with neuronal depressant and sleep-modulating properties. *Nature* 381:242–245.

de Lecea, L., T. S. Kilduff, C. Peyron, X. B. Gao, P. E. Foye, P. E. Danielson, C. Fukuhara, E. L. F. Battenberg, V. T. Gautvik, F. S. Bartlett II, W. N. Frankel, A. N. van den Pol, F. E. Bloom, K. M. Gautvik, and J. G. Sutcliffe. 1998. The hypocretins: hypothalamus-specific peptides with neuroexcitatory activity. *Proc Natl Acad Sci USA* 95:322–327.

Deboer, T., M. J. Vansteensel, L. Détári, and J. H. Meijer. 2003. Sleep states alter activity of suprachiasmatic nucleus neurons. *Nat Neurosci* 6:1086–1090.

DeMarco, G. J., H. A. Baghdoyan, and R. Lydic. 2003. Differential cholinergic activation of G proteins in rat and mouse brainstem: relevance for sleep and nociception. *J Comp Neurol* 457:175–184.

Dement, W. 1958. The occurrence of low voltage, fast electroencephalogram patterns during behavioral sleep in cat. *Electroenceph Clin Neurophysiol* 10:291–296.

Dement, W., and N. Kleitman. 1957. Cyclic variations in EEG during sleep and their relation to eye movements, body motility, and dreaming. *Electroenceph Clin Neurophysiol* 9: 673–690.

Dimsdale, J. E., D. Norman, D. DeJardin, and M. S. Wallace. 2007. The effect of opioids on sleep architecture. *J Clin Sleep Med* 15:33–36.

Dong, H.-L., S. Fukuda, E. Murata, Z. Zhu, and T. Higuchi. 2006. Orexins increase cortical acetylcholine release and electroencephalographic activation through orexin-1 receptor in the rat basal forebrain during isoflurane anesthesia. *Anesthesiology* 104:1023–1032.

Douglas, C. L., G. N. Bowman, H. A. Baghdoyan, and R. Lydic. 2005. C57BL/6J and B6.V-LEP[OB] mice differ in the cholinergic modulation of sleep and breathing. *J Appl Physiol* 98:918–929.

Drover, D. 2006. Patient State Index. *Best Pract Res Clin Anaesthesiol* 20:121–128.

Dzoljic, M. R., O. E. Ukponmwan, and P. R. Saxena. 1992. 5-HT1-like receptor agonists enhance wakefulness. *Neuropharmacology* 31:623–633.

Eckert, D. J., and A. Malhotra. 2008. Pathophysiology of adult obstructive sleep apnea. *Proc Am Thorac Soc* 5:144–153.

Eggermann, E., M. Serafin, L. Bayer, D. Machard, B. Saint-Mleux, B. E. Jones, and M. Muhlethaler. 2001. Orexins/hypocretins excite basal forebrain cholinergic neurones. *Neuroscience* 108:177–181.

Elazar, Z., and A. Berchanski. 2001. Glutamatergic-cholinergic synergistic interaction in the pontine reticular formation. Effects on catalepsy. *Naunyn Schemiedbergs Arch Pharmacol* 363:569–576.

Elmenhorst, D., R. Basheer, R. W. McCarley, and A. Bauer. 2009. Sleep deprivation increases A$_1$ adenosine receptor density in the rat brain. *Brain Res* 1258:53–58.

Eriksson, K. S., O. Sergeeva, R. E. Brown, and H. L. Haas. 2001. Orexin/hypocretin excites the histaminergic neurons of the tuberomammillary nucleus. *J Neurosci* 21:9273–9279.

España, R. A., B. A. Baldo, A. E. Kelley, and C. W. Berridge. 2001. Wake-promoting and sleep-suppressing actions of hypocretin (orexin): basal forebrain sites of action. *Neuroscience* 106:699–715.

España, R. A., and T. E. Scammell. 2004. Sleep neurobiology for the clinician. *Sleep* 27:811–820.

Fadel, J., R. Pasumarthi, and L. R. Reznikov. 2005. Stimulation of cortical acetylcholine release by orexin A. *Neuroscience* 130:541–547.

Fallon, J. H., and R. Y. Moore. 1978. Catecholamine innervation of the basal forebrain. IV. Topography of the dopamine projection to the basal forebrain and neostriatum. *J Comp Neurol* 180:545–580.

Faulhaber, J., A. Steiger, and M. Lancel. 1997. The GABA$_A$ agonist THIP produces slow wave sleep and reduces spindling activity in NREM sleep in humans. *Psychopharmacol (Berl)* 130:285–291.

Feldman, M. 1971. Physiological observations in a chronic case of "locked in" syndrome. *Neurology* 21:459–478.

Findlay, A. L., and J. N. Hayward. 1969. Spontaneous activity of single neurones in the hypothalamus of rabbits during sleep and waking. *J Physiol* 201:237–258.

Fink, K. B., and M. Gothert. 2007. 5-HT receptor regulation of neurotransmitter release. *Pharmacol Rev* 59:360–417.

Fiske, E., J. Grønli, B. Bjorvatn, R. Ursin, and C. M. Portas. 2006. The effect of GABAA antagonist bicuculline on dorsal raphé nucleus and frontal cortex extracellular serotonin: a window on SWS and REM sleep modulation. *Pharmacol Biochem Behav* 83:314–321.

Florin-Lechner, S. M., J. P. Druhan, G. Aston-Jones, and R. J. Valentino. 1996. Enhanced norepinephrine release in prefrontal cortex with burst stimulation of the locus coeruleus. *Brain Res* 742:89–97.

Foote, S. L., G. Aston-Jones, and F. E. Bloom. 1980. Impulse activity of locus coeruleus neurons in awake rats and monkeys is a function of sensory stimulation and arousal. *Proc Natl Acad Sci USA* 77:3033–3037.

Franks, N, P. 2008. General anaesthesia: from molecular targets to neuronal pathways of sleep and arousal. *Nat Rev Neurosci* 9:370–386.

Freeman, A., B. Ciliax, R. Bakay, J. Daley, R. D. Miller, G. Keating, A. Levey, and D. Rye. 2001. Nigrostriatal collaterals to thalamus degenerate in parkinsonian animal models. *Ann Neurol* 76:403–425.

George, R., W. L. Haslett, and D. J. Jenden. 1964. A cholinergic mechanism in the brainstem reticular formation: induction of paradoxical sleep. *Int J Neuropharmacol* 72:541–552.

Gerashchenko, D., C. Blanco-Centurian, M. A. Greco, and P. J. Shiromani. 2003. Effects of lateral hypothalamic lesion with the neurotoxin hypocretin-2-saporin on sleep in Long-Evans rats. *Neuroscience* 116:223–235.

Gerashchenko, D., and P. J. Shiromani. 2004. Effects of inflammation produced by chronic lipopolysaccharide administration on the survival of hypocretin neurons and sleep. *Brain Res* 1019:162–169.

Gerashchenko, D., J. P. Wisor, D. Burns, R. K. Reh, P. J. Shiromani, T. Sakurai, H. O. de la Iglesia, and T. S. Kilduff. 2008. Identification of a population of sleep-active cerebral cortex neurons. *Proc Natl Acad Sci USA* 105:10227–10232.

Giovannini, M. G., L. Bartolini, S. R. Kopf, and G. Pepeu. 1998. Acetylcholine release from the frontal cortex during exploratory activity. *Brain Res* 784:218–227.

Gironell, A., M. D. de la Calzada, T. Sagales, and L. Barraquer-Bordas. 1995. Absence of REM sleep and altered non-REM sleep caused by a haematoma in the pontine tegmentum. *J Neurol Neurosurg Psychiatry* 59:195–196.

Greco, M. A., P. M. Fuller, T. C. Jhou, S. Martin-Schild, J. E. Zadina, Z. Hu, P. Shiromani, and J. Lu. 2008. Opioidergic projections to sleep-active neurons in the ventrolateral preoptic nucleus. *Brain Res* 1245:96–107.

Greco, M. A., and P. J. Shiromani. 2001. Hypocretin receptor protein and mRNA expression in the dorsolateral pons of rats. *Mol Brain Res* 88:176–182.

Greene, R. W., and D. O. Carpenter. 1985. Actions of neurotransmitters on pontine medial reticular formation neurons of cat. *J Neurophysiol* 54:520–531.

Haas, H. L, and P. Panula. 2003. The role of histamine and the tuberomamillary nucleus in the nervous system. *Nat Rev Neurosci* 4:121–130.

Hagan, J. J., R. A. Leslie, S. Patel, M. L. Evans, T. A. Wattam, S. Holmes, C. D. Benham, S. G. Taylor, C. Routledge, P. Hemmati, R. P. Munton, T. E. Ashmeade, A. S. Shah, J. P. Hatcher, P. D. Hatcher, D. N. C. Jones, M. I. Smith, D. C. Piper, A. J. Hunter, R. A. Porter, and N. Upton. 1999. Orexin A activates locus coeruleus cell firing and increases arousal in the rat. *Proc Natl Acad Sci USA* 96:10911–10916.

Halassa, M. M., T. Fellin, and P. G. Haydon. 2007. The tripartite synapse: roles for gliotransmission in health and disease. *Trends Mol Med* 13:54–63.

Halassa, M. M., C. Florian, T. Fellin, J. R. Munoz, S.-Y. Lee, T. Abel, P. G. Haydon, and M. G. Frank. 2009. Astrocytic modulation of sleep homeostasis and cognitive consequences of sleep loss. *Neuron* 61:213–219.

Hara, J., C. T. Beuckmann, T. Nambu, J. T. Willie, R. M. Chemelli, C. M. Sinton, F. Sugiyama, K.-I. Yagami, K. Goto, M. Yanagisawa, and T. Sakurai. 2001. Genetic ablation of orexin neurons in mice results in narcolepsy, hypophagia, and obesity. *Neuron* 30:345–354.

Hascup, K. N., E. R. Hascup, F. Pomerleau, P. Huettl, and G. A. Gerhardt. 2008. Second-by-second measures of L-glutamate in the prefrontal cortex and striatum of freely moving mice. *J Pharmacol Exp Ther* 324:725–731.

Hein, L. 2006. Adrenoceptors and signal transduction in neurons. *Cell Tissue Res* 326:541–551.

Hermans, A., R. B. Keithley, J. M. Kita, L. A. Sombers, and R. M. Wightman. 2008. Dopamine detection with fast-scan cyclic voltammetry used with analog background subtraction. *Anal Chem* 80:4040–4048.

Hernandez-Peon, R., G. Chavez-Ibarra, P. J. Morgane, and C. Timo-Iaria. 1963. Limbic cholinergic pathways involved in sleep and emotional behavior. *Exp Neurol* 8:93–111.

Hervieu, G. J., J. E. Cluderay, D. C. Harrison, J. C. Roberts, and R. A. Leslie. 2001. Gene expression and protein distribution of the orexin-1 receptor in the rat brain and spinal cord. *Neuroscience* 103:777–792.

Himukashi, S., Y. Miyazaki, H. Takeshima, S. Koyanagi, K. Mukaida, T. Shichino, H. Uga, and K. Fukuda. 2005. Nociceptin system does not affect MAC of volatile anesthetics. *Acta Anaesthesiol Scand* 49:771–773.

Hobson, J. A. 1978. What is a behavioral state? Paper read at Neuroscience Symposia Vol. III: Aspects of behavioral neurobiology, at Bethesda, MD.

Hobson, J. A. 2002. *Dreaming: An Introduction to the Science of Sleep.* New York: Oxford University Press Inc.

Hobson, J. A., R. W. McCarley, and P. W. Wyzinski. 1975. Sleep cycle oscillation: reciprocal discharge by two brainstem neuronal groups. *Science* 189:55–58.

Hogg, R. C., M. Raggenbass, and D. Bertrand. 2003. Nicotinic acetylcholine receptors: from structure to function. *Rev Physiol Biochem Pharmacol* 147:1–46.

Hong, Z.-Y., Z.-L. Huang, W.-M. Qu, and N. Eguchi. 2005. Orexin A promotes histamine, but not norepinephrine or serotonin, release in frontal cortex of mice. *Acta Pharmacol Sin* 26:155–159.

Horvath, T. L., C. Peyron, S. Diano, A. Ivanov, G. Aston-Jones, T. S. Kilduff, and A. N. van den Pol. 1999. Hypocretin (orexin) activation and synaptic innervation of the locus coeruleus noradrenergic system. *J Comp Neurol* 415:145–159.

Houdouin, F., R. Cespuglio, and M. Jouvet. 1991. Effects induced by the electrical stimulation of the nucleus raphé dorsalis upon hypothalamic release of 5-hydroxyindole compounds and sleep parameters in the rat. *Brain Res* 565:48–56.

Hudetz, A. G. 2008. Are we unconscious during general anesthesia? *Int Anesthesiol Clin* 46:25–42.

Hudetz, A. G., J. D. Wood, and J. P. Kampine. 2003. Cholinergic reversal of isoflurane anesthesia in rats as measured by cross-approximate entropy of the electroencephalogram. *Anesthesiology* 99:1125–1131.

Huitron-Resendiz, S., M. P. Kristensen, M. Sánchez-Alavez, S. D. Clark, S. L. Grupke, C. Tyler, C. Suzuki, H.-P. Nothacker, O. Civelli, J. R. Criado, S. J. Henriksen, C. S. Leonard, and L. de Lecea. 2005. Urotensin II modulates rapid eye movement sleep through activation of brainstem cholinergic neurons. *J Neurosci* 25:5465–5474.

Iber, C., S. Ancoli-Israel, A. L. Chesson Jr, and S. F. Quan. 2007. *The AASM Manual for the Scoring of Sleep and Associated Events: Rules, Terminology, and Technical Specifications.* Westchester: American Academy of Sleep Medicine.

Inagaki, N., A. Yamatodani, M. Ando-Yamamoto, M. Tohyama, T. Watanabe, and H. Wada. 1988. Organization of histaminergic fibers in rat brain. *J Comp Neurol* 273:283–300.

Isaac, S. O., and C. W. Berridge. 2003. Wake-promoting actions of dopamine D1 and D2 receptor stimulation. *J Pharmacol Exp Ther* 307:386–394.

Ishii, M., and Y. Kurachi. 2006. Muscarinic acetylcholine receptors. *Curr Pharm Des* 12: 3573–3581.

Ishizuka, T., Y. Yamamoto, and A. Yamatodani. 2002. The effect of orexin-A and -B on the histamine release in the anterior hypothalamus in rats. *Neurosci Lett* 323:93–96.

Jacob, T. C., S. J. Moss, and R. Jurd. 2008. GABA$_A$ receptor trafficking and its role in the dynamic modulation of neuronal inhibition. *Nat Rev Neurosci* 9:331–343.

Jacobs, B. L. 1986. Single unit activity of locus coeruleus neurons in behaving animals. *Prog Neurobiol* 27:183–194.

Jasper, H. H., and J. Tessier. 1971. Acetylcholine liberation from cerebral cortex during paradoxical (REM) sleep. *Science* 172:601–602.

Johansen, J. W. 2006. Update on bispectral index monitoring. *Best Pract Res Clin Anaesthesiol* 20:81–99.

John, J., L. Ramanathan, and J. M. Siegel. 2008. Rapid changes in glutamate levels in the posterior hypothalamus across sleep-wake states in freely behaving rats. *Am J Physiol Regul Integr Comp Physiol* 295:R2041–R2049.

Johnson, M. D., R. K. Franklin, M. D. Gibson, R. B. Brown, and D. R. Kipke. 2008. Implantable microelectrode arrays for simultaneous electrophysiological and neurochemical recordings. *J Neurosci Methods* 174:62–70.

Jones, B. E. 2003. Arousal states. *Front Biosci* 8:S438–S451.

Jones, B. E., and A. Beaudet. 1987. Distribution of acetylcholine and catecholamine neurons in the cat brainstem: a choline acetyltransferase and tyrosine hydroxylase immunohistochemical study. *J Comp Neurol* 261:15–32.

Jones, B. E., and A. C. Cuello. 1989. Afferents to the basal forebrain cholinergic cell area from pontomesencephalic-catecholamine, serotonin, and acetylcholine-neurons. *Neuroscience* 31:37–61.

Jouvet, M. 1962. Recherches sur les structures nerveuses et les mécanisms responsables des différentes phases du sommeil physiologique. *Arch Ital Biol* 100:125–206.

Jouvet, M. 1965. Paradoxical sleep-a study of its nature and mechanisms. *Prog Brain Res* 18: 20–62.

Jouvet, M. 1969. Biogenic amines and the states of sleep. *Science* 163:32–41.

Jouvet, M., and F. Michel. 1959. Corrélations électromyographiques du sommeil chez le chat décortiqué et mésencéphalique chronique. *Comptes Rendus des Seances de la Societe de Biologie et de Ses Filiales* 153:422–425.

Jouvet, M., F. Michel, and J. Courjon. 1959. Sur la mise en jeu de deux mécaisemes á expression électro-encéphalographique différente au cours du sommeil physiologique chez le chat. *Comptes Rendus Hebdomadaires Des Seances de l Academie des Sciences* 248:3043–3045.

Jouvet, M, F. Michel, and J. Courjon. 1959. Sur un stade d'activité électrique rapide au cours du sommeil physiologique. *Comptes Rendus des Seances de la Societe de Biologie et de Ses Filiales* 153:1024–1028.

Kaitin, K. I. 1984. Preoptic area unit activity during sleep and wakefulness in the cat. *Exp Neurol* 83:347–357.

Kaneko, T., K. Itoh, R. Shigemoto, and N. Mizuno. 1989. Glutaminase-like immunoreactivity in the lower brainstem and cerebellum of the adult rat. *Neuroscience* 32:79–98.

Kaneko, Y., K. Shimada, K. Saitou, Y. Sugimoto, and C. Kamei. 2000. The mechanism responsible for the drowsiness caused by first generation H1 antagonists on the EEG pattern. *Methods Find Exp Clin Pharm* 22:163–168.

Kehr, J., and U. Ungerstedt. 1988. Fast HPLC estimation of gamma-aminobutyric acid in microdialysis perfusates: effects of nipecotic and 3-mecaptopropionic acids. *J Neurochem* 51:1308–1310.

Keifer, J. C., H. A. Baghdoyan, L. Becker, and R. Lydic. 1994. Halothane decreases pontine acetylcholine release and increases EEG spindles. *Neuroreport* 5:577–580.

Keifer, J. C., H. A. Baghdoyan, and R. Lydic. 1992. Sleep disruption and increased apneas after pontine microinjection of morphine. *Anesthesiology* 77:973–982.

Kelz, M. B., Y. Sun, J. Chen, Q. C. Meng, J. T. Moore, S. C. Veasey, S. Dixon, M. Thornton, H. Funato, and M. Yanagisawa. 2008. An essential role for orexins in emergence from general anesthesia. *Proc Natl Acad Sci USA* 105:1309–1314.

Kennedy, R. T., J. E. Thompson, and T. W. Vickroy. 2002. In vivo monitoring of amino acids by direct sampling of brain extracellular fluid at ultralow flow rates and capillary electrophoresis. *J Neurosci Methods* 114:39–49.

Kim, J., K. Nakajima, Y. Oomora, M. J. Wayner, and K. Sasaki. 2009. Electrophysiological effects of ghrelin on pedunculopontine tegmental neurons in rats: an in vitro study. *Peptides* 30:745–757.

Kimura, K., N. Tachibana, J. Kohyama, Y. Otsuka, S. Fukazawa, and R. Waki. 2000. A discrete pontine ischemic lesion could cause REM sleep behavior disorder. *Neurology* 55:894–895.

Kitsikis, A., and M. Steriade. 1981. Immediate behavioral effects of kainic acid injections into the midbrain reticular core. *Behav Brain Res* 3:361–380.

Kiyashchenko, L. I., B. Y. Mileykovskiy, N. Maidment, H. A. Lam, M.-F. Wu, J. John, J. Peever, and J. M. Siegel. 2002. Release of hypocretin (orexin) during waking and sleep states. *J Neurosci* 22:5282–5286.

Kiyono, S., M. L. Seo, M. Shibagaki, T. Watanabe, K. Maeyama, and H. Wada. 1985. Effects of α-fluoromethylhistidine on sleep-waking parameters in rats. *Physiol Behav* 34:615–617.

Ko, E. M., I. V. Estabrooke, M. McCarthy, and T. E. Scammell. 2003. Wake-related activity of tuberomammillary neurons in rats. *Brain Res* 992:220–226.

Kodama, T., Y. Takahashi, and Y. Honda. 1990. Enhancement of acetylcholine release during paradoxical sleep in the dorsal tegmental field of the cat brain stem. *Neurosci Lett* 114:277–282.

Kohlmeier, K. A., J. Burns, P. B. Reiner, and K. Semba. 2002. Substance P in the descending cholinergic projection to REM sleep-induction regions of the rat pontine reticular formation: anatomical and electrophysiological analyses. *Eur J Neurosci* 15:176–196.

Kottegota, S., I. Shaik, and S. A. Shippy. 2002. Demonstration of low flow push–pull perfusion. *J Neurosci Methods* 121:93–101.

Koyama, Y., and O. Hayaishi. 1994. Firing of neurons in the preoptic/anterior hypothalamic areas in rat: its possible involvement in slow wave sleep and paradoxical sleep. *Neurosci Res* 19: 31–38.

Koyama, Y., T. Kodama, K. Takahashi, K. Okai, and Y. Kayama. 2002. Firing properties of neurones in the laterodorsal hypothalamic area during sleep and wakefulness. *Psychiat Clin Neurosci* 56:339–340.

Kreuer, S. 2006. The Narcotrend monitor. *Best Pract Res Clin Anaesthesiol* 20:111–119.

Kshatri, A. M., H. A. Baghdoyan, and R. Lydic. 1998. Cholinomimetics, but not morphine, increase antinociceptive behavior from pontine reticular regions regulating rapid eye movement sleep. *Sleep* 21:677–685.

Kumar, V. M., S. Datta, G. S. Chhina, and B. Singh. 1986. Alpha adrenergic system in medial preoptic area in sleep-wakefulness in rats. *Brain Res Bull* 16:463–468.

Kurosawa, M., K. Okada, A. Sato, and S. Uchida. 1993. Extracellular release of acetylcholine, noradrenaline and serotonin increases in the cerebral cortex during walking in conscious rats. *Neurosci Lett* 161:73–76.

Kushikata, T., K. Hirota, H. Yoshida, M. Kudo, D. G. Lambert, D. Smart, J. C. Jerman, and A. Matsuki. 2003. Orexinergic neurons and barbiturate anesthesia. *Neuroscience* 121:855–863.

Lai, Y. Y., and J. M. Siegel. 1988. Medullary regions mediating atonia. *J Neurosci* 8:4790–4796.

Lai, Y. Y., and J. M. Siegel. 1991. Pontomedullary glutamate receptors mediating locomotion and muscle tone suppression. *J Neurosci* 11:2931–2937.

Lancel, M. 1997. The GABA$_A$ agonist THIP increases non-REM sleep and enhances non-REM sleep-specific delta activity in the rat during the dark period. *Sleep* 20:1099–1104.

Landolt, H. P., D. J. Dijk, S. E. Gaus, and A. A. Borbely. 1995. Caffeine reduces low-frequency delta activity in the human sleep EEG. *Neuropsychopharmacol* 12:229–238.

Landolt, H. P., E. Werth, A. A. Borbely, and D. J. Dijk. 1995. Caffeine intake (200 mg) in the morning affects human sleep and EEG power spectra at night. *Brain Res* 675:67–74.

Laposky, A. D., M. A. Bradley, D. L. Williams, J. Bass, and F. W. Turek. 2008. Sleep-wake regulation is altered in leptin-resistant (db/db) genetically obese and diabetic mice. *Am J Physiol Regul Integr Comp Physiol* 295:R2059–R2066.

Laposky, A. D., J. Shelton, J. Bass, C. Dugovic, N. Perrino, and F. W. Turek. 2006. Altered sleep regulation in leptin-deficient mice. *Am J Physiol Regul Integr Comp Physiol* 290:R894–R903.

Laureys, S., and G. Tononi, eds. 2009. *The Neurology of Consciousness: Cognitive Neuroscience and Neuropathology*. Burlington: Academic Press.

Lavie, P., H. Pratt, B. Scharf, R. Peled, and J. Brown. 1984. Localized pontine lesion: nearly total absence of REM sleep. *Neurology* 34:118–120.

Le, S., J. A. Gruner, J. R. Mathiasen, M. J. Marino, and H. Schaffhauser. 2008. Correlation between ex vivo receptor occupancy and wake-promoting activity of selective H$_3$ receptor antagonists. *J Pharmacol Exp Ther* 325:902–909.

Lee, M. G., O. K. Hassani, and B. E. Jones. 2005. Discharge of identified orexin/hypocretin neurons across the sleep-waking cycle. *J Neurosci* 25:6716–6720.

Leonard, T. O., and R. Lydic. 1995. Nitric oxide synthase inhibition decreases pontine acetylcholine release. *Neuroreport* 6:1525–1529.

Leonard, T. O., and R. Lydic. 1997. Pontine nitric oxide modulates acetylcholine release, rapid eye movement sleep generation, and respiratory rate. *J Neurosci* 17.774–785.

Li, Y., and A. N. van den Pol. 2008. μ-opioid receptor-mediated depression of the hypothalamic hypocretin/orexin arousal system. *J Neurosci* 28:2814–2819.

Ligneau, X., D. Perrin, L. Landais, J.-C. Camelin, T. P. G. Calmels, I. Berrebi-Bertrand, J.-M. Lecomte, R. Parmentier, C. Anaclet, J. S. Lin, V. Bertaina-Anglade, C. Drieu la Rochelle, F. d'Aniello, A. Rouleau, F. Gbahou, J. M. Arrang, C. R. Ganellin, H. Stark, W. Schunack, and J. C. Schwartz. 2007. BF2.649 [1-{3-[3-(4-Chlorophenyl)propoxy]propyl}piperidine, hydrochloride], a nonimidazole inverse agonist/antagonist at the human histamine H3 receptor: preclinical pharmacology. *J Pharmacol Exp Ther* 320:365–375.

Lin, J. S., K. Sakai, and M. Jouvet. 1988. Evidence for histaminergic arousal mechanisms in the hypothalamus of cats. *Neuropharmacology* 27:111–122.

Lin, J. S., K. Sakai, G. Vanni-Mercier, and M. Jouvet. 1989. A critical role of the posterior hypothalamus in the mechanisms of wakefulness determined by microinjection of muscimol in freely moving cats. *Brain Res* 479:225–240.

Lin, L., J. Faraco, R. Li, H. Kadotani, W. Rogers, X. Lin, X. Qiu, P. J. de Jong, S. Nishino, and E. Mignot. 1999. The sleep disorder canine narcolepsy is caused by a mutation in the hypocretin (orexin) receptor 2 gene. *Cell* 98:365–376.

Lincoln, D. W. 1969. Correlation of unit activity in the hypothalamus with EEG patterns associated with the sleep cycle. *Exp Neurol* 24:1–18.

Liu, R.-J., A. N. van den Pol, and G. K. Aghajanian. 2002. Hypocretins (orexins) regulate serotonin neurons in the dorsal raphé nucleus by excitatory direct and inhibitory indirect actions. *J Neurosci* 22:9453–9464.

Liu, Z.-W., and X. B. Gao. 2007. Adenosine inhibits activity of hypocretin/orexin neurons by the A1 receptor in the lateral hypothalamus: a possible sleep-promoting effect. *J Neurophysiol* 97:837–848.

Loomis, A. L., E. N. Harvey, and G. Hobart. 1935. Further observations on the potential rhythms of the cerebral cortex during sleep. *Science* 82:198–200.

Loomis, A. L., E. N. Harvey, and G. Hobart. 1935. Potential rhythms of the cerebral cortex during sleep. *Science* 81:597–598.

Loomis, A. L., E. N. Harvey, and G. Hobart. 1937. Cerebral states during sleep, as studied by human brain potentials. *J Exp Psych* 21:127–144.

Loomis, A. L., E. N. Harvey, and G. Hobart. 1938. Distribution of disturbance-patterns in the human electroencephalogram, with special reference to sleep. *J Neurophysiol* 1:413–430.

Lu, J., L. E. Nelson, N. Franks, M. Maze, N. L. Chamberlin, and C. B. Saper. 2008. Role of endogenous sleep-wake and analgesic systems in anesthesia. *J Comp Neurol* 508:648–662.

Lydic, R. 1996. Reticular modulation of breathing during sleep and anesthesia. *Curr Opin Pulm Med* 2:474–481.

Lydic, R., and H. A. Baghdoyan. 1993. Pedunculopontine stimulation alters respiration and increases ACh release in the pontine reticular formation. *Am J Physiol* 264:R544–R554.

Lydic, R., and H. A. Baghdoyan. 2002. Ketamine and MK-801 decrease acetylcholine release in the pontine reticular formation, slow breathing, and disrupt sleep. *Sleep* 25:615–620.

Lydic, R., and H. A. Baghdoyan. 2005. Sleep, anesthesiology, and the neurobiology of arousal state control. *Anesthesiology* 103:1268–1295.

Lydic, R., and H. A. Baghdoyan. 2007. Neurochemical mechanisms mediating opioid-induced REM sleep disruption. In *Sleep and Pain*, edited by G. Lavigne, B. Sessle, M. Choinière and P. Soja. Seattle: IASP Press.

Lydic, R., and J. F. Biebuyck. 1994. Sleep neurobiology: relevance for mechanistic studies of anaesthesia. *Br J Anaesth* 72:506–508.

Lydic, R., C. L. Douglas, and H. A. Baghdoyan. 2002. Microinjection of neostigmine into the pontine reticular formation of C57BL/6J mouse enhances rapid eye movement sleep and depresses breathing. *Sleep* 25:835–841.

Lydic, R., R. W. McCarley, and J. A. Hobson. 1983. The time-course of dorsal raphé discharge, PGO waves, and muscle tone averaged across multiple sleep cycles. *Brain Res* 274:365–370.

Manohar, S., H. Noda, and W. R. Adey. 1972. Behavior of mesencephalic reticular neurons in sleep and wakefulness. *Exp Neurol* 34:140–157.

Marcus, J. N., C. N. Aschenasi, C. E. Lee, R. M. Chemelli, C. B. Saper, M. Yanagisawa, and J. K. Elmquist. 2001. Differential expression of orexin receptors 1 and 2 in the rat brain. *J Comp Neurol* 435:6–25.

Markland, O. N., and M. L. Dyken. 1976. Sleep abnormalities in patients with brain stem lesions. *Neurology* 26:769–776.

Marks, G. A., O. W. Sachs, and C. G. Birabil. 2008. Blockade of GABA, type A, receptors in the rat pontine reticular formation induces rapid eye movement sleep that is dependent upon the cholinergic system. *Neuroscience* 156:1–10.

Marrosu, F., C. Portas, M. S. Mascia, M. A. Casu, M. Fa, M. Giagheddu, A. Imperatu, and G. L. Gessa. 1995. Microdialysis measurement of cortical and hippocampal acetylcholine release during sleep-wake cycle in freely moving cats. *Brain Res* 671:329–332.

Martinez-Mir, M. I., H. Pollard, J. Moreau, J. M. Arrang, M. Ruat, E. Traiffort, J. C. Schwartz, and J. M. Palacios. 1990. Three histamine receptors (H1, H2, and H3) visualized in the brain of human and non-human primates. *Brain Res* 526:322–327.

Mashour, G. A. 2006. Monitoring consciousness: EEG-based measures of anesthetic depth. *Semin Anesth Perioperat Med Pain* 25:205–210.

Mashour, G. A. Ed. 2008. Unconscious Processes. *Int Anesthesiol Clin* 46:1–205.

Materi, L. M., D. D. Rasmusson, and K. Semba. 2000. Inhibition of synaptically evoked cortical acetylcholine release by adenosine: an in vivo microdialysis study in the rat. *Neuroscience* 97:219–226.

McCarley, R. W. 2007. Neurobiology of REM and NREM sleep. *Sleep Med* 8:302–330.

McCarley, R. W., and J. A. Hobson. 1975. Neuronal excitability modulation over the sleep cycle: a structural and mathematical model. *Science* 189:58–60.

McGinty, D. J. 1969. Somnolence, recovery, and hyposomnia following ventro-medial diencephalic lesions in the rat. *Electroenceph Clin Neurophysiol* 26:70–79.

McGinty, D. J., and R. M. Harper. 1976. Dorsal raphé neurons: depression of firing during sleep in cats. *Brain Res* 101:569–575.

McGinty, D. J., and M. B. Sterman. 1968. Sleep suppression after basal forebrain lesions in the cat. *Science* 160:1253–1255.

Mendelson, W. B. 2001. The sleep-inducing effect of ethanol microinjection into the medial preoptic area is blocked by flumazenil. *Brain Res* 892:118–121.

Mendelson, W. B., J. V. Martin, M. Perlis, and R. Wagner. 1989. Enhancement of sleep by microinjection of triazolam into the medial preoptic area. *Neuropsychopharmacol* 2:61–66.

Menefee, L. A., M. J. M. Cohen, W. R. Anderson, K. Doghramji, E. D. Frank, and H. Lee. 2000. Sleep disturbance and nonmalignant chronic pain: a comprehensive review of the literature. *Pain Med* 1:156–172.

Methippara, M. M., M. N. Alam, R. Szymusiak, and D. J. McGinty. 2000. Effects of lateral preoptic area application of orexin-A on sleep-wakefulness. *Neuroreport* 11:3423–3426.

Meuret, P., S. B. Backman, V. Bonhomme, G. Plourde, and P. Fiset. 2000. Physostigmine reverses propofol-induced unconsciousness and attenuation of the auditory steady state response and bispectral index in human volunteers. *Anesthesiology* 93:708–717.

Mileykovskiy, B. Y., L. I. Kiyashchenko, and J. M. Siegel. 2005. Behavioral correlates of activity in identified hypocretin/orexin neurons. *Neuron* 46:787–798.

Miller, J. D., J. Farber, P. Gatz, H. Roffwarg, and D. C. German. 1983. Activity of mesencephalic dopamine and non-dopamine neurons across stages of sleep and waking in the rat. *Brain Res* 273:133–141.

Missale, C., S. R. Nash, S. W. Robinson, M. Jaber, and M. G. Caron. 1998. Dopamine receptors: from structure to function. *Physiol Rev* 78:189–225.

Mitani, A., K. Ito, A. E. Hallanger, B. H. Wainer, K. Kataoka, and R. W. McCarley. 1988. Cholinergic projections from the laterodorsal and pedunculopontine tegmental nuclei to the pontine gigantocellular field in the cat. *Brain Res* 451:397–402.

Mobarakeh, J. I., K. Takahashi, S. Sakurada, S. Nishino, T. Watanabe, M. Kato, and K. Yanai. 2005. Enhanced antinociception by intracerebroventricularly and intrathecally-administered orexin A and B (hypocretin-1 and -2) in mice. *Peptides* 26:767–777.

Moldofsky, H. 2001. Sleep and Pain. *Sleep Med Rev* 5:387–398.

Monti, J. M., L. D'Angelo, H. Jantos, and S. Pazos. 1988. Effects of a-fluoromethylhistidine on sleep and wakefulness in the rat. *J Neural Transm* 72:141–145.

Monti, J. M., M. Fernandez, and H. Jantos. 1990. Sleep during acute dopamine D1 agonist SKF 38393 or D1 antagonist SCH 23390 administration in rats. *Neuropsychopharmacol* 3:153–162.

Monti, J. M., M. Hawkins, H. Jantos, L. D'Angelo, and M. Fernandez. 1988. Biphasic effects of dopamine D-2 receptor agonists on sleep and wakefulness in the rat. *Psychopharmacol (Berl)* 95:395–400.

Monti, J. M., and H. Jantos. 1992. Dose-dependent effects of the 5-HT1A receptor agonist 8-OH-DPAT on sleep and wakefulness in the rat. *J Sleep Res* 1:169–175.

Monti, J. M., and H. Jantos. 2006a. Effects of activation and blockade of 5-HT$_{2A/2C}$ receptors in the dorsal raphé nucleus on sleep and waking in the rat. *Prog Neuropsychopharmacol Biol Psychiatry* 30:1189–1195.

Monti, J. M., and H. Jantos. 2006b. Effects of the serotonin 5-HT$_{2A/2C}$ receptor agonist DOI and of the selective 5-HT$_{2A}$ or 5-HT$_{2C}$ receptor antagonists EMD 281014 and SB-243213, respectively, on sleep and waking in the rat. *Eur J Pharmacol* 553:163–170.

Monti, J. M., H. Jantos, and M. Fernandez. 1989. Effects of the selective dopamine D-2 receptor agonist, quinpirole on sleep and wakefulness in the rat. *Eur J Pharmacol* 169:61–66.

Monti, J. M., S. R. Pandi-Perumal, and C. M. Sinton, eds. 2008. *Neurochemistry of Sleep and Wakefulness*. New York: Cambridge University Press.

Mooney, M. E., D. V. Herin, J. M. Schmitz, N. Moukaddam, C. E. Green, and J. Grabowski. 2009. Effects of oral methamphetamine on cocaine use: a randomized, double blind, placebo-controlled trial. *Drug Alcohol Depend* 101:34–41.

Moreno-Balandran, M. E., M. Garzon, C. Bodalo, F. Reinoso-Suárez, and I. de Andrés. 2008. Sleep-wakefulness effects after microinjections of hypocretin-1 (orexin A) in cholinoceptive areas of the cat oral pontine tegmentum. *Eur J Neurosci* 28:331–341.

Moriarty, S., L. Hedley, J. Flores, R. Martin, and T. S. Kilduff. 2008. Selective 5-HT$_{2A}$ and 5-HT$_6$ receptor antagonists promote sleep in rats. *Sleep* 31:34–44.

Morrow, J. D., S. Vikraman, L. Imeri, and M. R. Opp. 2008. Effects of serotonergic activation by 5-hydroxytryptophan on sleep and body temperature of C57BL/6J and interleukin-6-deficient mice are dose and time related. *Sleep* 31:21–33.

Moruzzi, G. 1964. Reticular influences on the EEG. *Clin Neurophysiol* 16:2–17.

Moruzzi, G., and H. W. Magoun. 1949. Brain stem reticular formation and activation of the EEG. *Electroenceph Clin Neurophysiol* 1:455–473.

Murillo-Rodriguez, E., M. Liu, C. Blanco-Centurian, and P. J. Shiromani. 2008. Effects of hypocretin (orexin) neuronal loss on sleep and extracellular adenosine levels in the rat basal forebrain. *Eur J Neurosci* 28:1191–1198.

Nambu, T., T. Sakurai, K. Mizukami, Y. Hosoya, M. Yanagisawa, and K. Goto. 1999. Distribution of orexin neurons in the adult rat brain. *Brain Res* 827:243–260.

Nauta, W. J. H. 1946. Hypothalamic regulation of sleep in rats. An experimental study. *J Neurophysiol* 9:285–316.

Nicholson, A. N., and B. M. Stine. 1986. Antihistamines: impaired performance and the tendency to sleep. *Eur J Clin Pharmacol* 30:27–32.

Nishino, S., and T. Kanbayashi. 2005. Symptomatic narcolepsy, cataplexy, and hypersomnia, and their implications in the hypothalamic hypocretin/orexin system. *Sleep Med Rev* 9:269–310.

Nishino, S., B. Ripley, S. Overeem, G. J. Lammers, and E. Mignot. 2000. Hypocretin (orexin) deficiency in human narcolepsy. *Lancet* 355:39–40.

Nishino, S., and T. Sakurai, eds. 2005. *The Orexin/Hypocretin System: Physiology and Pathophysiology*. Totowa, NJ: Humana Press Inc.

Nitz, D., and J. M. Siegel. 1996. GABA release in posterior hypothalamus across sleep-wake cycle. *Am J Physiol* 271:R1707–R1712.

Núñez, A., W. Buño, and F. Reinoso-Suárez. 1998. Neurotransmitter actions on oral pontine tegmental neurons of the rat: an in vitro study. *Brain Res* 804:144–148.

Núñez, A., M. E. Moreno-Balandran, M. L. Rodrigo-Angulo, M. Garzon, and I. de Andrés. 2006. Relationship between the perifornical hypothalamic area and oral pontine reticular nucleus in the rat. Possible implication of the hypocretinergic projection in the control of rapid eye movement sleep. *Eur J Neurosci* 24:2834–2842.

Oishi, Y., Z.-L. Huang, B. B. Fredholm, Y. Urade, and O. Hayaishi. 2008. Adenosine in the tubero-mammillary nucleus inhibits the histaminergic system via A1 receptors and promotes non-rapid eye movement sleep. *Proc Natl Acad Sci USA* 105:19992–19997.

Onoe, H., and K. Sakai. 1995. Kainate receptors: a novel mechanism in paradoxical (REM) sleep generation. *Neuroreport* 6:353–356.

Orser, B. A. 2006. Extrasynaptic GABA$_A$ receptors are critical targets for sedative-hypnotic drugs. *J Clin Sleep Med* 2:S12–S18.

Orser, B. A. 2007. Lifting the fog around anesthesia. *Sci Am* 296:54–61.

Osman, N. I., H. A. Baghdoyan, and R. Lydic. 2005. Morphine inhibits acetylcholine release in rat prefrontal cortex when delivered systemically or by microdialysis to basal forebrain. *Anesthesiology* 103:779–787.

Pal, D., and B. N. Mallick. 2006. Role of noradrenergic and GABA-ergic inputs in pedunculo-pontine tegmentum for regulation of rapid eye movement sleep in rats. *Neuropharmacology* 51:1–11.

Panula, P., S. Pirvola, S. Auvinen, and M. S. Airaksinen. 1989. Histamine-immunoreactive nerve fibers in the rat brain. *Neuroscience* 28:585–610.

Panula, P., H.-Y. T. Yang, and E. Costa. 1984. Histamine-containing neurons in the rat hypothalamus. *Proc Natl Acad Sci USA* 81:2572–2576.

Parmentier, R., C. Anaclet, C. Guhennec, E. Brousseau, D. Bricout, T. Giboulot, D. Bozyczko-Coyne, K. Spiegel, H. Ohtsu, M. Williams, and J. S. Lin. 2007. The brain $H_3$-receptor as a novel therapeutic target for vigilance and sleep-wake disorders. *Biochem Pharmacol* 73:1157–1171.

Paxinos, G., and C. Watson. 2007. *The Rat Brain in Stereotaxic Coordinates*, 6th ed. New York: Academic Press.

Peyron, C., J. Faraco, W. Rogers, B. Ripley, S. Overeem, Y. Charnay, S. Nevsimalova, M. Aldrich, D. Reynolds, R. Albin, R. Li, M. Hungs, M. Pedrazzoli, M. Padigaru, M. Kucherlapati, J. Fan, R. Maki, G. J. Lammers, C. Bouras, R. Kucherlapati, S. Nishino, and E. Mignot. 2000. A mutation in a case of early onset narcolepsy and a generalized absence of hypocretin peptides in human narcoleptic brains. *Nat Med* 6:991–997.

Peyron, C., D. K. Tighe, A. N. van den Pol, L. de Lecea, H. C. Heller, J. G. Sutcliffe, and T. S. Kilduff. 1998. Neurons containing hypocretin (orexin) project to multiple neuronal systems. *J Neurosci* 18:9996–10015.

Phillips, P. E. M., and R. M. Wightman. 2003. Critical guidelines for validation of the selectivity of in-vivo chemical microsensors. *Trends Anal Chem* 22:509–514.

Piper, D. C., N. Upton, M. I. Smith, and A. J. Hunter. 2000. The novel brain neuropeptide, orexin-A, modulates the sleep-wake cycle of rats. *Eur J Neurosci* 12:726–730.

Plourde, G., D. Chartrand, P. Fiset, S. Font, and S. B. Backman. 2003. Antagonism of sevoflurane anaesthesia by physostigmine: effects on the auditory steady-state response and bispectral index. *Br J Anaesth* 91:583–586.

Porkka-Heiskanen, T., R. E. Strecker, and R. W. McCarley. 2000. Brain site-specificity of extracellular adenosine concentration changes during sleep deprivation and spontaneous sleep: an in vivo microdialysis study. *Neuroscience* 99:507–517.

Porkka-Heiskanen, T., R. E. Strecker, M. Thakkar, A. A. Bjorkum, R. W. Greene, and R. W. McCarley. 1997. Adenosine: a mediator of the sleep-inducing effects of prolonged wakefulness. *Science* 276:1265–1268.

Portas, C. M., B. Bjorvatn, S. Fagerland, J. Gronli, V. Mundal, E. Sorensen, and R. Ursin. 1998. On-line detection of extracellular levels of serotonin in dorsal raphé nucleus and frontal cortex over the sleep/wake cycle in the freely moving rat. *Neuroscience* 83:807–814.

Portas, C. M., M. Thakkar, D. Rainnie, and R. W. McCarley. 1996. Microdialysis perfusion of 8-hydroxy-2-(di-n-propylamino)tetralin (8-OH-DPAT) in the dorsal raphé nucleus decreases serotonin release and increases rapid eye movement sleep in the freely moving cat. *J Neurosci* 16:2820–2828.

Punjabi, N. M. 2008. The epidemiology of adult obstructive sleep apnea. *Proc Am Thorac Soc* 5:136–143.

Python, A., T. Steimer, Z. de Saint Hilaire, R. Mikolajewski, and S. Nicolaidis. 2001. Extracellular serotonin variations during vigilance states in the preoptic area of rats: a microdialysis study. *Brain Res* 910:49–54.

Ranson, S. W. 1939. Somnolence caused by hypothalamic lesions in the monkey. *Arch Neurol Psychiatry* 41:1–23.

Rechtschaffen, A., and A. Kales. 1968. *A Manual of Standardized Terminology: Techniques and Scoring System for Sleep Stages of Human Subjects*. Washington, DC: US Government Printing Office.

Regoli, D., G. Drapeau, S. Dion, and P. D'Orleans-Juste. 1989. Receptors for substance P and related neurokinins. *Pharamacology* 38:1–15.

Reinoso-Suarez, F., I. de Andrés, M. L. Rodrigo-Angulo, and M. Garzón. 2001. Brain structures and mechanisms involved in the generation of REM sleep. *Sleep Med Rev* 5:63–77.

Robinson, D. L., A. Hermans, A. T. Seipel, and R. M. Wightman. 2008. Monitoring rapid chemical communication in the brain. *Chem Rev* 108:2554–2584.

Roehrs, T., and T. Roth. 2005. Sleep and pain: interaction of two vital functions. *Semin Neurol* 25:106–116.

Roth, T., R. Rogowski, S. Hull, H. Schwartz, G. Koshorek, B. Corser, and D. Seiden. 2007. Efficacy and safety of doxepin 1 mg, 3 mg, and 6 mg in adults with primary insomnia. *Sleep* 30: 1555–1561.

Russell, I. J., A. T. Perkins, J. E. Michalek, and Oxybate SXB-26 Fibromyalgia Syndrome Study Group. 2009. Sodium oxybate relieves pain and improves function in fibromyalgia syndrome: a randomized, double-blind, placebo-controlled, multicenter clinical trial. *Arthritis Rheum* 60:299–309.

Sagales, T., S. Erill, and E. F. Domino. 1969. Differential effects of scopolamine and chlorpromazine on REM and NREM sleep in normal male subjects. *Clin Pharmacol Ther* 10: 522–529.

Saitou, K., T. Kaneko, Y. Sugimoto, Z. Chen, and C. Kamei. 1999. Slow wave sleep-inducing effects of first generation H1-antagonists. *Biol Pharm Bull* 22:1079–1082.

Sakai, K., D. Salvert, M. Touret, and M. Jouvet. 1977. Afferent connections of the nucleus raphé dorsalis in the cat as visualized by the horseradish peroxidase technique. *Brain Res* 137:11–35.

Sakai, K., Y. Yoshimoto, P. H. Luppi, P. Fort, M. El Mansari, D. Salvert, and M. Jouvet. 1990. Lower brainstem afferents to the cat posterior hypothalamus: a double-labeling study. *Brain Res Bull* 24:437–455.

Sakurai, T., A. Amemiya, M. Ishii, I. Matsuzaki, R. M. Chemelli, H. Tanaka, S. C. Williams, J. A. Richardson, G. P. Kozlowski, S. Wilson, J. R. S. Arch, R. E. Buckingham, A. C. Haynes, S. A. Carr, R. S. Annan, D. E. McNulty, W.-S. Liu, J. A. Terrett, N. A. Elshourbagy, D. J. Bergsma, and M. Yanagisawa. 1998. Orexin and orexin receptors: a family of hypothalamic neuropeptides and G protein-coupled receptors that regulate feeding behavior. *Cell* 92:572–595.

Sallanon, M., M. Denoyer, K. Kitahama, C. Aubert, N. Gay, and M. Jouvet. 1989. Long-lasting insomnia induced by neuron lesions and its transient reversal by muscimol injection into the posterior hypothalamus in the cat. *Neuroscience* 32:669–683.

Sanford, L. D., X. Tang, J. Xiao, R. J. Ross, and A. R. Morrison. 2003. GABAergic regulation of REM sleep in reticularis pontis oralis and caudalis in rats. *J Neurophysiol* 90: 938–945.

Sarter, M., and J. P. Bruno. 2000. Cortical cholinergic inputs mediating arousal, attentional processing and dreaming: differential afferent regulation of the basal forebrain by telencephalic and brainstem afferents. *Neuroscience* 95:933–952.

Schwierin, B., A. A. Borbely, and I. Tobler. 1996. Effects of $N^6$-cyclopentyladenosine and caffeine on sleep regulation in the rat. *Eur J Pharmacol* 300:163–171.

Shafer, A. 1995. Metaphor and anesthesia. *Anesthesiology* 83:1331–1342.

Shaw, I. R., G. Lavigne, P. Mayer, and M. Choinière. 2005. Acute intravenous administration of morphine perturbs sleep architecture in healthy pain-free young adults: a preliminary study. *Sleep* 28:677–682.

Sherin, J. E., J. K. Elmquist, F. Torrealba, and C. B. Saper. 1998. Innervation of histaminergic tuberomamillary neurons by GABAergic and galaninergic neurons in the ventrolateral preoptic nucleus of the rat. *J Neurosci* 18:4705–4721.

Sherin, J. E., P. J. Shiromani, R. W. McCarley, and C. B. Saper. 1996. Activation of ventrolateral preoptic neurons during sleep. *Science* 271:216–219.

Shiromani, P. J., D. M. Armstrong, and J. C. Gillin. 1988. Cholinergic neurons from the dorsolateral pons project to the medial pons: a WGA-HRP and choline acetyltransferase immunohistochemical study. *Neurosci Lett* 95:19–23.

Shou, M., C. R. Ferrario, K. N. Schultz, T. E. Robinson, and R. T. Kennedy. 2006. Monitoring dopamine in vivo by microdialysis sampling and on-line CE-laser-induced fluorescence. *Anal Chem* 78:6717–6725.

Silber, M. H., S. Ancoli-Israel, M. H. Bonnet, S. Chokroverty, M. M. Grigg-Damberger, M. Hirshkowitz, S. Kapen, S. A. Keenan, M. H. Kryger, T. Penzel, M. R. Pressman, and C. Iber. 2007. The visual scoring of sleep in adults. *J Clin Sleep Med* 3:121–131.

Silva, A. P., C. Cavadas, and E. Grouzmann. 2002. Neuropeptide Y and its receptors as potential therapeutic drug targets. *Clin Chim Acta* 326:3–25.

Simon, H., M. Le Moal, L. Stinus, and A. Calas. 1979. Anatomical relationships between the ventral mesencephalic tegmentum-A10 region and the locus coeruleus as demonstrated by anterograde and retrograde tracing techniques. *J Neural Transm* 44:77–86.

Sitaram, N., A. M. Moore, and J. C. Gillin. 1978. Experimental acceleration and slowing of REM sleep ultradian rhythm by cholinergic agonist and antagonist. *Nature* 274:490–492.

Sitaram, N., R. J. Wyatt, S. Dawson, and J. C. Gillin. 1976. REM sleep induction by physostigmine infusion during sleep. *Science* 191:1281–1283.

Smith, A., C. J. Watson, K. J. Frantz, B. Eppler, R. T. Kennedy, and J. Peris. 2004. Differential increase in taurine levels by low-dose ethanol in the dorsal and ventral striatum revealed by microdialysis with on-line capillary electrophoresis. *Alcohol Clin Exp Res* 28:1028–1038.

Smith, M. T., and J. A. Haythornthwaite. 2004. How do sleep disturbance and chronic pain inter-relate? Insights from the longitudinal and cognitive-behavioral clinical trials literature. *Sleep Med Rev* 8:119–132.

Soffin, E. M., C. H. Gill, S. J. Brough, J. C. Jerman, and C. H. Davies. 2004. Pharmacological characterization of the orexin receptor subtype mediating postsynaptic excitation in the rat dorsal raphé nucleus. *Neuropharmacology* 46:1168–1176.

Sood, S., J. K. Dhawan, V. Ramesh, J. John, G. Gopinath, and V. M. Kumar. 1997. Role of medial preoptic area beta adrenoceptors in the regulation of sleep-wakefulness. *Pharmacol Biochem Behav* 57:1–5.

Spiegel, K., E. Tasali, P. Penev, and E. Van Cauter. 2004. Brief communication: sleep curtailment in healthy young men is associated with decreased leptin levels, elevated ghrelin levels, and increased hunger and appetite. *Ann Intern Med* 141:846–850.

Steinfels, G. F., J. Heym, R. E. Strecker, and B. L. Jacobs. 1983. Behavioral correlates of dopaminergic unit activity in freely moving cats. *Brain Res* 258:217–228.

Steininger, T. L., M. N. Alam, H. Gong, R. Szymusiak, and D. J. McGinty. 1999. Sleep-waking discharge of neurons in the posterior lateral hypothalamus of the albino rat. *Brain Res* 840:138–147.

Steininger, T. L., H. Gong, D. J. McGinty, and R. Szymusiak. 2001. Subregional organization of preoptic area/anterior hypothalamic projections to arousal-related monoaminergic cell groups. *J Comp Neurol* 429:638–653.

Steriade, M., and R. W. McCarley, eds. 2005. *Brain Control of Wakefulness and Sleep*. New York: Kluwer Academic/Plenum Publishers.

Steriade, M., G. Oakson, and N. Ropert. 1982. Firing rates and patterns of midbrain reticular neurons during steady and transitional states of the sleep-waking cycle. *Exp Brain Res* 46:37–51.

Stevens, D. R., R. W. McCarley, and R. W. Greene. 1992. Excitatory amino acid-mediated responses and synaptic potentials in medial pontine reticular formation neurons of the rat in vitro. *J Neurosci* 12:4188–4194.

Strecker, R. E., S. Moriarty, M. M. Thakkar, T. Porkka-Heiskanen, R. Basheer, L. J. Dauphin, D. G. Rainnie, C. M. Portas, R. W. Greene, and R. W. McCarley. 2000. Adenosinergic modulation of basal forebrain and preoptic/anterior hypothalamic neuronal activity in the control of behavioral state. *Behav Brain Res* 115:183–204.

Suntsova, N., R. Szymusiak, M. N. Alam, R. Guzman-Marin, and D. J. McGinty. 2002. Sleep-waking discharge patterns of median preoptic nucleus neurons in rats. *J Physiol* 543:665–677.

Sweet, C. P., and J. A. Hobson. 1968. The effects of posterior hypothalamic lesions on behavioral and electrographic manifestations of sleep and waking in cats. *Arch Ital Biol* 106:283–293.

Szentirmai, E., L. Kapás, and J. M. Krueger. 2007. Ghrelin microinjection into forebrain sites induces wakefulness and feeding in rats. *Am J Physiol Regul Integr Comp Physiol* 292:R575–R585.

Szentirmai, E., and J. M. Krueger. 2006. Central administration of neuropeptide Y induces wakefulness in rats. *Am J Physiol Regul Integr Comp Physiol* 291:R473–R480.

Szymusiak, R., M. N. Alam, T. L. Steininger, and D. J. McGinty. 1998. Sleep-waking discharge patterns of ventrolateral preoptic/anterior hypothalamic neurons in rats. *Brain Res* 803:178–188.

Szymusiak, R., and D. J. McGinty. 1986. Sleep-related neuronal discharge in the basal forebrain of cats. *Brain Res* 370:82–92.

Taheri, S., L. Lin, D. Austin, T. Young, and E. Mignot. 2004. Short sleep duration is associated with reduced leptin, elevated ghrelin, and increased body mass index. *PLoS Med* 1(e62):210–217.

Taheri, S., M. Mahmoodi, J. Opacka-Juffry, M. A. Ghatei, and S. R. Bloom. 1999. Distribution and quantification of immunoreactive orexin A in rat tissues. *FEBS Lett* 457:157–161.

Takahashi, K., Y. Koyama, Y. Kayama, and M. Yamamoto. 2002. Effects of orexin on the laterodorsal tegmental neurones. *Psychiat Clin Neurosci* 56:335–336.

Takahashi, K., J. S. Lin, and K. Sakai. 2006. Neuronal activity of histaminergic tuberomammillary neurons during wake-sleep states in the mouse. *J Neurosci* 26:10292–10298.

Takahashi, K., J. S. Lin, and K. Sakai. 2008. Neuronal activity of orexin and non-orexin waking-active neurons during wake-sleep states in the mouse. *Neuroscience* 153:860–870.

Tanase, D., W. A. Martin, H. A. Baghdoyan, and R. Lydic. 2001. G protein activation in rat ponto-mesencephalic nuclei is enhanced by combined treatment with a mu opioid and an adenosine $A_1$ receptor agonist. *Sleep* 24:1–11.

Tao, R., Z. Ma, J. T. McKenna, M. Thakkar, S. Winston, R. E. Strecker, and R. W. McCarley. 2006. Differential effect of orexins (hypocretins) on serotonin release in the dorsal and median raphé nuclei of freely behaving rats. *Neuroscience* 141:1101–1105.

Thakkar, M., C. Portas, and R. W. McCarley. 1996. Chronic low-amplitude electrical stimulation of the laterodorsal tegmental nucleus of freely moving cats increases REM sleep. *Brain Res* 723:223–227.

Thakkar, M. M., V. Ramesh, E. G. Cape, S. Winston, R. E. Strecker, and R. W. McCarley. 1999. REM sleep enhancement and behavioral cataplexy following orexin (hypocretin)-II receptor antisense perfusion in the pontine reticular formation. *Sleep Res Online* 2:113–120.

Thakkar, M. M., V. Ramesh, R. E. Strecker, and R. W. McCarley. 2001. Microdialysis perfusion of orexin-A in the basal forebrain increases wakefulness in freely behaving rat. *Arch Ital Biol* 139:313–328.

Thakkar, M. M., R. E. Strecker, and R. W. McCarley. 1998. Behavioral state control through differential serotonergic inhibition in the mesopontine cholinergic nuclei: a simultaneous unit recording and microdialysis study. *J Neurosci* 18:5490–5497.

Thakkar, M. M., S. C. Engemann, K. M. Walsh, and P. K. Sahota. 2008. Adenosine and the homeostatic control of sleep: effects of A1 receptor blockade in the perifornical lateral hypothalamus. *Neuroscience* 153:875–880.

Thakkar, M. M., S. Winston, and R. W. McCarley. 2003. $A_1$ receptor and adenosine homeostatic regulation of sleep-wakefulness: effects of antisense to the $A_1$ receptor in the cholinergic basal forebrain. *J Neurosci* 23:4278–4287.

Thakkar, M. M., S. Winston, and R. W. McCarley. 2008. Effect of microdialysis perfusion of 4,5,6,7-tetrahydroisoxazolo-[5,4-C]pyridine-3-ol in the perifornical hypothalamus on sleep-wakefulness: role of δ-subunit containing extrasynaptic $GABA_A$ receptors. *Neuroscience* 153:551–555.

Thannickal, T. C., R. Y. Moore, R. Nienhuis, L. Ramanathan, S. Gulyani, M. Aldrich, M. Cornford, and J. M. Siegel. 2000. Reduced number of hypocretin neurons in human narcolepsy. *Neuron* 27:469–474.

Ticho, S. R., and M. Radulovacki. 1991. Role of adenosine in sleep and temperature regulation in the preoptic area of rats. *Pharmacol Biochem Behav* 40:33–40.

Tokunaga, S., Y. Takeda, K. Shinomiya, M. Hirase, and C. Kamei. 2007. Effects of some $H_1$-antagonists on the sleep-wake cycle in sleep-disturbed rats. *J Pharmacol Sci* 103:201–206.

Tose, R., T. Kushikata, H. Yoshida, M. Kudo, K. Furukawa, S. Ueno, and K. Hirota. 2009. Orexin A decreases ketamine-induced anesthesia time in the rat: the relevance to brain noradrenergic neuronal activity. *Anesth Analg* 108:491–495.

Tóth, A., T. Hajnik, L. Záborszky, and L. Détári. 2007. Effect of basal forebrain neuropeptide Y administration on sleep and spontaneous behavior in freely moving rats. *Brain Res Bull* 72:293–301.

Trescot, A. M., S. Datta, M. G. Lee, and H. Hansen. 2008. Opioid Pharmacology. *Pain Physician* 11:S133–S153.

Trulson, M. E., and B. L. Jacobs. 1979. Raphé unit activity in freely moving cats: correlation with level of behavioral arousal. *Brain Res* 163:135–150.

Tung, A., B. Bluhm, and W. B. Mendelson. 2001. Sleep inducing effects of propofol microinjection into the medial preoptic area are blocked by flumazenil. *Brain Res* 908:155–160.

Urbain, N., K. Creamer, and G. Debonnel. 2006. Electrophysical diversity of the dorsal raphé cells across the sleep-wake cycle of the rat. *J Physiol* 573.3:679–695.

Valldeoriola, F., J. Santamaria, F. Graus, and E. Tolosa. 1993. Absence of REM sleep, altered NREM sleep and supranuclear horizontal gaze palsy caused by a lesion of the pontine tegmentum. *Sleep* 16:184–188.

Van Dort, C. J., H. A. Baghdoyan, and R. Lydic. 2008. Neurochemical modulators of sleep and anesthetic states. *Int Anesthesiol Clin* 46:75–104.

Van Dort, C. J., H. A. Baghdoyan, and R. Lydic. 2009. Adenosine $A_1$ and $A_{2A}$ receptors in mouse prefrontal cortex modulate acetylcholine release and behavioral arousal. *J Neurosci* 29:871–881.

Vanini, G., P. Torterolo, R. McGregor, M. H. Chase, and F. R. Morales. 2007. GABAergic processes in the mesencephalic tegmentum modulate the occurrence of active (rapid eye movement) sleep in guinea pigs. *Neuroscience* 145:1157–1167.

Vanini, G., C. J. Watson, R. Lydic, and H. A. Baghdoyan. 2008. γ-aminobutyric acid-mediated neurotransmission in the pontine reticular formation modulates hypnosis, immobility, and breathing during isoflurane anesthesia. *Anesthesiology* 109:978–988.

Vanni-Mercier, G., K. Sakai, and M. Jouvet. 1984. Neurones spécifiques de l'éveil dans l'hypothalamus postérieur du chat. *Comptes Rendus de l Acad des Sciences Serie III* 298:195–200.

Vazquez, J., and H. A. Baghdoyan. 2001. Basal forebrain acetylcholine release during REM sleep is significantly greater than during waking. *Am J Physiol Regul Integr Comp Physiol* 280:R598–R601.

Virus, R. M., M. Djuricic-Nedelson, M. Radulovacki, and R. D. Green. 1983. The effects of adenosine and $2'$-deoxycoformycin on sleep and wakefulness in rats. *Neuropharmacology* 22:1401–1404.

Vittoz, N. M., and C. W. Berridge. 2006. Hypocretin/orexin selectively increases dopamine efflux within the prefrontal cortex: involvement of the ventral tegmental area. *Neuropsychopharmacol* 31:384–395.

von Economo, C. V. 1930. Sleep as a problem of localization. *J Nerv Ment Dis* 72:249–259.

Walker, E. A. 1998. Clinical and pathological examination of patients with neurological disorders. In *The Genesis of Neuroscience*, edited by E. R. Laws Jr. and G. B. Udvarhelyi. Park Ridge: AANS.

Walsh, J. K., S. Deacon, D.-J. Dijk, and J. Lundahl. 2007. The selective extrasynaptic $GABA_A$ agonist, gaboxadol, improves traditional hypnotic efficacy measures and enhances slow wave activity in a model of transient insomnia. *Sleep* 30:593–602.

Walsh, J. K., D. Mayleben, C. Guico-Pabia, K. Vandormael, R. Martinez, and S. Deacon. 2008. Efficacy of the selective extrasynaptic GABAA agonist, gaboxadol, in a model of transient insomnia: a randomized, controlled clinical trial. *Sleep Med* 9:393–402.

Walsh, J. K., G. Zammit, P. K. Schweitzer, J. Ondrasik, and T. Roth. 2006. Tiagabine enhances slow wave sleep and sleep maintenance in primary insomnia. *Sleep Med* 7:155–161.

Wang, M., G. T. Roman, K. Schultz, C. Jennings, and R. T. Kennedy. 2008. Improved temporal resolution for in vivo microdialysis by using segmented flow. *Anal Chem* 80:5607–5615.

Wang, W., H. A. Baghdoyan, and R. Lydic. 2009. Leptin replacement restores supraspinal cholinergic antinociception in leptin deficient obese mice. *J Pain* 10:836–843.

Watanabe, S., T. Kuwaki, M. Yanagisawa, Y. Fokuda, and M. Shimoyama. 2005. Persistent pain and stress activate pain-inhibitory orexin pathways. *Neuroreport* 16:5–8.

Watanabe, T., Y. Taguchi, S. Shiosaka, J. Tanaka, H. Kubota, Y. Terano, M. Tohyama, and H. Wada. 1984. Distribution of the histaminergic neuron system in the central nervous system of rats; a fluorescent immunohistochemical analysis with histidine decarboxylase as a marker. *Brain Res* 295:13–25.

Watson, C. J., R. Lydic, and H. A. Baghdoyan. 2007. Sleep and GABA levels in the oral part of rat pontine reticular formation are decreased by local and systemic administration of morphine. *Neuroscience* 144:375–386.

Watson, C. J., R. Lydic, and H. A. Baghdoyan. 2008. Pontine reticular formation (PnO) administration of hypocretin-1 increases PnO GABA levels and wakefulness. *Sleep* 31:453–464.

Watson, C. J., B. J. Venton, and R. T. Kennedy. 2006. In vivo measurements of neurotransmitters using microdialysis sampling. *Anal Chem* 78:1391–1399.

Webster, H. H., and B. E. Jones. 1988. Neurotoxic lesions of the dorsolateral pontomesencephalic tegmentum-cholinergic cell area in the cat. II. Effects upon sleep-waking states. *Brain Res* 458:285–302.

Willie, J. T., R. M. Chemelli, C. M. Sinton, S. Tokita, S. C. Williams, Y. Y. Kisanuki, J. N. Marcus, C. Lee, J. K. Elmquist, K. A. Kohlmeier, C. S. Leonard, J. A. Richardson, R. E. Hammer, and M. Yanagisawa. 2003. Distinct narcolepsy syndromes in orexin receptor-2 and orexin null mice: molecular genetic dissection of non-REM and REM sleep regulatory processes. *Neuron* 38:715–730.

Wilson, S. J., J. E. Bailey, A. S. Rich, J. Nash, M. Adrover, A. Tournoux, and D. J. Nutt. 2005. The use of sleep measures to compare a new 5HT$_{1A}$ agonist with buspirone in humans. *J Psychopharmacol* 19:609–613.

Winsky-Sommerer, R., V. V. Vyazovskiy, G. E. Homanics, and I. Tobler. 2007. The EEG effects of THIP (gaboxadol) on sleep and waking are mediated by the GABA$_A$ δ-subunit-containing receptors. *Eur J Neurosci* 25:1893–1899.

Wisor, J. P., S. Nishino, I. Sora, G. H. Uhl, E. Mignot, and D. M. Edgar. 2001. Dopaminergic role in stimulant-induced wakefulness. *J Neurosci* 21:1787–1794.

Woodbridge, P. D. 1957. Changing concepts concerning depth of anesthesia. *Anesthesiology* 18:536–550.

Woolf, N. J., and L. L. Butcher. 1986. Cholinergic systems in the rat brain: III. Projections from the pontomesencephalic tegmentum to the thalamus, tectum, basal ganglia, and basal forebrain. *Brain Res Bull* 16:603–637.

Woolf, N. J., and L. L. Butcher. 1989. Cholinergic systems in the rat brain: IV. Descending projections of the pontomesencephalic tegmentum. *Brain Res Bull* 23:519–540.

Woolf, N. J., F. Eckenstein, and L. L. Butcher. 1984. Cholinergic systems in the rat brain: I. Projections to the limbic telencephalon. *Brain Res Bull* 13:751–784.

Xi, M.-C., and M. H. Chase. 2008. Effects of eszopiclone and zolpidem on sleep and waking states in the adult guinea pig. *Sleep* 31:1043–1051.

Xi, M.-C., S. J. Fung, J. Yamuy, F. R. Morales, and M. H. Chase. 2002. Induction of active (REM) sleep and motor inhibition by hypocretin in the nucleus pontis oralis of cat. *J Neurophysiol* 87:2880–2888.

Xi, M.-C., F. R. Morales, and M. H. Chase. 1999. Evidence that wakefulness and REM sleep are controlled by a GABAergic pontine mechanism. *J Neurophysiol* 82:2015–2019.

Xi, M.-C., F. R. Morales, and M. H. Chase. 2001. Effects on sleep and wakefulness of the injection of hypocretin-1 (orexin-A) into the laterodorsal tegmental nucleus of the cat. *Brain Res* 901:259–264.

Xu, Y.-L., R. K. Reinscheid, S. Huitron-Resendiz, S. D. Clark, Z. Wang, S. H. Lin, F. A. Brucher, J. Zeng, N. K. Ly, S. J. Henriksen, L. de Lecea, and O. Civelli. 2004. Neuropeptide S: a neuropeptide promoting arousal and anxiolytic effects. *Neuron* 43:487–497.

Yamamoto, T., N. Nozaki-Taguchi, and T. Chiba. 2002. Analgesic effect of intrathecally administered orexin-A in the rat formalin test and in the rat hot plate test. *Br J Pharmacol* 137:170–176.

Zaborszky, L., B. E. Cullinan, and A. Braun. 1991. Afferents to basal forebrain cholinergic neurons: an update. *Adv Exp Med Biol* 295:43–100.

Zhang, G., L. Wang, H. Liu, and J. Zhang. 2004. Substance P promotes sleep in the ventrolateral preoptic area of rats. *Brain Res* 1028:225–232.

Zhang, J.-H., S. Sampogna, F. R. Morales, and M. H. Chase. 2004. Distribution of hypocretin (orexin) immunoreactivity in the feline pons and medulla. *Brain Res* 995:205–217.

# Chapter 4
# Anesthetic Modulation of Auditory Perception: Linking Cellular, Circuit, and Behavioral Effects

**Matthew I. Banks**

**Abstract** The cortical effects of general anesthetics are well characterized at the molecular, synaptic, cellular, and network levels, and several prominent endpoints of anesthetic agents are likely mediated by actions in cerebral cortex, including amnesia and loss of consciousness. However, the links between these different levels of effects and the behavioral endpoints to which they likely contribute are not well understood. Here, we present work from our lab and others' on the effects of the general anesthetic isoflurane on auditory cortex and sensory processing. Our goal is to create an experimental and theoretical framework for understanding how the effects of isoflurane on synaptic physiology, network firing activity, and sensory processing may be related and may yield insights into the normal functioning of the brain.

**Keywords** Auditory cortex · isoflurane · synaptic inhibition · neural networks · neuronal synchrony · sensory discrimination · auditory-evoked responses · electrophysiology

## Introduction

In recent years, research in the fields of anesthetic mechanisms and the neural basis of consciousness have begun to converge (John and Prichep 2005; Hameroff 2006; Mashour 2006; Alkire, Hudetz, and Tononi 2008). General anesthetics are useful tools for providing controlled, reversible changes in awareness, whose underlying mechanisms can be explored with electrophysiological and imaging techniques. The effects of general anesthetic agents on neocortical circuits and cortically mediated behaviors are varied and profound, with multiple protein targets, cell type and region specificity, and endpoint-specific dose dependencies. Sensory processing is a readily observed target for anesthetic modulation, and this is the basis of promising technologies for perioperative monitoring of anesthetic depth (Drummond 2000). The auditory cortex is an attractive model system for studying the link between

M.I. Banks (✉)
Department of Anesthesiology, University of Wisconsin, Madison, WI, 53706, USA

A. Hudetz, R. Pearce (eds.), *Suppressing the Mind*, Contemporary Clinical Neuroscience, DOI 10.1007/978-1-60761-462-3_4,

81

cellular, network, and behavioral effects of anesthetic agents. Cortical auditory processing has been studied extensively in patients, volunteers, and animals, and the extent of modulation of cortical auditory-evoked responses by general anesthetics has been shown to be correlated with level of awareness and intraoperative memory (Schwender et al. 1993). Loss of sensory awareness and, more generally, loss of consciousness are likely mediated at least in part by cortical actions of anesthetics (John and Prichep 2005; Alkire, Hudetz, and Tononi 2008). Prominent network level effects measured electrophysiologically in cortical population activity include slowing of cortical oscillations (Whittington, Jefferys, and Traub 1996; Dickinson et al. 2003), increased amplitude (and thus synchrony) of low-frequency oscillations at concentrations leading to loss of consciousness (MacIver, Mandema et al. 1996; John et al. 2001), and suppression of spontaneous and evoked activity at higher concentrations (Schwender et al. 1993; MacIver, Makinen et al. 1996; Mandema et al. 1996). At the cellular level, general anesthetics modulate the time course and amplitude of synaptic responses via actions on presynaptic voltage-gated channels and postsynaptic ligand-gated channels (Jones and Harrison 1993; Franks and Lieb 1994; Perouansky et al. 1995; Lukatch and MacIver 1996; Banks and Pearce 1999; Hemmings 2009). Although these various effects are well described in isolation, our understanding of the links between molecular, cellular, circuit, and behavioral effects is in its infancy. Here, we review work from our lab and others' related to the effects of the general anesthetic isoflurane on auditory cortex in rodents and suggest models for how isoflurane could impair sensory processing and sensory awareness.

## Modulation of Synaptic Responses by Isoflurane

The cellular and molecular targets of general anesthetics in neocortical circuits have been studied extensively (Rudolph and Antkowiak 2004; Franks 2008). Some agents such as etomidate and benzodiazepines act on $\gamma$-aminobutyric acid type A ($GABA_A$) receptors exclusively, while others such as the volatile gas agents (e.g., isoflurane) act on both $GABA_A$ receptors and other targets including presynaptic voltage-gated sodium and calcium channels (Jones and Harrison 1993; Schlame and Hemmings 1995; Banks and Pearce 1999). Although volatile anesthetics lack specificity in their molecular targets, the importance of these agents' actions on $GABA_A$ receptors for producing anesthesia is suggested by the similar behavioral effects produced by these agents and the more selective agents (Franks and Lieb 1994; Grasshoff, Rudolph, and Antkowiak 2005). The majority of general anesthetic agents prolong $GABA_A$ receptor-mediated synaptic inhibition (Otis and Mody 1992; Jones and Harrison 1993; Banks and Pearce 1999; Verbny, Merriam, and Banks 2005) by binding to specific sites on the receptor and modulating its affinity for GABA and/or its channel-gating kinetics allosterically (Belelli et al. 1997; Mihic et al. 1997; Jenkins et al. 2001). The similarity of the effects of such chemically diverse molecules on inhibitory postsynaptic currents (IPSCs) suggests that prolongation of inhibition plays an essential role in the neural basis of anesthesia. In pyramidal cells and GABAergic interneurons of murine auditory cortex, isoflurane prolongs IPSCs by >2-fold at 1 MAC and >3-fold at 2 MAC (Fig. 4.1) (Merriam, Netoff, and Banks 2005; Verbny, Merriam, and Banks 2005). The peak amplitudes of IPSCs in cortical cells are suppressed by volatile agents, especially at high concentrations (Tanelian et al. 1993; Zimmerman, Jones, and Harrison 1994; Antkowiak 2001; Verbny, Merriam, and Banks 2005). Increases in tonic inhibition likely due to direct activation of postsynaptic $GABA_A$ receptors or enhanced sensitivity of these receptors to ambient GABA are also observed (Yeung et al. 2003). Such profound effects on synaptic inhibition are likely to underlie the dramatic effects on network activity we will describe below. Interestingly, lower (i.e., subhypnotic) concentrations still produce substantial effects on IPSCs, e.g., $\sim$1.5 fold increase in $\tau_{Decay}$ at 0.3 MAC (Fig. 4.1). In the hippocampus, actions on $GABA_A$ receptors, including prolongation of IPSCs and enhancement of tonic inhibition at subhypnotic doses of anesthetic agents, have been associated with these agents' amnestic properties (Banks and Pearce 1999; Caraiscos et al. 2004). Across its clinically relevant concentration range, isoflurane's effects on multiple molecular targets appear to act synergistically, with prolongation of synaptic inhibition coupled with presynaptic suppression of glutamate release (Perouansky et al. 1995; MacIver, Mikulec et al. 1996). Interestingly, GABA release does not appear to be suppressed, indicating that anesthetic agents act selectively at glutamatergic versus GABAergic presynaptic terminals (Peters et al. 2008). The functional implications of these electrophysiological observations in the primary sensory cortex are still unclear, but in experiments outlined below we suggest that they may underlie impairments in sensory processing. First, however, we will address the circuit-level implications of prolonged inhibition in neocortical networks.

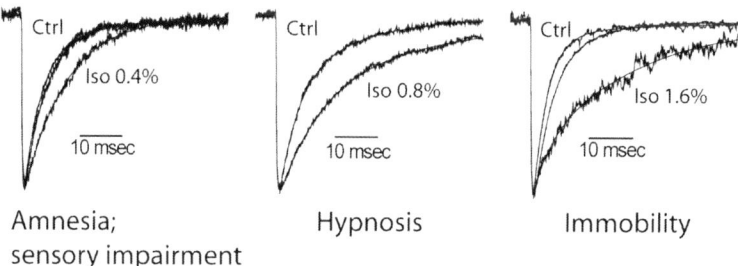

**Fig. 4.1** Dose-dependent prolongation of IPSCs by isoflurane. Shown are averaged, normalized spontaneous IPSCs recorded under whole-cell voltage clamp from three different cells treated with 0.1 mM isoflurane (*left*), 0.2 mM isoflurane (*center*), and 0.4 mM isoflurane (*right*). Weighted decay time constants were as follows: *Left*, Ctrl 7.3 ms, Iso 9.9 ms, and Wash 8.2 ms; *center*, Ctrl 9.5 ms, Iso 21.4 ms, and Wash not recorded; *right*, Ctrl 5.0 ms, Iso 19.3 ms, and Wash 5.9 ms. Reprinted with permission from Merriam, Netoff, and Banks (2005)

## Prolonged Inhibition Promotes Network Synchrony

Investigation of the effects of anesthetic agents on synaptic currents in isolation is an important step in understanding the link between these agents' molecular targets and their behavioral effects. However, cortical neurons receive input from thousands of neurons and in turn project to dozens or hundreds of target cells. By virtue of these connections, firing times of cells in these networks are influenced by the firing times of their inputs. It is in these spike time influences that the predominant effects of anesthetic agents are manifested in cortical network behavior. To understand how modulation of synaptic connectivity parameters impacts on the dynamic firing behavior of neuronal networks, two complimentary approaches are typically followed. First, mathematical models of neural networks can be used to investigate how changes in coupling parameters influence network firing behavior. Neurons themselves are nonlinear oscillators, and synchronous oscillations at a variety of frequencies are observed in cortical circuits in vivo (Buzsaki 2006). Networks of GABAergic neurons interconnected via chemical inhibitory and electrical synapses are capable of generating synchronous oscillations in cortex (Whittington, Traub, and Jefferys 1995; Beierlein, Gibson, and Connors 2000). Thus, these modeling efforts often focus on measures of network synchrony, such as oscillation frequency and coherence, as a function of the strength and duration of IPSCs in the network. In small, deterministic models, prolonged inhibition leads to greater coherence and slower oscillation frequency (Van Vreeswijk, Abbott, and Ermentrout 1994; Di Garbo, Barbi, and Chillemi 2002; Lewis and Rinzel 2003). This result is also observed in larger networks, but within a restricted range of IPSC duration; slowing IPSCs beyond this range leads to decreases in coherence and breakdown in synchrony (Wang and Buzsaki 1996; White et al. 1998).

A second approach to investigating the influence of modulation of synaptic coupling on network dynamics is to measure network behavior in vivo or in brain slice preparations in the absence and presence of modulators such as anesthetic agents. The effects of general anesthetic agents on neural activity as measured using EEG are consistent across agent and species (MacIver, Mandema et al. 1996; Makinen et al. 1996; John et al. 2001; John and Prichep 2005). High-frequency, low-amplitude activity shifts to lower frequencies and higher amplitudes, indicating increased synchronization of neural activity (but on a slower time scale). At higher concentrations of anesthetic agents, a temporal structure known as "burst suppression" emerges, with periods of flat EEG interrupted by synchronous bursts of activity. As described below, increased synchrony in the presence of anesthetic agents may be related to the effects of these agents on synaptic inhibition (Merriam, Netoff, and Banks 2005). The phenomenon of burst suppression indicates that reduced ongoing (spontaneous) cortical activity coincides with synchronous bursts of activity. There is evidence that cortical responsiveness changes dynamically during bursts and interburst isoelectric periods (Kisley and Gerstein 1999; Detsch et al. 2002; Hudetz and Imas 2007). As in large-scale neural network models, drugs that are reported to prolong IPSCs typically slow oscillations observed under control conditions (Whittington, Traub, and Jefferys 1995; Antkowiak and Hentschke 1997;

Fisahn et al. 1998; Dickinson et al. 2003). Within a certain range of drug concentrations, these drugs increase the amplitude of these population responses as well, indicating greater network synchrony (MacIver, Mandema et al. 1996). Low-frequency oscillations tend to exhibit greater coherence locally and across certain brain regions (e.g., between frontal and prefrontal regions), but interhemispheric and anterior–posterior couplings are disrupted (John et al. 2001; Imas et al. 2006). Similarly, long-range coordination of faster (e.g., gamma) rhythms is disrupted by general anesthetics (Faulkner, Traub, and Whittington 1998; Imas et al. 2006).

These two approaches are complimentary in terms of the insight they provide on a mechanistic level, on the one hand, and on the level of realistic network behavior on the other, but still leave unexplained the relationship between modulation of synaptic coupling and the behavior of physiological networks replete with stochastic and cell-type-specific firing behavior. We chose a hybrid approach that sought to balance control over coupling parameters seen in network models with realistic firing behavior observed in real neural networks. In this approach, we recorded from pairs of neurons in auditory cortex coupled via dynamic clamp (Sharp et al. 1993) (Fig. 4.2A), in which virtual synapses are inserted between neurons and the amplitude and kinetics of synaptic conductances can be varied to investigate how their modulation can influence spiking behavior in cortical networks (Merriam, Netoff, and Banks 2005). Layer 1 GABAergic interneurons in auditory cortex of juvenile mice form sparsely connected networks in which cells are coupled bidirectionally via both chemical inhibitory and electrical synapses with about 10% probability. The time constant of decay of IPSCs under control conditions is ~7 ms at 34°C, and isoflurane prolongs this decay by several fold depending on the concentration (as for pyramidal neurons). We investigated spiking behavior in these two-cell networks by measuring the phase relationships of spikes fired in response to tonic excitation (Fig 4.2B), as would occur, for example, upon activation of slow metabotropic receptors that tonically depolarize interneurons in cortical circuits (Bergles et al. 1996; Beierlein, Gibson, and Connors 2000). Under control conditions, the phase distribution was heavily biased toward antisynchronous spiking, i.e., cell 1 would fire, inhibit cell 2, which would then recover, fire, and inhibit cell 1. Occasionally, the cells would fire in phase, i.e., synchronously. Interestingly, when IPSCs were prolonged by 2- or 3-fold, as would be the case under hypnotic and surgical doses of isoflurane, respectively, this bistable firing behavior would become stronger, as the tendency for the network to dwell in the synchronous firing state increased with IPSC duration (Fig. 4.2B). In fact, it was the ratio of IPSC duration to the network firing interval that was most highly correlated with synchronous spiking: the greater the portion of the interspike interval that was taken up by the IPSC, the more the cells tended to synchronize rather than alternate spiking (Fig. 4.2C). When gap junction synapses were included in addition to the chemical inhibitory synapses, this tendency was even more pronounced (Fig. 4.2D).

These results suggest a mechanism for the tendency of anesthetic agents at hypnotic doses to increase low-frequency synchronization in cortical networks. On the scale of local circuits, prolonging inhibition in interneuronal networks leads to greater synchronous firing in these networks, which in turn project this synchronized

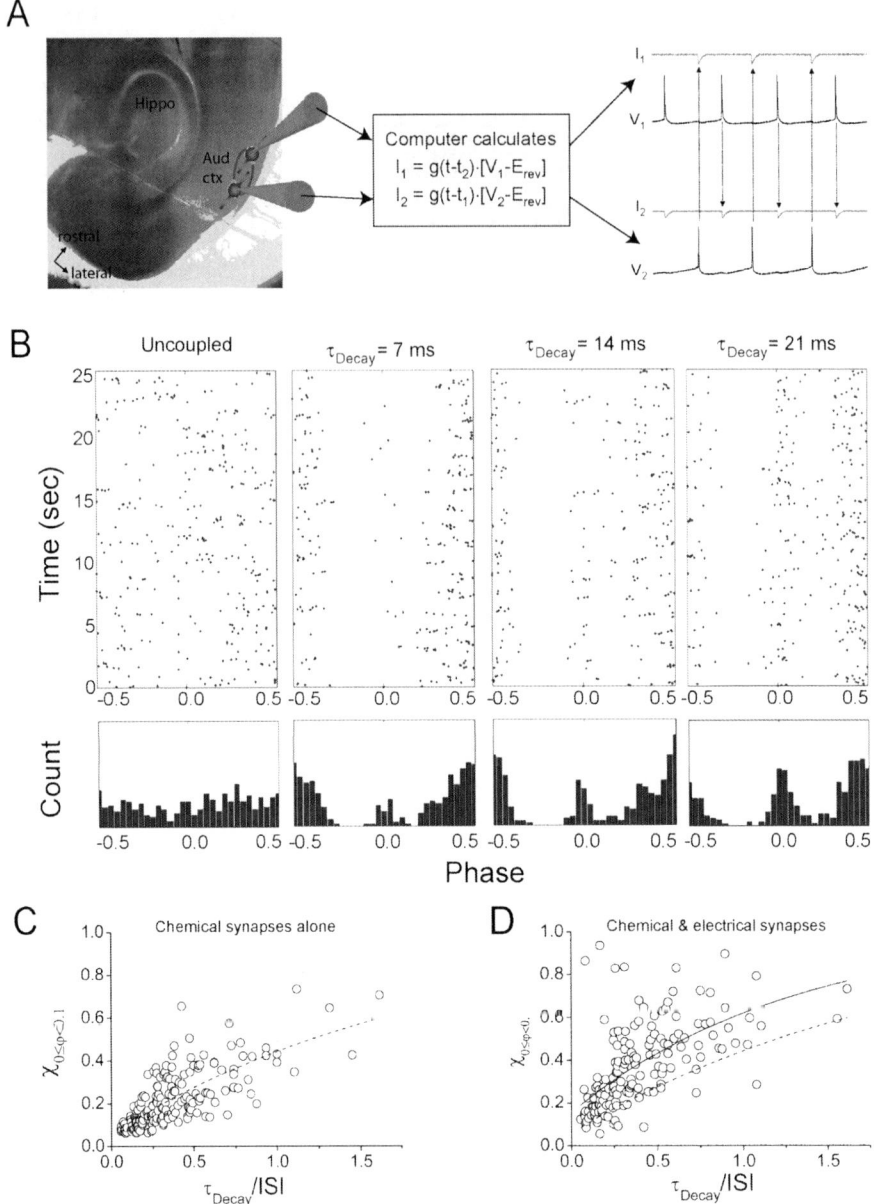

**Fig. 4.2** (continued)

inhibition onto their pyramidal cell targets with great efficacy. On the scale of cortical columns and hypercolumns, layer 1 networks themselves are poised to coordinate activity across wide cortical areas by virtue of their long horizontal projections within layer 1 that terminate on the apical dendrites of layer 2/3 and layer

5 pyramidal cells (Mitani et al. 1985; Zhu and Zhu 2004). Thus, synchronization of interneuronal networks in general, and layer 1 networks specifically, may underlie widespread synchronization observed under anesthesia. The role this synchronization plays in disrupting cortical information processing remains to be determined, but it is interesting that similar synchronization is also observed in slow-wave sleep (Steriade, McCormick, and Sejnowski 1993).

## Isoflurane Modulates Cortical Sensory-Evoked Responses

Measurement of sensory-evoked responses is a simple and effective means of testing brain responsiveness to sensory stimuli and the maintenance of sensory processing under different drug conditions. To assay cortical processing specifically, cortical-evoked responses are frequently measured using surface electrodes in humans or via more invasive recording techniques in animals. Cortical auditory-evoked responses have been studied extensively in the context of anesthetic depth, memory formation, and sensory processing (Pockett 1999). These responses are modulated in humans and experimental animals by a wide array of general anesthetic agents, including volatile anesthetics (Schwender et al. 1993; Santarelli et al. 2003). In humans, anesthetic effects on auditory-evoked responses are observed even at subhypnotic doses of the drug (Madler et al. 1991), and in some studies evoked responses are virtually abolished at surgical levels of anesthesia (Schwender et al. 1993). These results suggest that information flow through the auditory cortex is severely reduced at moderate to high levels of anesthesia. However, more recent studies suggest that

---

**Fig. 4.2** Prolonged inhibition promotes synchrony in interneuronal networks. **A** Dynamic clamp recording technique used to study the effect of prolonged inhibition on network firing behavior. Two layer 1 GABAergic interneurons in a murine brain slice containing auditory cortex were recorded under whole-cell current clamp (*left*). A computer system monitored the membrane potential and calculated the current that would flow through a synapse connecting the two cells (*middle*), injecting that current into the "postsynaptic" cell every time an action potential occurs in the "presynaptic" cell (*right*). The time course and amplitude of the virtual synapses can be specified by the experimenter. **B** Spike rasters (*top*) and corresponding histograms (*bottom*) recorded from two cells under four different coupling conditions as specified. Cells are firing quasi-periodically due to injection of depolarizing currents. Spike times in the cell are referenced to the spike times in the other cell and presented as relative phase in fractions of the unit circle ($\pm 0.5$). Uncoupled cells fire independently (*left*). Cells coupled with IPSCs fire antisynchronously when coupled with control IPSCs ($\tau_{Decay} = 7$ ms) but tend to fire more and more synchronous spikes as inhibition is prolonged. **C** Fraction of total spikes fired in two-cell networks coupled only with GABAergic synapses that are synchronous ($\chi_0 < \phi < 0.1$) as a function of the ratio of IPSC $\tau_{Decay}$ to network firing interval. As the duration of inhibition becomes longer relative to the mean interspike interval in the network, more and more spikes tend to be synchronous. *Dashed line* is the best polynomial fit to the data. **D** Same as **C**, but for two-cell networks coupled bidirectionally with both gap junctions and GABAergic synapses. *Solid line* is the best polynomial fit to the data. Note that the tendency to fire synchronous spikes is greater in these networks compared to networks coupled only with GABAergic synapses (*dashed line* replotted from **C**). **B–D** Replotted with permission from (Merriam, Netoff, and Banks 2005)

substantial activation of auditory cortex remains at hypnotic doses of general anesthetic agents. For example, imaging techniques have shown that stimulus-evoked
activity in auditory cortex is maintained during anesthesia with propofol (Plourde
et al. 2006). Similarly, in animals there are reports showing profound depression
of auditory-evoked responses during anesthesia, but paradoxically in some animals
these responses are enhanced (Rabe et al. 1980; Santarelli et al. 2003), similar
to the enhancement observed in a recent study on visual-evoked responses (Imas
et al. 2005).

We investigated the effects of isoflurane on rats using chronically implanted
epidural electrodes (Fig. 4.3). At subhypnotic doses, isoflurane has little effect on

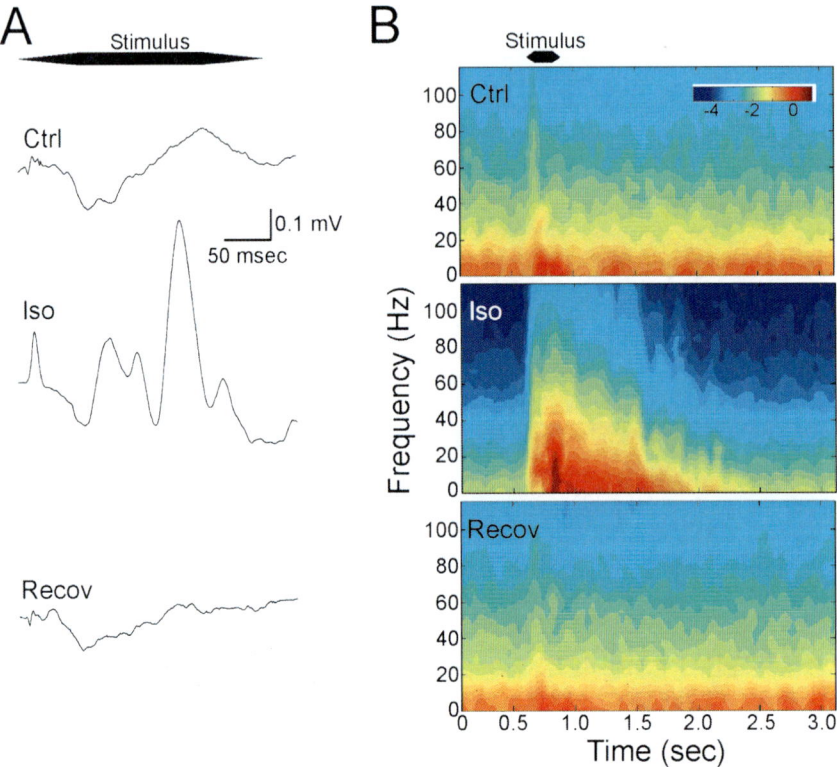

**Fig. 4.3** Enhanced auditory-evoked responses at hypnotic doses of isoflurane. **A** Mean auditory-
evoked responses recorded from a rat in response to a 250 ms 10–20 kHz FM sweep under control
conditions (*top trace*), in the presence of 1.6% isoflurane (*middle trace*), and 1 h after cessation
of drug application (*bottom trace*). Duration of the stimulus is indicated by the *black trapezoid*.
Note the enhanced response, especially at the end of the stimulus, in isoflurane. **B** Spectral analysis derived from the same data as in **A**. Short-term Fourier transforms were applied to successive
overlapping 256 ms segments of Hamming-windowed single-trial time domain responses and averaged to yield the spectrograms shown. Note the different time scale in these plots compared to those
in **A**. Note also the suppression of prestimulus spontaneous activity and the dramatic enhancement
of the response amplitude and duration under isoflurane. *Color map* depicts log power, in mV$^2$/Hz

the overall amplitude of the auditory-evoked response, but significantly suppresses high-frequency (>80 Hz; "high gamma") components of both spontaneous and evoked activity (Rummel and Banks 2007). At hypnotic doses, this suppression of high-frequency activity is even more pronounced, consistent with previous reports showing that the amplitudes of high-gamma (80–200 Hz) components are good predictors of arousal level in anesthetized patients undergoing surgery (Scheller et al. 2005). The suppression also extends to lower frequency components as well (Fig. 4.3). Paradoxically, the overall magnitude and especially low-frequency components of the evoked responses are enhanced dramatically (Fig. 4.3). This increase in response magnitude is directly coupled to the broadband suppression of spontaneous activity, suggesting that under isoflurane, sensory stimuli presented to a quiescent cortical circuit elicit a burst response that is stereotyped (i.e., invariant) and thus likely information-poor (Alkire, Hudetz, and Tononi 2008).

## Isoflurane Impairs Sensory Processing

The data reviewed above indicate that general anesthetics have multiple, interconnected effects on neural activity in sensory neocortex. Modulation by isoflurane of its molecular targets produces changes in spontaneous and evoked network activity whose details vary with drug concentration. Two critical and related issues in understanding the cortical effects of these agents are (1) the impact of subhypnotic doses of general anesthetics on cortical information processing and (2) the contribution of this modulation to loss of consciousness.

There have been surprisingly few studies addressing the first issue. In the visual system, low doses of isoflurane are reported to reduce contrast sensitivity in human volunteers (Taylor et al. 1998). However, slightly higher concentrations of isoflurane had little effect on accuracy in a visual search task in spite of significant effects on neural activity in visual cortical areas (Heinke and Schwarzbauer 2001). We studied the effects of 0.2 and 0.4% isoflurane on auditory discrimination in rats (Fig. 4.4) (Burlingame et al. 2007). In this study, we trained rats to discriminate upward- from downward-going frequency-modulated swept tones using a two-alternative forced choice, positive reinforcement training strategy. For pairs of stimuli with broad frequency range (10 kHz) and long duration (250 ms), animals readily learned the task to 80–90% accuracy over the course of several weeks. After training the rats on the "easy" stimulus pair, we tested them successively on more difficult pairs, i.e., stimuli with smaller frequency ranges and shorter durations. Performance decreased substantially with these changes in stimulus parameters. We chose three stimulus pairs ("easy", "medium," and "hard") and tested the animals under three drug conditions (0, 0.2, and 0.4% isoflurane; Fig. 4.4). Remarkably, the effects of isoflurane depended on the task difficulty: for the "easy" stimulus pair, neither 0.2 nor 0.4% isoflurane had an effect on performance, whereas for the "medium" stimulus pair 0.4% but not 0.2% impaired performance. For the "hard" stimulus pair, both 0.2 and 0.4% isoflurane had significant effects. The absence of effect for the "easy"

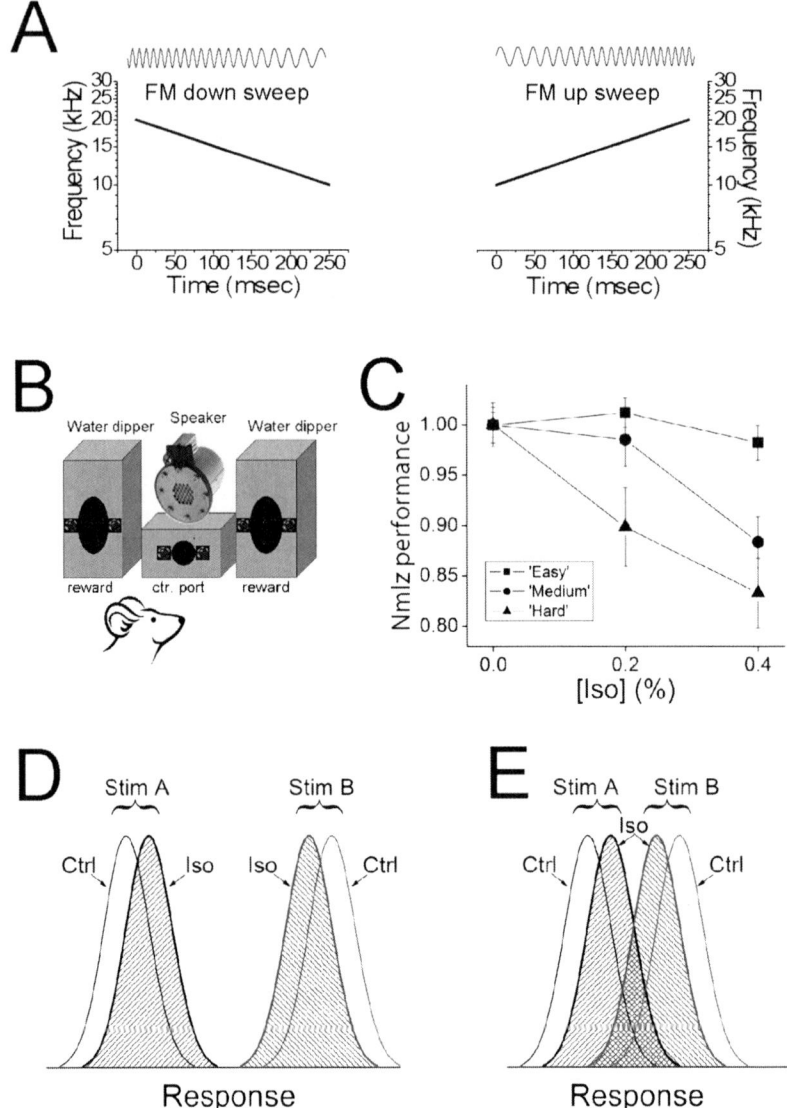

**Fig. 4.4** (continued)

stimulus pair indicates that the effects of isoflurane are unrelated to effects on motor control or motivation and are most likely explained by effects on sensory processing directly or on attention. The latter might be more critical for more demanding tasks, and its effects are likely to be manifested in the evoked responses of auditory cortical neurons (Alain and Izenberg 2003).

The task dependence of the effects of isoflurane on sensory discrimination can be explored using a simple model (Fig. 4.4D,E). In this model, successful sensory discrimination requires adequate separation of population responses. These responses are stochastic and are described by probability distributions; these distributions for two stimuli being discriminated are illustrated in Fig. 4.4 for the simple case of the response being represented by a single parameter (e.g., mean firing rate during the stimulus). Stimulus discriminability is an inverse function of the overlap of the two distributions. A modest dose of isoflurane is postulated to shift these distributions slightly in opposite directions for the two stimuli, for example, as would occur if isoflurane tended to make varied stimuli produce more uniform responses. When stimuli are easily discriminated, probability distributions of responses exhibit little overlap, so modest effects of anesthetics on those responses do not affect overlap and discriminability (Fig. 4.4D). More difficult discriminations involve more overlapping distributions, and modest cellular effects of anesthetics may have measurable effects on overlap and thus discriminability (Fig. 4.4E).

## Selective Effects of Isoflurane on Intracortical Processing

It is clear from the experiments described above that our understanding of the links between anesthetic actions measured electrophysiologically and those measured behaviorally is limited by our lack of understanding of the processes of sensory coding and sensory awareness in the absence of anesthetic agents. Here, we propose a specific hypothesis on how anesthetic agents impair sensory awareness based on the model of distributed representation of sensory awareness across the cortical hierarchy. In this model, the neural basis of sensory awareness lies in the activation of one or more activity traces that span multiple levels of the cortical hierarchy, with

---

**Fig. 4.4** Subhypnotic doses of isoflurane impair sensory discrimination. **A,B** Auditory discrimination task used to test the effect of subhypnotic doses of isoflurane on sensory processing in rats. Animals were trained to initiate a trial by placing their snouts in the center port (**B**) and to respond to a randomly presented up (**A**, *right*) or down (**A**, *left*) FM sweep by proceeding to one of two reward stations. **C** Mean normalized performance on the sensory discrimination task in **A** for three stimulus pairs and three drug conditions. Plotted are the normalized performance data averaged across seven animals for the "easy" stimulus pair (*squares*; 250 ms, 10 kHz sweep range), "medium" stimulus pair (*circles*; 125 ms, 5 kHz range), and "hard" stimulus pair (*triangles*; 16 ms, 630 Hz range) as a function of isoflurane concentration (mean ± SD). **D,E** Depiction of a model to explain the results in **C**. Neural responses are parameterized by one random variable, whose distributions for two presented stimuli are depicted by the *thin* curves. The difficulty of the task is correlated with the degree of overlap of the neural responses. Low doses of isoflurane are hypothesized to shift the distributions along the stimulus parameter axis by a small degree (*thick, hatch-filled curves*, respectively). In the case of the "easy" stimulus pair (**D**), the small shift attributed to isoflurane has little effect on the overlap of the distributions and thus little effect on discrimination. In the case of the "hard" stimulus pair (**E**), the small shift significantly increases overlap and thus degrades performance. Data in **C** adapted from (Burlingame et al. 2007)

information flowing bidirectionally between different levels in the sensory and association cortical hierarchy. Although sensory processing is classically described as a feedforward process, feedback afferents in neocortex provide comparable numbers of synaptic connections (Rockland and Virga 1989; Salin, Kennedy, and Bullier 1995; Budd 1998) and are likely to modulate feedforward responses (Sandell and Schiller 1982). In some cases, feedback also drives columnar activity prior to, or in the absence of, ascending input (Cauller and Kulics 1991; Mignard and Malpeli 1991). Under conscious conditions, sensory stimuli always occur within a context of memory- and behavioral state-dependent expectations whose neural bases lie primarily in top-down cortical information flow. Psychophysical studies suggest that "top-down" processes such as priming, context, expectation, and attention influence behavioral responses to sensory stimuli (Warren 1970; Haist et al. 2001; Alain and Izenberg 2003; Alain 2007; Davis and Johnsrude 2007; Fritz, Elhilali, and Shamma 2007). Influences of expectation and other top-down processes on cortical activity are likely to include modulated activation of infragranular and supragranular pyramidal cells due to the concentration of descending excitatory inputs to layer 1 and infragranular layers (Zeki and Shipp 1988; Felleman and Van Essen 1991; Cauller 1995). Changes in response latency on behavioral tasks that may be related to top-down influences such as multimodal convergence and attention (Fritz et al. 2007; Hecht, Reiner, and Karni 2008) are mirrored in changes in synchrony and relative spike timing measured in neocortical circuits that are likely due to modulatory effects of descending inputs on cortical activation patterns (Engel, Fries, and Singer 2001; Krupa et al. 2004). These considerations lead to the postulation that the interaction, or some would say comparison (John and Prichep 2005), between these expectation-based top-down signals and bottom-up signals triggered by environmental stimuli constitutes, or is critical for, sensory awareness. Ongoing ("spontaneous") cortical neural activity and substantial trial-by-trial variability in responses to sensory stimuli, both of which are suppressed by hypnotic doses of general anesthetics (Makinen et al. 1996; Kisley and Gerstein 1999), likely reflect intracortically driven activation patterns and the interaction of these patterns with sensory perturbations from the outside world.

We suggest that the absence of top-down information flow is a hallmark of unconscious sensory processing and further suggest that general anesthetics trigger loss of consciousness by interfering with these top-down processes. Traditionally, descriptions of the actions of general anesthetics on cortical circuits focused on the dramatic reduction in neural activity observed at hypnotic doses (Schwender et al. 1993). However, more recent studies have shown that stimulus-related activity in A1 is present under general anesthesia, albeit at a reduced level. This reduction in activity is present even in the absence of loss of consciousness (Dueck et al. 2005; Kerssens et al. 2005; Plourde et al. 2006). Although important aspects of cortical responses are altered by general anesthesia (Wang 2007), electrophysiological studies have shown that sensory-evoked responses can be enhanced dramatically under anesthesia compared to awake conditions (Imas et al. 2005). Similarly, during slow-wave sleep, cortical responses to transcranial magnetic stimulation are enhanced locally, but the spread of activity due to cortico-cortical interactions is reduced (Massimini et al. 2005). These data contradict the model that general anesthetics cause loss of consciousness by simply turning off neocortical circuits and are consistent with more subtle effects of these agents that may involve endogenous mechanisms underlying sensory awareness.

Data from experiments on sensory cortex in vivo are consistent with a model in which bottom-up processes are relatively unaffected by hypnotic doses of general anesthetics, while top-down modulation of these responses due to context, expectation, salience, and integration of local receptive field information is disrupted. We suggest that differential sensitivity to general anesthetics of ascending versus descending inputs underlies these effects in vivo and that these agents depress selectively top-down inputs to a cortical column while leaving responses to ascending stimuli relatively unaffected. Studies in visual cortex indicate that general anesthetics have far greater effects on cortico-cortical projections in the form of descending contextual modulation and global integration of local motion cues than on responses to ascending sensory inputs (Lamme, Zipser, and Spekreijse 1998; Pack, Berezovskii, and Born 2001). These studies suggest that responses to initial, ascending information in cortical circuits remain relatively unaffected under anesthesia, while descending modulation of these responses is dramatically altered. This pathway-specific suppression of excitatory neurotransmission by isoflurane may be mediated by synapse-specific modulation of voltage-gated ion channels and/or proteins involved in exocytosis and endocytosis, as suggested for selective effects observed in glutamatergic versus GABAergic synaptic terminals (Peters et al. 2008).

**Acknowledgements** Supported by National Institutes of Health (Bethesda, MD, USA) DC006013 (to M. I. Banks) and the Department of Anesthesiology, University of Wisconsin, Madison, WI, USA.

# References

Alain, C. 2007. Breaking the wave: effects of attention and learning on concurrent sound perception. *Hear Res* 229(1–2):225–236.

Alain, C., and A. Izenberg. 2003. Effects of attentional load on auditory scene analysis. *J Cogn Neurosci* 15(7):1063–1073.

Alkire, M. T., A. G. Hudetz, and G. Tononi. 2008. Consciousness and anesthesia. *Science* 322(5903):876–880.

Antkowiak, B. 2001. How do general anaesthetics work? *Naturwissenschaften* 88(5):201–213.

Antkowiak, B., and H. Hentschke. 1997. Cellular mechanisms of gamma rhythms in rat neocortical brain slices probed by the volatile anaesthetic isoflurane. *Neurosci Lett* 231(2):87–90.

Banks, M. I., and R. A. Pearce. 1999. Dual actions of volatile anesthetics on GABA(A) IPSCs: dissociation of blocking and prolonging effects. *Anesthesiology* 90(1):120–134.

Beierlein, M., J. R. Gibson, and B. W. Connors. 2000. A network of electrically coupled interneurons drives synchronized inhibition in neocortex. *Nat Neurosci* 3(9):904–910.

Belelli, D., J. J. Lambert, J. A. Peters, K. Wafford, and P. J. Whiting. 1997. The interaction of the general anesthetic etomidate with the gamma-aminobutyric acid type A receptor is influenced by a single amino acid. *Proc Natl Acad Sci USA* 94(20):11031–11036.

Bergles, D. E., V. A. Doze, D. V. Madison, and S. J. Smith. 1996. Excitatory actions of norepinephrine on multiple classes of hippocampal CA1 interneurons. *J Neurosci* 16(2):572–585.

Budd, J. M. 1998. Extrastriate feedback to primary visual cortex in primates: a quantitative analysis of connectivity. *Proc Biol Sci* 265(1400):1037–1044.

Burlingame, R. H., S. Shrestha, M. R. Rummel, and M. I. Banks. 2007. Subhypnotic doses of isoflurane impair auditory discrimination in rats. *Anesthesiology* 106(4):754–762.

Buzsaki, G. 2006. *Rhythms of the brain.* Oxford: Oxford University Press.

Caraiscos, V. B., E. M. Elliott, K. E. You-Ten, V. Y. Cheng, D. Belelli, J. G. Newell, M. F. Jackson, J. J. Lambert, T. W. Rosahl, K. A. Wafford, J. F. MacDonald, and B. A. Orser. 2004. Tonic inhibition in mouse hippocampal CA1 pyramidal neurons is mediated by alpha5 subunit-containing gamma-aminobutyric acid type A receptors. *Proc Natl Acad Sci USA* 101(10): 3662–3667.

Cauller, L. 1995. Layer I of primary sensory neocortex: where top-down converges upon bottom-up. *Behav Brain Res* 71(1–2):163–170.

Cauller, L. J., and A. T. Kulics. 1991. The neural basis of the behaviorally relevant N1 component of the somatosensory-evoked potential in SI cortex of awake monkeys: evidence that backward cortical projections signal conscious touch sensation. *Exp Brain Res* 84(3): 607–619.

Davis, M. H., and I. S. Johnsrude. 2007. Hearing speech sounds: top-down influences on the interface between audition and speech perception. *Hear Res* 229(1–2):132–147.

Detsch, O., E. Kochs, M. Siemers, B. Bromm, and C. Vahle-Hinz. 2002. Increased responsiveness of cortical neurons in contrast to thalamic neurons during isoflurane-induced EEG bursts in rats. *Neurosci Lett* 317(1):9–12.

Di Garbo, A., M. Barbi, and S. Chillemi. 2002. Synchronization in a network of fast-spiking interneurons. *Biosystems* 67(1–3):45–53.

Dickinson, R., S. Awaiz, M. A. Whittington, W. R. Lieb, and N. P. Franks. 2003. The effects of general anaesthetics on carbachol-evoked gamma oscillations in the rat hippocampus *in vitro. Neuropharmacology* 44(7):864–872.

Drummond, J. C. 2000. Monitoring depth of anesthesia: with emphasis on the application of the bispectral index and the middle latency auditory evoked response to the prevention of recall. *Anesthesiology* 93(3):876–882.

Dueck, M. H., F. Petzke, H. J. Gerbershagen, M. Paul, V. Hesselmann, B. Girnus, B. Krug, B. Sorger, R. Goebel, R. Lehrke, V. Sturm, and U. Boerner. 2005. Propofol attenuates responses of the auditory cortex to acoustic stimulation in a dose-dependent manner: a FMRI study. *Acta Anaesthesiol Scand* 49(6):784–791.

Engel, A. K., P. Fries, and W. Singer. 2001. Dynamic predictions: oscillations and synchrony in top-down processing. *Nat Rev Neurosci* 2(10):704–716.

Faulkner, H. J., R. D. Traub, and M. A. Whittington. 1998. Disruption of synchronous gamma oscillations in the rat hippocampal slice: a common mechanism of anaesthetic drug action. *Br J Pharmacol* 125(3):483–492.

Felleman, D. J., and D. C. Van Essen. 1991. Distributed hierarchical processing in the primate cerebral cortex. *Cereb Cortex* 1(1):1–47.

Fisahn, A., F. G. Pike, E. H. Buhl, and O. Paulsen. 1998. Cholinergic induction of network oscillations at 40 Hz in the hippocampus *in vitro. Nature* 394(6689):186189.

Franks, N. P. 2008. General anaesthesia: from molecular targets to neuronal pathways of sleep and arousal. *Nat Rev Neurosci* 9(5):370–386.

Franks, N. P., and W. R. Lieb. 1994. Molecular and cellular mechanisms of general anaesthesia. *Nature* 367(6464):607–614.

Fritz, J. B., M. Elhilali, S. V. David, and S. A. Shamma. 2007. Auditory attention–focusing the searchlight on sound. *Curr Opin Neurobiol* 17(4):437–455.

Fritz, J. B., M. Elhilali, and S. A. Shamma. 2007. Adaptive changes in cortical receptive fields induced by attention to complex sounds. *J Neurophysiol* 98(4):2337–2346.

Grasshoff, C., U. Rudolph, and B. Antkowiak. 2005. Molecular and systemic mechanisms of general anaesthesia: the 'multi-site and multiple mechanisms' concept. *Curr Opin Anaesthesiol* 18(4):386–391.

Haist, F., A. W. Song, K. Wild, T. L. Faber, C. A. Popp, and R. D. Morris. 2001. Linking sight and sound: fMRI evidence of primary auditory cortex activation during visual word recognition. *Brain Lang* 76(3):340–350.

Hameroff, S. R. 2006. The entwined mysteries of anesthesia and consciousness: is there a common underlying mechanism? *Anesthesiology* 105(2):400–412.

Hecht, D., M. Reiner, and A. Karni. 2008. Enhancement of response times to bi- and tri-modal sensory stimuli during active movements. *Exp Brain Res* 185(4):655–665.

Heinke, W., and C. Schwarzbauer. 2001. Subanesthetic isoflurane affects task-induced brain activation in a highly specific manner: a functional magnetic resonance imaging study. *Anesthesiology* 94(6):973–981.

Hemmings, H. C. 2009. Molecular targets of general anesthetics in the nervous system. In *Suppressing the mind: anesthetic modulation of memory and consciousness*, edited by A. G. Hudetz and R. A. Pearce. New York City: Springer.

Hudetz, A. G., and O. A. Imas. 2007. Burst activation of the cerebral cortex by flash stimuli during isoflurane anesthesia in rats. *Anesthesiology* 107(6):983–991.

Imas, O. A., K. M. Ropella, B. D. Ward, J. D. Wood, and A. G. Hudetz. 2005. Volatile anesthetics enhance flash-induced gamma oscillations in rat visual cortex. *Anesthesiology* 102(5):937–947.

Imas, O. A., K. M. Ropella, J. D. Wood, and A. G. Hudetz. 2006. Isoflurane disrupts anterio-posterior phase synchronization of flash-induced field potentials in the rat. *Neurosci Lett* 402(3):216–221.

Jenkins, A., E. P. Greenblatt, H. J. Faulkner, E. Bertaccini, A. Light, A. Lin, A. Andreasen, A. Viner, J. R. Trudell, and N. L. Harrison. 2001. Evidence for a common binding cavity for three general anesthetics within the GABAA receptor. *J Neurosci* 21(6):RC136.

John, E. R., and L. S. Prichep. 2005. The anesthetic cascade: a theory of how anesthesia suppresses consciousness. *Anesthesiology* 102(2):447–471.

John, E. R., L. S. Prichep, W. Kox, P. Valdes-Sosa, J. Bosch-Bayard, M. Aubert, M. Tom, F. di Michele, and L. D. Gugino. 2001. Invariant reversible QEEG effects of anesthetics. *Conscious Cogn* 10(2):165–183.

Jones, M. V., and N. L. Harrison. 1993. Effects of volatile anesthetics on the kinetics of inhibitory postsynaptic currents in cultured rat hippocampal neurons. *J Neurophysiol* 70(4):1339–1349.

Kerssens, C., S. Hamann, S. Peltier, X. P. Hu, M. G. Byas-Smith, and P. S. Sebel. 2005. Attenuated brain response to auditory word stimulation with sevoflurane: a functional magnetic resonance imaging study in humans. *Anesthesiology* 103(1):11–19.

Kisley, M. A., and G. L. Gerstein. 1999. Trial-to-trial variability and state-dependent modulation of auditory-evoked responses in cortex. *J Neurosci* 19(23):10451–10460.

Krupa, D. J., M. C. Wiest, M. G. Shuler, M. Laubach, and M. A. Nicolelis. 2004. Layer-specific somatosensory cortical activation during active tactile discrimination. *Science* 304(5679):1989–1992.

Lamme, V. A., K. Zipser, and H. Spekreijse. 1998. Figure-ground activity in primary visual cortex is suppressed by anesthesia. *Proc Natl Acad Sci USA* 95(6):3263–3268.

Lewis, T. J., and J. Rinzel. 2003. Dynamics of spiking neurons connected by both inhibitory and electrical coupling. *J Comput Neurosci* 14(3):283–309.

Lukatch, H. S., and M. B. MacIver. 1996. Synaptic mechanisms of thiopental-induced alterations in synchronized cortical activity. *Anesthesiology* 84(6):1425–1434.

MacIver, M. B., J. W. Mandema, D. R. Stanski, and B. H. Bland. 1996. Thiopental uncouples hippocampal and cortical synchronized electroencephalographic activity. *Anesthesiology* 84(6):1411–1424.

MacIver, M. B., A. A. Mikulec, S. M. Amagasu, and F. A. Monroe. 1996. Volatile anesthetics depress glutamate transmission via presynaptic actions. *Anesthesiology* 85(6):823–834.

Madler, C., I. Keller, D. Schwender, and E. Poppel. 1991. Sensory information processing during general anaesthesia: effect of isoflurane on auditory evoked neuronal oscillations. *Br J Anaesth* 66(1):81–87.

Makinen, S., K. Hartikainen, J. T. Eriksson, and V. Jantti. 1996. Spontaneous and evoked cortical dynamics during deep anaesthesia. *Int J Neural Syst* 7(4):481–487.

Mashour, G. A. 2006. Integrating the science of consciousness and anesthesia. *Anesth Analg* 103(4):975–982.

Massimini, M., F. Ferrarelli, R. Huber, S. K. Esser, H. Singh, and G. Tononi. 2005. Breakdown of cortical effective connectivity during sleep. *Science* 309 (5744):2228–2232.

Merriam, E. B., T. I. Netoff, and M. I. Banks. 2005. Bistable network behavior of layer I interneurons in auditory cortex. *J Neurosci* 25(26):61756186.

Mignard, M., and J. G. Malpeli. 1991. Paths of information flow through visual cortex. *Science* 251(4998):1249–1251.

Mihic, S. J., Q. Ye, M. J. Wick, V. V. Koltchine, M. D. Krasowski, S. E. Finn, M. P. Mascia, C. F. Valenzuela, K. K. Hanson, E. P. Greenblatt, R. A. Harris, and N. L. Harrison. 1997. Sites of alcohol and volatile anaesthetic action on GABA(A) and glycine receptors. *Nature* 389(6649):385389.

Mitani, A., M. Shimokouchi, K. Itoh, S. Nomura, M. Kudo, and N. Mizuno. 1985. Morphology and laminar organization of electrophysiologically identified neurons in the primary auditory cortex in the cat. *J Comp Neurol* 235(4):430–447.

Otis, T. S., and I. Mody. 1992. Modulation of decay kinetics and frequency of GABAA receptor-mediated spontaneous inhibitory postsynaptic currents in hippocampal neurons. *Neuroscience* 49(1):13–32.

Pack, C. C., V. K. Berezovskii, and R. T. Born. 2001. Dynamic properties of neurons in cortical area MT in alert and anaesthetized macaque monkeys. *Nature* 414(6866):905–908.

Perouansky, M., D. Baranov, M. Salman, and Y. Yaari. 1995. Effects of halothane on glutamate receptor-mediated excitatory postsynaptic currents. A patch-clamp study in adult mouse hippocampal slices. *Anesthesiology* 83(1):109–119.

Peters, J. H., S. J. McDougall, D. Mendelowitz, D. R. Koop, and M. C. Andresen. 2008. Isoflurane differentially modulates inhibitory and excitatory synaptic transmission to the solitary tract nucleus. *Anesthesiology* 108(4):675–683.

Plourde, G., P. Belin, D. Chartrand, P. Fiset, S. B. Backman, G. Xie, and R. J. Zatorre. 2006. Cortical processing of complex auditory stimuli during alterations of consciousness with the general anesthetic propofol. *Anesthesiology* 104(3):448–457.

Pockett, S. 1999. Anesthesia and the electrophysiology of auditory consciousness. *Conscious Cogn* 8(1):45–61.

Rabe, L. S., L. Moreno, B. M. Rigor, and N. Dafny. 1980. Effects of halothane on evoked field potentials recorded from cortical and subcortical nuclei. *Neuropharmacology* 19(9): 813–825.

Rockland, K. S., and A. Virga. 1989. Terminal arbors of individual "feedback" axons projecting from area V2 to V1 in the macaque monkey: a study using immunohistochemistry of anterogradely transported Phaseolus vulgaris-leucoagglutinin. *J Comp Neurol* 285(1):54–72.

Rudolph, U., and B. Antkowiak. 2004. Molecular and neuronal substrates for general anaesthetics. *Nat Rev Neurosci* 5(9):709–720.

Rummel, M. R., Banks, M. I. 2007. Within-trial correlations across time and frequency in cortical evoked potentials are altered abruptly at anesthetic doses causing loss of consciousness. *Soc Neurosci Abstr* 33:505.2.

Salin, P. A., H. Kennedy, and J. Bullier. 1995. Spatial reciprocity of connections between areas 17 and 18 in the cat. *Can J Physiol Pharmacol* 73(9):1339–1347.

Sandell, J. H., and P. H. Schiller. 1982. Effect of cooling area 18 on striate cortex cells in the squirrel monkey. *J Neurophysiol* 48(1):38–48.

Santarelli, R., E. Arslan, L. Carraro, G. Conti, M. Capello, and G. Plourde. 2003. Effects of isoflurane on the auditory brainstem responses and middle latency responses of rats. *Acta Otolaryngol* 123(2):176–181.

Scheller, B., G. Schneider, M. Daunderer, E. F. Kochs, and B. Zwissler. 2005. High-frequency components of auditory evoked potentials are detected in responsive but not in unconscious patients. *Anesthesiology* 103(5):944–950.

Schlame, M., and H. C. Hemmings, Jr. 1995. Inhibition by volatile anesthetics of endogenous glutamate release from synaptosomes by a presynaptic mechanism. *Anesthesiology* 82(6):1406–1016.

Schwender, D., S. Klasing, C. Madler, E. Poppel, and K. Peter. 1993. Depth of anesthesia. Midlatency auditory evoked potentials and cognitive function during general anesthesia. *Int Anesthesiol Clin* 31(4):89–106.

Sharp, A. A., M. B. O'Neil, L. F. Abbott, and E. Marder. 1993. The dynamic clamp: artificial conductances in biological neurons. *Trends Neurosci* 16(10):389–394.

Steriade, M., D. A. McCormick, and T. J. Sejnowski. 1993. Thalamocortical oscillations in the sleeping and aroused brain. *Science* 262(5134):679–685.

Tanelian, D. L., P. Kosek, I. Mody, and M. B. MacIver. 1993. The role of the GABAA receptor/chloride channel complex in anesthesia. *Anesthesiology* 78(4):757–776.

Taylor, S. R., O. A. Khan, M. L. Swart, G. G. Lockwood, and J. G. Jones. 1998. Effects of a low concentration of isoflurane on contrast sensitivity in volunteers. *Br J Anaesth* 81(2):176–179.

Van Vreeswijk, C., L. F. Abbott, and G. B. Ermentrout. 1994. When inhibition not excitation synchronizes neural firing. *J Comput Neurosci* 1(4):313–321.

Verbny, Y. I., E. B. Merriam, and M. I. Banks. 2005. Modulation of gamma-aminobutyric acid type A receptor-mediated spontaneous inhibitory postsynaptic currents in auditory cortex by midazolam and isoflurane. *Anesthesiology* 102(5):962–969.

Wang, X. 2007. Neural coding strategies in auditory cortex. *Hear Res* 229(1–2):81–93.

Wang, X. J., and G. Buzsaki. 1996. Gamma oscillation by synaptic inhibition in a hippocampal interneuronal network model. *J Neurosci* 16(20):6402–6413.

Warren, R. M. 1970. Perceptual restoration of missing speech sounds. *Science* 167(917):392–393.

White, J. A., C. C. Chow, J. Ritt, C. Soto-Trevino, and N. Kopell. 1998. Synchronization and oscillatory dynamics in heterogeneous, mutually inhibited neurons. *J Comput Neurosci* 5(1):5–16.

Whittington, M. A., R. D. Traub, and J. G. Jefferys. 1995. Synchronized oscillations in interneuron networks driven by metabotropic glutamate receptor activation. *Nature* 373(6515):612–615.

Whittington, M. A., J. G. Jefferys, and R. D. Traub. 1996. Effects of intravenous anaesthetic agents on fast inhibitory oscillations in the rat hippocampus *in vitro*. *Br J Pharmacol* 118(8):1977–1986.

Yeung, J. Y., K. J. Canning, G. Zhu, P. Pennefather, J. F. MacDonald, and B. A. Orser. 2003. Tonically activated GABAA receptors in hippocampal neurons are high-affinity, low-conductance sensors for extracellular GABA. *Mol Pharmacol* 63(1):2–8.

Zeki, S., and S. Shipp. 1988. The functional logic of cortical connections. *Nature* 335(6188):311–317.

Zhu, Y., and J. J. Zhu. 2004. Rapid arrival and integration of ascending sensory information in layer 1 nonpyramidal neurons and tuft dendrites of layer 5 pyramidal neurons of the neocortex. *J Neurosci* 24(6):1272–1279.

Zimmerman, S. A., M. V. Jones, and N. L. Harrison. 1994. Potentiation of gamma-aminobutyric acidA receptor Cl- current correlates with *in vivo* anesthetic potency. *J Pharmacol Exp Ther* 270(3):987–991.

# Chapter 5
# Cortical Disintegration Mechanism of Anesthetic-Induced Unconsciousness

Anthony Hudetz

**Abstract** The fundamental thesis of this paper is that under general anesthesia, the brain preserves its reactivity to sensory stimulation, but the information conveyed by these stimuli is not integrated – therefore, it is not consciously perceived. This happens at anesthetic depths associated with behavioral unresponsiveness from which unconsciousness is inferred. A further hypothesis is that general anesthetic agents interfere with the integration of sensory information encoded in cortical neuronal activity in the gamma frequency band. A critical region whose connectivity with the rest of the brain may be disrupted in anesthesia is the posterior parietal association cortex. While it is true that most anesthetic agents utilize different cellular and molecular pathways and neuron groups to produce their pharmacological effects, a common systems level action that may lead to unconsciousness by all anesthetics is the disruption of cortical information processing. The plausibility of these hypotheses and supporting experimental evidence is discussed in the context of theoretical considerations for defining consciousness and inferring the presence or absence of consciousness in an anesthetized subject.

**Keywords** Consciousness · gamma oscillations · neuronal synchrony · visual evoked response · information exchange · entropy · neuronal networks · functional connectivity

## Definition and Assessment of Consciousness in Anesthetized Subjects

There is no accepted objective definition of consciousness. Although all people think they know well what consciousness is, when asked, they report vastly different conceptualizations of consciousness (Zeman 2002). Proposed definitions of consciousness are either too general or too restrictive. Consciousness is best

A. Hudetz (✉)
Department of Anesthesiology, Medical College of Wisconsin, Milwaukee, WI, USA
e-mail: ahudetz@mcw.edu

A. Hudetz, R. Pearce (eds.), *Suppressing the Mind*, Contemporary Clinical Neuroscience, DOI 10.1007/978-1-60761-462-3_5,
© Humana Press, a part of Springer Science+Business Media, LLC 2010

defined as *subjective experience,* which emphasizes its qualitative, first-person character (Chalmers 1996). The experiences include sensations, feelings, perceptions, emotions, thoughts, beliefs, desires, volitions, images including inner speech, and possibly also objectless experiences. In our vocabulary, awareness and consciousness are used interchangeably. One cannot define one in terms of the other; the Jamesian definition of consciousness (James 1890) as the "awareness of oneself and the environment" (Young and Pigott 1999) is circular. Movement does not imply consciousness, although in anesthesia spontaneous movement warns of the possibility of consciousness. However, it is difficult to determine if these movements are voluntary, i.e., consciously willed or not. Eger and Sonner (2006) suggests that only immobility and amnesia define the anesthetic state; unconsciousness or analgesia are not essential to its definition. However, we obviously care about the *experience* of pain, or "discomfort," and this cannot be addressed without a reference to consciousness.

Moreover, wakefulness and consciousness are not the same phenomena. Although a continuum of states – from wakefulness through drowsiness to deep sleep or anesthesia – seems intuitive, and in fact this concept is used in anesthesia depth monitors like BIS, such a one-dimensional model of states of consciousness is an oversimplification. Wakefulness is a behavioral indication of central nervous system arousal, whereas consciousness assumes subjective experience. A creature having only subliminal sensations of any kind may not be conscious, although it could be considered to be awake. Patients in a vegetative state become periodically awake, while in all likelihood they remain unconscious at all times. In neurology, the common distinction between wakefulness and awareness as two constitutive dimensions of consciousness is often made (Laureys 2005). In this two-dimensional model, sleepwalking should properly be called *awake unconscious.* In this condition, complex sensorimotor functions are present, but awareness is absent. Only parts of the brain are active due to an incomplete, differential activation relative to slow-wave sleep state. Conversely, dreams represent subjective experience that implies awareness without wakefulness. In normal, healthy individuals, arousal enables the conscious state, but in other cases, such as subjects under the influence of hallucinogenic agents like ketamine, experience can occur with a limited degree of arousal. While the two-dimensional model of consciousness describes neurological cases well, it is possible that a distinction between wakefulness and awareness will not be sufficient to characterize all states of consciousness that may, in fact, be multidimensional. A three-dimensional model of sleep–wake states was developed by Hobson and Pace-Schott (2002). This model may have to be further extended in the future to properly account for anesthetized states that are characterized by changes in multiple dimensions including sensation, awareness, explicit memory, implicit memory, voluntary movement, etc.

While we all know about our own consciousness, an objective assessment of consciousness from a second- or third-person point of view is, strictly speaking, impossible. However, for practical purposes, one has to operationalize consciousness by interrogating a subject using verbal commands – essentially, by applying a simplified Turing test (Stins 2009). Arguing by analogy, if others systematically behave as we do when we are conscious, then it is plausible to conclude that they are conscious as well. The limitation of this approach becomes evident when we try to apply the principle to noncommunicating, paralyzed subjects. Here one has to resort to surrogate measures, such as neuronal evoked responses measured with electrophysiological or brain imaging methods. However, it remains disputable whether a positive evoked neuronal response that resembles the normal response in the conscious subject allows one to infer the presence of consciousness (i.e., subjective experience). These responses may reflect implicit information processing without experience. Until we know the necessary and sufficient neural events underlying conscious information processing, we will not be able to ascertain the presence of consciousness from an observed brain response. The neurobiological correlates of consciousness may also lay beyond the reach of our current biological measurement modalities. These arguments illuminate the need for continuing research into the neurobiological basis of consciousness and unconsciousness.

A philosophical problem also arises with respect to what constitutes subjective experience. In particular, does experience have to be reportable? Block (2005) distinguishes two forms of consciousness: phenomenal and access consciousness, thereby distinguishing primary (phenomenal) experience from its availability for cognitive operations. Events outside one's immediate attention may be phenomenally conscious, but not access conscious. Each type may have its own neural correlate. If this distinction holds, then it may be possible for a lightly anesthetized subject to have phenomenal experience of pain of some form, but not have the ability to consciously reflect on it or to report this experience. This is similar to the condition of blindsight, the phenomenon that occurs when patients sometimes sense the presence of a stimulus in their blind visual field while they deny actually seeing it and cannot describe it in anyway (Weiskrantz 1990).

A possible reason for failed access consciousness may be inattention or amnesia. It is obviously important to conceptually distinguish between consciousness and memory during anesthesia. It is well known that patients who can communicate under an appropriate level of sedation may not recall these events after emergence. Unfortunately, the highly publicized practice advisory of the American Society of Anesthesiology Task Force on Intraoperative Awareness codified a confusing definition: "Intraoperative awareness occurs when a patient becomes conscious during a procedure performed under general anesthesia and subsequently has recall of these events." However, it is clear that intraoperative awareness can occur either with or without subsequent recall. An experience may occur over a brief amount of time, but may not be encoded or the memory trace may later be erased.

On the other hand, consciousness at its roots may in fact depend on some form of very short-term memory. James (1890) noted that without memory there would be no conscious sensation. If sensory data were not stored for even a brief duration of time, they would flee before having a chance to be analyzed. Could it be that anesthetics suppress consciousness by blocking memory? Evidence suggests that a minimum duration of undisturbed processing time of 200–300 ms is required for conscious perception (Libet 1982; Libet et al. 1991). A sensory frame is 80–100 ms long (Koenig et al. 2002), and two or three consecutive frames may be necessary for the temporal-contextual interpretation of raw sense data (John 2002). The necessity for such a short, iconic memory for conscious cognition was recently discussed (Erdelyi 2004; Robbins 2004). One could argue that at an appropriate level of anesthesia, sensory data may not be interpreted consciously if they cannot be held in iconic memory. This could happen if anesthetics slow down processing times by prolonging synaptic inhibition and nerve conduction times or truncating the neuronal responses to stimuli. We will present data in support of this effect. Without iconic memory there would be no conscious perception and, as we noted before, unconsciousness under anesthesia could be characterized as the *"forgotten present"* (Hudetz 2006, 2008).

Some hold that consciousness is an irreducible, fundamental entity and thus cannot be explained in reductionist or materialist terms (Chalmers 1996). We would like to believe that consciousness is an emergent phenomenon of a neurobiological process. While fundamental philosophical difficulties in defining consciousness remain, to study the neural correlates of anesthetic-induced unconsciousness there is no alternative to adopting a mechanistic working hypothesis, similar to what was suggested for studying the neural correlates of consciousness itself (Crick and Koch 2003). In the past, consciousness has been linked to a multitude of brain regions, neuronal processes, receptors, molecules, and quantum phenomena (Hameroff, Kaszniak, and Scott 1996, 1998; Metzinger 2000; Baars, Banks, and Newman 2003; Velmans and Schneider 2007). Few theories promoted an objective definition of consciousness. For example, Shulman, Hyder, and Rothman (2003) suggested that consciousness is directly related to global brain energy. Baars (2005) defined consciousness as the dissemination of information across a "global workspace" of the brain. Tononi (2004) defined a conscious system by its capacity to integrate information. The latter definition implies that consciousness is a graded phenomenon, with more integrated information comes, in some sense, more consciousness. Conversely, consciousness is reduced either when the available information is reduced or when the capacity of the system to integrate is reduced. This model was recently applied to illustrate how anesthetic agents may suppress consciousness (Alkire, Hudetz, and Tononi 2008). The current treatment of the subject is consistent with this model, and some of the supporting data are presented in greater detail.

## Anesthetic Targets in the Brain That Produce Unconsciousness

General anesthetic agents affect virtually every cell in the body and every part of the brain. Anesthetics decrease brain metabolism in an essentially global manner (Alkire et al. 1997, 1999). Nevertheless, there are well-defined subcortical sites whose pharmacological inactivation produces sedation, atonia, analgesia, and unconsciousness, and which, therefore, have been proposed as key anesthetic targets. These sites include the thalamus (Miller et al. 1989; Angel 1991; Alkire, Haier, and Fallon 2000), mesopontine tegmentum (Devor and Zalkind 2001), septohippocampal system (Ma et al. 2002), and hypothalamus (Nelson et al. 2002). Most of these sites participate in the ascending arousal system that enables the cortex to process information. The basal forebrain is a key source of cholinergic cortical arousal and may be another critical anesthetic target (Dong et al. 2006). Numerous brainstem regions are likely involved as well (Lydic 1996; Lydic and Biebuyck 1994). The thalamus has been considered to be of central importance in anesthesia (Angel 1991; Alkire, Haier, and Fallon 2000) because its metabolism and blood flow is consistently suppressed by various anesthetic agents (Alkire and Miller 2005). Recently, studies have begun to view the cerebral cortex as the primary target of general anesthetics when applied through conventional routes (Hentschke, Schwarz, and Antkowiak 2005; Velly et al. 2007; Alkire, Hudetz, and Tononi 2008). It is possible that all interventions that cause unconsciousness ultimately produce their effect by a disruption of cortical integrative function. In this chapter, we consider cortical sensory integration as a common mechanism of anesthetic suppression of conscious experience.

Specific brain sites and mechanisms participating in the regulation of the state of consciousness have also been delineated by drug microinjection studies in an attempt to reverse the state of anesthesia. Thus, Alkire et al. (2007) injected nicotine into the central median nucleus of the thalamus of sevoflurane-anesthetized rats and was able to "wake up" the animals in the continued presence of the anesthetic. We injected rats with the acetylcholinesterase inhibitor neostigmine via the intracerebroventricular route and saw EEG activation together with the return of spontaneous movements (Hudetz, Wood, and Kampine 2003). The muscarinic $m_2$ receptor agonist oxotremorine produced the same effect. The degree and certainty that these behavioral activations indicate the return of consciousness in an animal model remains to be determined. After all, the observed behavior may reflect a sleepwalking, "zombie" like state not accompanied by conscious experience. Decerebrate rats recover and are able to right themselves, groom, and move around, although they are unable to forage food for themselves (Woods 1964). Nevertheless, human anesthesia reversal studies with physostigmine in propofol-anesthetized patients suggest that the emergence to consciousness is real (Fiset et al. 1999; Meuret et al. 2000).

## Reactivity of the Anesthetized Brain to Sensory Stimuli

An important step toward understanding how anesthetics may produce unconsciousness is to recognize that the anesthetized brain is not oblivious to sensory stimulation. Earlier neurophysiological studies suggested that anesthesia may result from a disconnection of sensory cortex from peripheral input at the level of thalamocortical relay neurons (Angel 1991; Detsch et al. 1999; Ries and Puil 1999a,b). Yet, other observations have shown that cortical reactivity is preserved, or even augmented, at anesthetic depths associated with loss of consciousness (Rabe et al. 1980; Ogawa et al. 1992; Erchova, Lebedev, and Diamond 2002). We studied the effects of halothane, isoflurane, and desflurane on cortical visual evoked potentials in a rat model (Imas et al. 2004; Imas et al. 2005a,b). The flash-evoked middle latency response shows up as a brief, stimulus-locked oscillation at gamma frequency (30–40 Hz, maximum amplitude at 40 ms). This appears to be a resonance-type phenomenon, as cortical circuits preferentially respond to stimulation at their proper frequency. Sometimes, the response is prolonged with a dispersed phase – we will say more about gamma oscillations later. In brief, we found that the visual evoked gamma response was not attenuated at the time the rats lost their righting reflex – a putative index of loss of consciousness (Franks 2008).

These results contrast with the observation that in human patients, general anesthetics decrease the amplitude and prolong the latency of sensory evoked potentials (Banoub, Tetzlaff, and Schubert 2003). The difference is most likely due to the fact that evoked potentials are usually tested with stimuli delivered at relatively high frequency (10–40 Hz). Anesthesia prolongs the recovery of neuronal excitability after each stimulus, which leads to frequency-dependent attenuation of the averaged evoked response (Detsch et al. 1999). The same effect is seen with steady-state evoked responses with numerous anesthetics (Munglani et al. 1993; Andrade et al. 1996; Plourde and Villemure 1996; Plourde et al. 1998; Bonhomme et al. 2000). With long stimulus intervals such as those we used (5 s), there is ample time for neuronal excitability to recover after each stimulus, and thus the direct cortical response is preserved. These findings suggest that under general anesthesia, cortical processing circuits are reactive, but cannot be driven at as high a frequency as during wakefulness. Indeed, Osa, Ando, and Adachi-Usami (1989) found that halothane did not attenuate human visual evoked potentials when tested at low-frequency (0.5 or 1 Hz) stimuli.

Cortical single-unit responses to sensory stimulation are usually attenuated under anesthesia with volatile agents (Ikeda and Wright 1974a,b; Villeneuve and Casanova 2003), although Tigwell and Sauter (1992) found reliable neuron responses in the monkey striate cortex at isoflurane concentrations between 0.5 and 0.9%. Isoflurane/nitrous oxide produced greater suppression of visual evoked firing than did halothane/nitrous oxide (Villeneuve and Casanova 2003). We recently found that both isoflurane and desflurane increased the peak reactivity of cortical units to flash stimulation in the early poststimulus period (0–100 ms). We will return to this finding later in this chapter.

Interestingly, at deep levels of anesthesia, characterized by partially or fully suppressed spontaneous EEG, the brain becomes hyperexcitable. In this state, sensory stimuli elicit brief bursts of cortical activity from an isoelectric baseline. This is surprising because in states with burst-suppressed EEG, patients are always nonresponsive and are considered totally unconscious. Nevertheless, such EEG bursts have been elicited by visual, auditory, and micromechanical stimuli (Hartikainen, Rorarius, Makela et al. 1995; Hartikainen, Rorarius, Perakyla et al. 1995; Hartikainen and Rorarius 1999; Hudetz and Imas 2007; Kroeger and Amzica 2007; Rojas et al. 2008). Since the bursts exhibit long latencies (100–300 ms), they are unlikely to be mediated by thalamocortical relays. It is more likely that the bursts are produced by transiently increased neuronal excitability through polysynaptic or nonsynaptic reticular, basal forebrain or nonspecific thalamic pathways (Hudetz and Imas 2007; Kroeger and Amzica 2007).

Taken together, these results suggest that under anesthesia sensory stimuli continue to excite the cerebral cortex. Therefore, anesthetic-induced unconsciousness cannot be explained by cortical deafferentation or a lack of cortical responsiveness. Instead, we entertain the hypothesis that unconsciousness during anesthesia is due to the inability of the brain to integrate or interpret the sensory information it receives (Hudetz 2006).

## Effect of Anesthesia on Gamma Oscillations

Gamma oscillations are central for our hypothesis because they are thought to play an important role in the encoding and transfer of information in the brain. Spontaneous and evoked gamma oscillations (typically 30–80 Hz) in the EEG, local field potentials, and neuronal firing rates are involved in various cognitive functions, including attention (Bouyer et al. 1980; Tiitinen et al. 1993; Muller, Gruber, and Keil 2000; Steinmetz et al. 2000; Fries et al. 2001), short-term memory (Tallon-Baudry, Kreiter, and Bertrand 1999), feature binding (Treisman 1996), conscious perception (Keil et al. 1999; Srinivasan et al. 1999; Tallon-Baudry and Bertrand 1999), and voluntary action (Pfurtscheller, Flotzinger, and Neuper 1994; Donoghue et al. 1998; Riehle et al. 2000). Moreover, gamma oscillations in field potentials and unit discharges tend to synchronize each other in a network of brain regions participating in a given perceptual or motor task (Engel et al. 1991; Bressler, Coppola, and Nakamura 1993; Desmedt and Tomberg 1994; Munk et al. 1996; Roelfsema et al. 1997; Maloney et al. 1997; Sarnthein et al. 1998; Rodriguez et al. 1999; von Stein et al. 1999; von Stein, Chiang, and Konig 2000). While transient increases in gamma oscillations accompany both consciously perceived and unperceived stimuli, only the consciously perceived stimuli produce long-distance gamma synchronization (Melloni et al. 2007). Furthermore, synchronization at different frequencies in the gamma range mediates different cognitive functions or subsystems participating in the overall task (Knight 2007).

It was proposed nearly 20 years ago that a suppression of 40 Hz gamma oscillations may underlie anesthetic-induced loss of consciousness (Kulli and Koch 1991; Schwender et al. 1995; Plourde and Villemure 1996; Schwender et al. 1996; Schwender, Daunderer, Mulzer et al. 1997; Schwender, Daunderer, Schnatmann et al. 1997; Plourde et al. 1998). However, experimental studies in animals have not revealed a consistent effect of anesthetics on gamma oscillations. Several investigators noted that spontaneous gamma or beta band power can be either decreased or increased in the anesthetized state (Buzsaki, Leung, and Vanderwolf 1983; Steriade 1998; Kral et al. 1999; Vanderwolf 2000; Imas et al. 2004, 2005a,b; Fell et al. 2005). This discrepancy may be due in part to the technique of recording (large scalp electrodes in humans versus small intracortical electrodes in animals), the recorded cortical region, the method of spectral analysis (e.g., FFT vs. wavelet), the type and depth of anesthesia, or to a difference in the nature of gamma oscillations in various species.

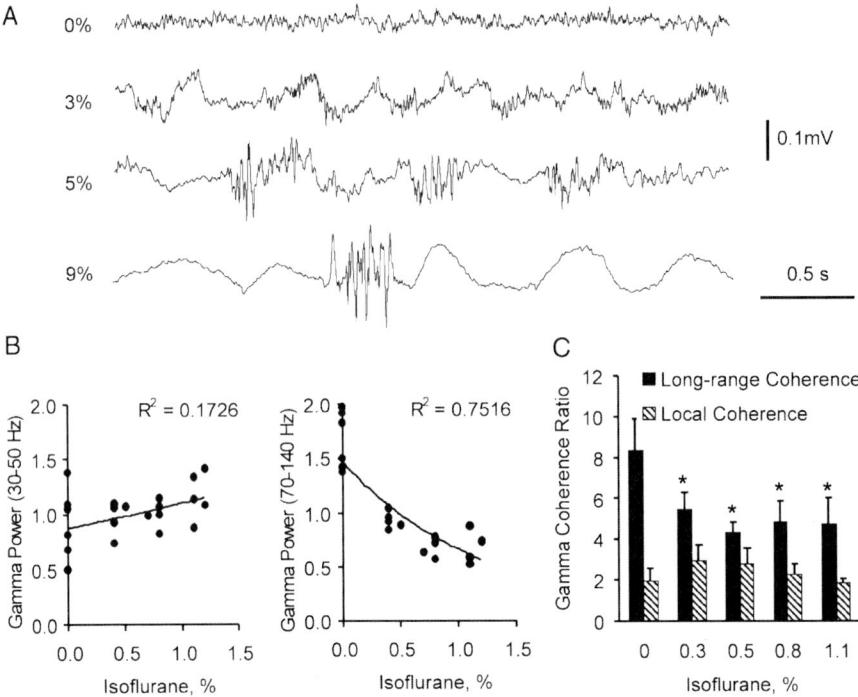

**Fig. 5.1** Effect of volatile anesthetics on gamma oscillations in the rat cerebral cortex. **A:** Spontaneous gamma bursts under anesthesia. EEG traces were recorded in the occipital cortex at four concentrations of desflurane (indicated to the left of each trace) in one rat. Traces were not filtered. Low-amplitude high-frequency activity is continuous in the conscious animal. With increasing desflurane concentration, gamma activity takes the form of bursts with increasing amplitude at a dominant frequency of 35–40 Hz. Consciousness is presumed lost when the rats lose their righting reflex between 3 and 5% desflurane (average: 4.3%). **B:** Effect of isoflurane on spontaneous gamma power in two frequency bands. EEG in the frontal cortex was recorded for several minutes at steady-state isoflurane concentrations, and band-limited power was calculated from power spectral density. Only the power of high-frequency (70–140 Hz) gamma oscillations is suppressed by isoflurane. Note that rats lose their righting reflex at 0.8% isoflurane. **C:** Effect of isoflurane on flash-induced gamma coherence. Coherence was calculated using the wavelet method from field potentials measured with multielectrode arrays in motor cortex and visual cortex. Long-range coherence between motor cortex and visual cortex was significantly decreased by isoflurane, while local coherence within the motor cortex was unchanged. *p<0.05 versus 0% isoflurane

A property of gamma oscillations that has received little attention is their temporal structure at different anesthetic levels. While gamma activity is relatively stationary in the wakeful state, it becomes frequently interrupted and occurs in bursts with increasing anesthetic depth. At the same time, the amplitude of gamma bursts increases. An example of EEG gamma bursts during desflurane anesthesia is shown in Fig. 5.1A. Obviously, the effect of anesthesia on the average gamma power would depend on both the amplitude and the duration of the gamma bursts. Of note is that these gamma bursts occur at moderate depths of anesthesia. Thus, they are not identical to the well-known activity bursts during burst-suppressed EEG. Wide-frequency bursts reaching into the lower gamma band were recently reported (Hermer-Vazquez, Hermer-Vazquez, and Srinivasan 2009). These bursts became infrequent above 1% isoflurane.

Our earlier studies on gamma power (Imas et al. 2004, 2005a,b) were conducted in the presence of intermittent visual stimulation, which raised the possibility that cortical arousal in response to the stimulus is responsible for the augmentation of gamma oscillations. We recently reexamined the effects of volatile anesthetics on spontaneous cortical gamma activity in the rat. As illustrated in Fig. 5.1B for isoflurane, there was no significant change in the power of low-frequency (30–50 Hz) gamma oscillations; however, the power of high-frequency (70–140 Hz) gamma oscillations was attenuated by the anesthetic in a concentration-dependent manner. This observation suggests that a distinction between various gamma frequency bands is important when evaluating the effect of anesthesia.

John et al. (2001) found that the critical change in the EEG that correlated with the loss and return of consciousness was a decrease in gamma coherence between the frontal hemispheres as well as between the frontal and occipital regions of the same hemisphere. Based on these findings, Mashour (2004) and Mashour, Forman, and Campagna (2005) proposed that the mechanism of anesthetic-induced loss of consciousness is "cognitive unbinding" of conscious sensory representations that result from the suppression of gamma coherence at 40 Hz. Hameroff (2006) also emphasized the importance of gamma synchrony for consciousness. To determine the concentration-dependent effect of anesthetics on long-range gamma synchronization, we examined anterior-posterior EEG coherence in rats with various volatile anesthetics (Imas et al. 2006). Multiple microwire electrodes were implanted in the frontal cortex (primary and secondary motor areas) and in the primary visual cortex. We recorded local field potentials before and after flash stimulation and calculated the flash-induced increase in gamma coherence relative to the prestimulus baseline. We found that volatile anesthetics attenuated the evoked increase in long-range gamma coherence between frontal motor and occipital visual areas, but local gamma coherence within the frontal cortex was unchanged (Fig. 5.1C). This result was consistent with the observed suppression of gamma coherence in anesthetized patients (John et al. 2001).

## Effect of Anesthetics on Cortical Information Integration

Tononi (2004) suggested that consciousness derives from the brain's capacity to integrate information. At any moment, the brain can be in one of a great many configurational states; the realization of a specific state represents a reduction in prior uncertainty and, therefore, equates to a large amount of information. At the same time, components of the brain as a system at any level (functional regions, neurons, synapses, molecules, etc) are not independent but closely linked (integrated). Consciousness requires both a large repertoire of brain states (information) and their availability to the system as a whole (integration). This conclusion leads to the hypothesis that anesthetics might disrupt consciousness in two principle ways: by reducing the repertoire of available brain states (reducing the amount of information) or by disrupting the communication among the system components (reducing integration) (Alkire, Hudetz, and Tononi 2008).

We have made three relevant observations that are consistent with this hypothesis. First, we studied the concentration-dependent effects of halothane or isoflurane anesthesia on interhemispheric cross-approximate entropy of the EEG in rats. This quantity measures the statistical dissimilarity or independence of neuronal activity of two brain regions – in this case, the frontal cortices of the two hemispheres in a temporal window of 100 ms. We found that each anesthetic decreased cross-approximate entropy with a significant change occurring at the concentration that produced the loss of righting reflex – a putative behavioral index of loss of consciousness in the rat. An example with isoflurane is shown in Fig. 5.2A. This result suggests that the two hemispheres became more interdependent, equivalent to a reduction in the repertoire of available independent brain states. One does not need both hemispheres to be conscious. Therefore, these findings should be replicated for other brain regions, and perhaps for selected components of specific neurofunctional systems. Nevertheless, the principle is clear: Anesthesia is accompanied by a reduction in the repertoire of neuronal states due to global hypersynchronization. We should be quick to note that this hypersynchronization is different from the above-mentioned gamma synchronization, as the latter involves a specific coordinated network of processing regions or neurons, whereas global hypersynchronization is nonspecific.

---

**Fig. 5.2** Effect of anesthetics on sensory integration in the rat cerebral cortex. **A**: Isoflurane diminishes the probability of independent states of two hemispheres as indicated by the concentration-dependent reduction of cross-approximate entropy of bifrontal EEG. **B**: Locations of bipolar semi-micro electrodes chronically implanted in the rat cerebral cortex to study the effect of anesthetics on functional connectivity. V1M and V1B: primary visual cortex monocular and binocular regions, respectively; V2LA and V2LP: secondary visual cortex lateral anterior and lateral posterior regions, respectively; V2MM and V2ML: secondary visual cortex medial and medial lateral region, respectively; PtA: parietal association cortex; TeA: temporal association cortex; PF: pre-frontal cortex; and FEF: frontal eye field. **C**: Example of the effect of isoflurane anesthesia on flash-evoked intracortical field potentials filtered to the gamma band (20–60 Hz) in frontal (PF), parietal (PtA), and occipital (V1M) sites. Note that the early (middle latency) response at 0–100 ms in visual cortex is preserved in anesthesia, although the rest of the response beyond 100 ms is truncated. Flash-evoked responses in parietal and occipital cortex are attenuated in anesthesia. **D**: Schematic summary of the effect of volatile anesthetics on directional information exchange among three principal cortical regions as measured by transfer entropy. Transfer entropy was calculated from flash-evoked gamma field potential responses similar to those shown in panel (**C**). Numbers by the arrows indicate the percentage change in transfer entropy from the conscious to the unconscious state (loss of righting reflex) along each direction as calculated for the narrow gamma band of 40–50 Hz. Data represent an average from multiple experiments with halothane, isoflurane, and desflurane. A significant reduction in transfer entropy in the feedback direction but not in the feedforward direction occurs upon loss of consciousness

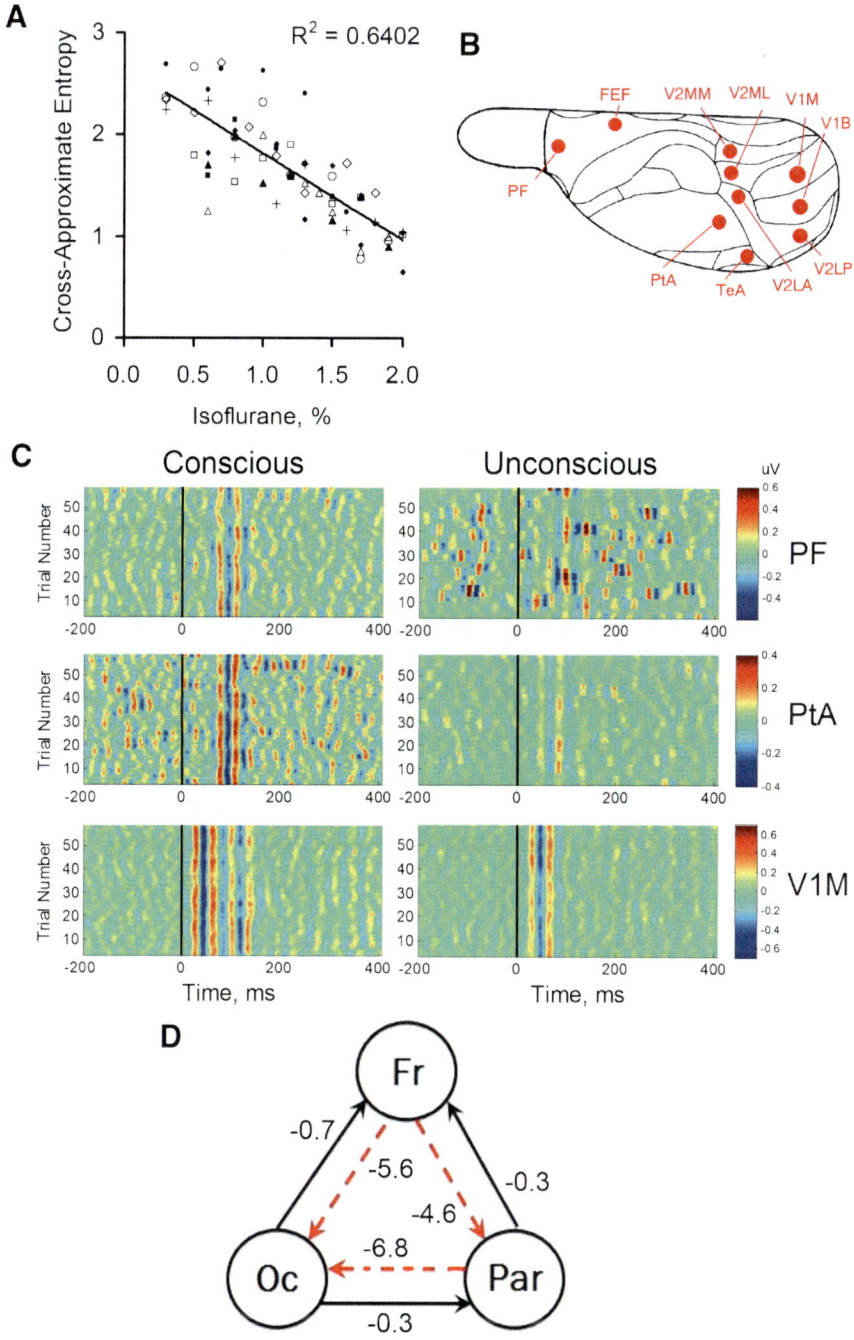

**Fig. 5.2** (continued)

In the second experiment, we estimated the information exchange between distant cortical sites using *transfer entropy* (Imas et al. 2005a,b). Transfer entropy offers a nonlinear, model-free estimation of directional information flow. Since the brain is highly active at rest and the EEG patterns repeatedly shift from one network to another (Koenig et al. 2002), we found it advantageous to transiently augment the network connectivity in a specific neurofunctional system by sensory stimulation. In our study, multiple recording electrodes were chronically implanted in various regions of the visual system, in the adjoining parietal and temporal association areas, and in the frontal region at various locations (shown in Fig. 5.2B). The visual system was stimulated with discrete light flashes, and intracortical field potentials were recorded in each region. An example is shown in Fig. 5.2C. Transfer entropy was then calculated from the poststimulus time-frequency power of the field potentials in the gamma band obtained by wavelet transformation. The analysis showed that the effect of volatile anesthetics on transfer entropy was asymmetrical. At critical anesthetic concentrations that produced unconsciousness, feedback transfer entropies in the frontoparietal, fronto-occipital, and parieto-occipital directions were significantly reduced. Moreover, these reductions were significantly larger than the change in transfer entropies in the corresponding feedforward directions, suggesting that the critical anesthetic effect associated with unconsciousness (loss of the righting reflex) was the preferential reduction in feedback information transfer as carried by gamma oscillations. The findings are summarized schematically in Fig. 5.2D.

These results were originally interpreted in light of the investigations that suggested a role for cortico-cortical feedback or recurrent processing in conscious perception. Cortico-cortical associational projections throughout the hierarchy of modality-specific, multimodal, and supramodal regions are bidirectional (Coogan and Burkhalter 1990). The feedback projections greatly outnumber the feedforward projections, suggesting their functional importance. This reciprocal connectivity has been proposed as an essential substrate for conscious perception and conscious behavior (Cauller and Kulics 1991; Cauller 1995; Jackson and Cauller 1998). As suggested, the forward connections represent and analyze incoming sensory data, whereas the feedback projections play a modulating role in the selection and contextual interpretation of information (Shao and Burkhalter 1996; Lamme and Roelfsema 2000; Pascual-Leone and Walsh 2001).

Feedforward and feedback processes are also thought to segregate temporally. It was suggested that neural activities in primary visual cortex within the first 100 ms after stimulus presentation indicate stimulus registration, whereas those that follow the stimulus by more than 100 ms reflect feedback signaling from higher visual processing regions (Super, Spekreijse, and Lamme 2001; Del Cul, Baillet, and Dehaene 2007; Garrido et al. 2007). The latter components have been correlated with conscious perception (Del Cul, Baillet, and Dehaene 2007) and differentiated between explicit and implicit retrieval (Badgaiyan 2005). Gamma oscillations associated with consciously perceived versus unperceived stimuli were shown to diverge at 40–180 ms and remain different up to 300 ms (Melloni et al. 2007).

To examine if anesthetics exert specific effects on feedforward and feedback components of visual evoked neuronal activity, we recently examined the

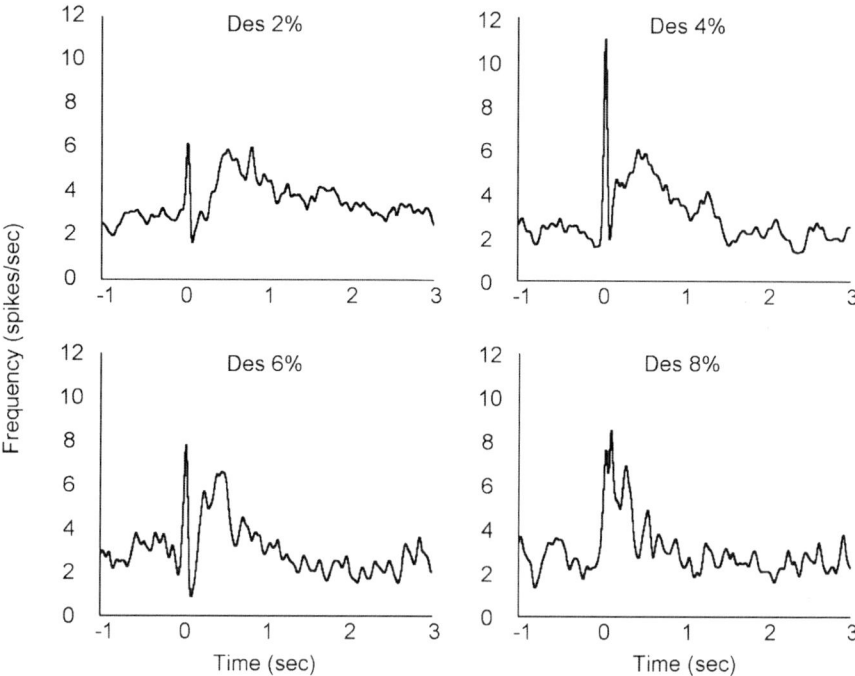

**Fig. 5.3** Effect of desflurane on unit response to visual stimulation in one rat. Light flashes produced by a blue photodiode were delivered to the left eye, and extracellular spikes were recorded in the primary visual cortex of the right hemisphere. Individual units were separated and the spikes trains were averaged to obtain a population response. Desflurane augments the early increase in spike rate usually peaking at 40 ms; flash is at time zero. In contrast, desflurane attenuates the subsequent long-latency response that normally peaks in the 0.2–1.0 s range

concentration-dependent effect of isoflurane and desflurane on flash-induced unit firing in visual cortex of rats. Neuronal firing rates were obtained from extracellular recordings with implanted electrode arrays after spike sorting with principal components analysis. As mentioned before, anesthesia did not attenuate the peak increase in firing rate within the first 100 ms after flash. On the other hand, as illustrated in Fig. 5.3, the long-latency unit response that followed at >150 ms was significantly diminished. Suppression of the long-latency response was paralleled by a decrease in spontaneous firing rate consistent with the suggested general decrease in neuronal excitability (Berg-Johnsen and Langmoen 1990; Hentschke, Schwarz, and Antkowiak 2005). The sensitivity of long-latency (200–500 ms) unit responses to volatile anesthetics was previously observed in cat visual cortex and rat somatosensory cortex at deep anesthesia only (Robson 1967; Ikeda and Wright 1974a,b; Chapin, Waterhouse, and Woodward 1981).

Assuming that the long-latency response component represents recurrent processing by an interaction with higher cortical regions, these results are consistent with the hypothesized selective effect of anesthetics on feedback signaling.

However, the long-latency response may also depend on numerous polysynaptic pathways, which could explain their enhanced sensitivity to anesthetics (Banoub, Tetzlaff, and Schubert 2003). Anesthetic depression of synaptic transmission may result in a cumulative loss of signaling such that the more synapses involved, the greater overall suppression is produced. In turn, complex information processing that relies on high synaptic connectivity would be affected the most. In addition to their obvious effect on synaptic communication, general anesthetics may impede axonal conduction along unmyelinated fibers (Berg-Johnsen and Langmoen 1986; Swindale 2003). As these effects may not occur uniformly in the brain, they may disrupt the timing and synchronization of neuronal messages, thus impeding communication within functional neuronal assemblies that integrate sensory information.

Our findings are consistent with the results of human neuroimaging studies that found attenuated auditory and somatosensory activation in the primary sensory regions under general anesthesia (Heinke et al. 2004; Hofbauer et al. 2004; Heinke and Koelsch 2005; Kerssens et al. 2005; Veselis et al. 2005; Plourde et al. 2006). This is because a smaller hemodynamic response may be explained by a shorter duration of the electrophysiological response (Pawela, Hudetz et al. 2008). In addition, Heinke et al. (2004) found that propofol suppresses frontal lobe activity before it suppresses temporal lobe activity during a language task when subjects became unresponsive. This observation may suggest that frontoparietal communication is impaired because frontal activation is reduced under anesthesia.

Recently, important experimental support was obtained for the frontoparietal feedback hypothesis of anesthesia from human subjects undergoing general anesthesia. Lee, Kim et al. (2009) showed that directed feedback connectivity derived from multichannel EEG measurements was diminished upon the loss of consciousness in patients anesthetized with propofol. The effect was reversed when patients began to respond to verbal commands. In a parallel paper, Lee, Mashour et al. (2009) calculated the capacity of the cortex for neural information integration based on Tononi's information theory. They found that propofol significantly reduced information integration capacity most prominently when calculated from the EEG gamma band. The spatiotemporal organization of gamma activity was also disrupted. These studies emphasize the importance of gamma oscillations in large-scale information integration in the brain and implicate their disruption as a possible mechanism of general anesthesia.

## Cortical Regions Responsible for Unconsciousness

Consciousness survives injuries to large portions of the cerebral cortex, although specific functional deficits would obviously be present. This phenomenon supports the concept of "mass effect" of consciousness consistent with Tononi's information integration theory (Tononi 2004), which implies that consciousness is a graded phenomenon. Indeed, if the plane of anesthesia were deepened slowly and gradually, one would experience a gradual diminution of consciousness. On the other hand, lesions to specific subcortical structures, particularly those along the midline ponto-thalamic axis, produce an immediate and full loss of consciousness (Schiff and Plum 2000). The efficacy of these lesions may lie in their unimpeded and complete disruption of global cortical integrative function due to a diminution of cortical arousal that normally enables consciousness. The question remains whether there are any critical cortical processing regions or networks whose elimination may invariably confer unconsciousness.

It has been suggested that lateral frontoparietal association networks play a critical role in conscious perception, attention, working memory, and episodic retrieval (Rees, Kreiman, and Koch 2002; Naghavi and Nyberg 2005; Del Cul, Baillet, and Dehaene 2007). Frontoparietal regions were preferentially deactivated in patients in a vegetative state (Laureys 2005) and during propofol anesthesia (Fiset et al. 1999). Alkire (2008) recently described a hypothesized "consciousness circuit" consisting of the implicated frontoparietal regions and the posterior cingulate cortex. This system shows a certain degree of homology with the so-called "default network" of the brain (Raichle et al. 2001), which is highly active in the unstimulated brain but exhibits a decrease in activity during goal-directed behavior. Observations from epilepsy, stroke, vegetative state, and anesthesia converge suggesting that a common cortical area, the region of posterior cingulate/retrosplenial/precuneus, has a critical role in consciousness (Vogt and Laureys 2005).

In spite of their membership in the core functional networks of the brain, the role of frontal cortical sites in consciousness remains questionable (Wheeler and Stuss 2003). Stuss, Picton, and Alexander (2001) suggested that the frontal lobes are critical for self-awareness. However, frontal injury from stroke may produce an akinetic syndrome rather than loss of consciousness (Lipschutz et al. 1991). It is possible that the critical change in cortical information transfer for loss of consciousness occurs in the local feedback loops of the posterior parietal cortex, including sensory and association regions, leading to a generalized failure of information integration. Furthermore, the suppression of voluntary movements under anesthesia may be an independent effect of anesthetics acting on the motor cortex in addition to their effects on the spinal cord (Rampil and Laster 1992; Rampil, Mason, and Singh 1993; Rampil and King 1996; Antognini, Carstens, and Buzin 1999; Antognini and Carstens 2002; Antognini, Barter, and Carstens 2005; Jugovac, Imas, and Hudetz 2006). With some anesthetics, such as ketamine, motor suppression may be the primary effect, dissociated from the smaller influence on sensory functions (Plourde, Baribeau, and Bonhomme 1997; Alkire, Haier, and Fallon 2000). An agent-invariant "final common pathway" to unconsciousness may reside with the disruption of information integration in a network of the posterior cortex.

## Network Structure and the Vulnerability of Conscious Sensory Integration

The topological structure of functional networks of the brain at both cellular and regional scales may be a significant determinant of when and how anesthetic agents disrupt information processing necessary for consciousness. We address two possible mechanisms: the role of specific long-range connections and the vulnerability of global connectivity due to network topology. Much of this work is in progress, but data already suggest the potential relevance of this topic for anesthesia research.

Since the discovery of so-called "resting state functional connectivity" (RSFC) of the brain subtended by low-frequency correlations of the MRI BOLD signals among specific brain regions recorded under task-free conditions (Biswal et al. 1995, 1997), functional networks of the brain have been identified that consist of major segregated processing modules linked with long-range fiber connections. This work has now been complemented with diffusion imaging-based tractography, providing further anatomical delineation of these networks. Thus, from RSFC, Meunier et al. (2009) found three major segregated functional modules with frontal, central, and posterior locations. One region, the dorsal fronto-cingulo-parietal module, showed extensive intermodular connections that appeared to mediate interactions between anterior motor and posterior sensory regions. Using similar methodology, He, Wang et al. (2009) identified five major modules that included motor- and sensory-specific systems as well as a medial frontoparietal and a lateral frontoparietal system. Studies with diffusion spectrum imaging tractography revealed two major bilaterally linked fiber backbones. One connects the medial prefrontal cortex through the cingulate cortex to the precuneus, and the other connects the lateral prefrontal cortex to parietal cortex (Hagmann et al. 2008; Honey et al. 2009). The medially positioned modules, together with the lateral parietal module, appear to overlap with the previously mentioned default network of the brain (Gusnard and Raichle 2001; Raichle et al. 2001). The activity of the default system is reduced, although not eliminated, under anesthesia (Vincent et al. 2007; Pawela, Biswal et al. 2008). This finding, for the time being, leaves the question of its role in consciousness unanswered. This suggests that RSFC reflects to a large part anatomical connectivity. On the other hand, Peltier et al. (2005) found a significant reduction in functional connectivity in patients anesthetized with sevoflurane. Because of the well-known vasodilator effect of most inhalational anesthetics, the specificity of these findings with respect to neuronal activity will have to be confirmed.

A number of recent investigations reveal that regional architecture of the brain follows the so called "small-world" pattern. Small world is an economical structure that is characterized by local clusters with high connectivity surrounding principal hubs. Sparse long-range connectivity connects the relatively segregated clusters. This topology is economical in the sense that it maximizes the efficiency of information processing while minimizing its cost in wiring (Bassett and Bullmore 2006). Thus, small-world topology is thought to confer the brain high efficiency of information exchange at relatively low cost, i.e., fewer nerve fibers and synaptic relays.

Networks of the brain can be defined at the neuronal level (neuronal connectivity) or regional level based on anatomical parcellation of the brain to nearly equivalent volumes (regional connectivity). A common graph theoretical measure of small-world topology is the ratio that compares the tendency of connections to cluster locally to the minimum path length between all pairs of regions. Data on connectivity have been obtained from anatomical tract tracing, MRI diffusion tensor imaging, diffusion spectrum imaging, EEG, MEG, and MRI RSFC. From these measures, principal hubs have been found throughout unimodal and heteromodal association cortex, including posterior cingulate, lateral temporal, lateral parietal, and medial and lateral prefrontal regions (Salvador et al. 2005; Achard et al. 2006; Hagmann et al. 2007; Bassett et al. 2006; Buckner et al. 2009; Gong et al. 2009; Honey et al. 2009; Wang et al. 2009). The medially positioned cortical hubs resemble the corresponding components of the default network. Moreover, the medial parietal region (posterior cingulate and precuneus) appears to be the principal hub in all studies conducted so far. Functional connectivity from MRI is in general consistent with, although more variable than, anatomical connectivity obtained with diffusion tractography (Honey et al. 2009).

The very design of small-world topology, which makes the brain so effectively connected, may also make the brain particularly vulnerable to anesthesia. We hypothesize that as anesthetics in increasing doses suppress synaptic transmission in more and more circuits, small regions and subnets become more and more isolated until, at some critical level, connectivity across the whole system is disrupted. Consequently, unconsciousness ensues. We hypothesize that an abrupt transition from consciousness to unconsciousness is a direct consequence of an abrupt disintegration of the cortical information processing network. In fact, Steyn-Ross, Steyn-Ross, Wilcocks et al. (2001) and Steyn-Ross, Steyn-Ross, Sleigh et al. (2001) presented a number of computer simulations based on regular cortical neuronal network anatomy that demonstrated a rapid state transition at a critical anesthetic dose. Global disconnection has been predicted by mathematical models of idealized network models (Cammarota and Onaral 1998; Rozenfeld and Ben-Avraham 2007).

Computational modeling has also been used to estimate the resilience or vulnerability of reconstructed human and animal brain networks (Bullmore and Sporns 2009). Networks with small-world architecture may be assumed to be particularly vulnerable to lesions because of the sparseness of their long-range connections. Computational lesioning based on anatomical connectivity data of the macaque monkey showed that lesions of parietal regions had the greatest potential to disrupt the integrative aspects of neocortical function (Honey and Sporns 2008). In human RSFC-based networks, the removal of hubs conferred a significant vulnerability (Gong et al. 2009); a selective attack on global hubs had a particularly dramatic effect (He, Wang et al. 2009). Achard et al. (2006) found that the RSFC-based human network was almost equally resilient to random attack and targeted attack of hubs. The network disintegrated when 40% of its most connected nodes were removed. However, in the former study, network connectivity decreased abruptly when a relatively small fraction of connections were removed (<20%), implying a critical level of tolerance at which the network becomes globally disconnected (He, Wang et al. 2009). Computational models have been applied to examine network connectivity changes in Alzheimer's disease (He, Chen et al. 2009) and epilepsy (Dyhrfjeld-Johnsen et al. 2007), but have not been applied to anesthesia. In a single patient so far, long-range intrahemispheric and interhemispheric connectivity was found to be absent in a minimally conscious patient (Salvador et al. 2005).

It is clear that both the network structure and the cellular/regional probability of lesions will significantly influence the critical transition at which networks may functionally disintegrate. Thus, the resilience of a brain network in the face anesthesia will depend on the specificity and nature of synaptic transmission failure. A detailed knowledge of the cellular targets and effects of anesthetics across all brain regions will be essential to estimate this effect. Understanding the mechanisms of anesthesia may also require investigating and modeling the brain's vulnerability at neuronal-synaptic circuit level rather than at regional network level.

We endorse the view that there may be a critical anesthetic level at which consciousness is abruptly lost, as indicated by an abrupt change in network connectivity. However, under certain conditions, consciousness may also be reduced gradually if no abrupt change in network connectivity occurs. The state of consciousness in anesthesia is not steady but shifts periodically; the temporal dynamics of these state changes will have to be factored into a more thorough understanding of their dependence on network connectivity. Finally, integrative capacities of the brain can be lost not only by functional disconnection but also by hypersynchronization of neuronal activity, such as that seen during bursting activity. Conscious information processing would be incompatible with either condition. Figure 5.4 illustrates these effects schematically.

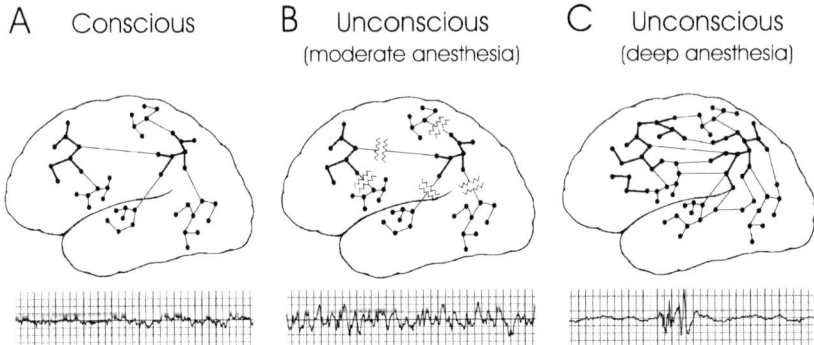

**Fig. 5.4** Schematic summary of the postulated effect of general anesthesia on information processing in cortical networks. The network is assumed to be functionally structured and according to small-world topology. In this structure, principal processing modules have strong intrinsic local connectivity while the modules are coupled by sparse long-range connections. **A**: In the conscious brain, frontal and parietal association areas (indicated by heavy lines) play a central role in information integration from the satellite sensory areas (indicated by intermediate lines). **B**: When consciousness is suppressed by general anesthetics by breaking or corrupting information transfer between the association regions and between the association and sensory regions (moderate anesthesia). **C**: Under deep anesthesia, such as in states characterized by burst-suppressed EEG, cortical modules can become hypersynchronized and this, too, prevents meaningful information processing. In all cases, primary reactivity of the early sensory areas is preserved; however, higher integration of sensory information is disrupted. Typical EEG patterns corresponding to each state are shown in the last row

**Disclosure Statement**  This work was supported by the grant R01 GM-56398 from the Institute of General Medical Sciences, National Institutes of Health, Bethesda, Maryland, USA. This work was not an industry-supported study and the author has no financial conflicts of interest. The comments of Drs. Zeljko Bosnjak, Quinn Hogan, Thomas Stekiel, and Anna Stadnicka to this manuscript are appreciated. The contribution of graduate students Jeannette Vizuete and Sivesh Pillay to the experiments from which results are presented here is appreciated.

# References

Achard, S., R. Salvador, B. Whitcher, J. Suckling, and E. Bullmore. 2006. A resilient, low-frequency, small-world human brain functional network with highly connected association cortical hubs. *J Neurosci* 26(1):63–72.

Alkire, M. T. 2008. Loss of effective connectivity during general anesthesia. *Int Anesthesiol Clin* 46(3):55–73.

Alkire, M. T., and J. Miller. 2005. General anesthesia and the neural correlates of consciousness. *Prog Brain Res* 150:229–244.

Alkire, M. T., R. J. Haier, and J. H. Fallon. 2000. Toward a unified theory of narcosis: brain imaging evidence for a thalamocortical switch as the neurophysiologic basis of anesthetic- induced unconsciousness. *Conscious Cogn* 9(3):370–286.

Alkire, M. T., A. G. Hudetz, and G. Tononi. 2008. Consciousness and anesthesia. *Science* 322(5903):876–880.

Alkire, M. T., R. J. Haier, N. K. Shah, and C. T. Anderson. 1997. Positron emission tomography study of regional cerebral metabolism in humans during isoflurane anesthesia. *Anesthesiology* 86(3):549–557.

Alkire, M. T., C. J. Pomfrett, R. J. Haier, M. V. Gianzero, C. M. Chan, B. P. Jacobsen, and J. H. Fallon. 1999. Functional brain imaging during anesthesia in humans: effects of halothane on global and regional cerebral glucose metabolism. *Anesthesiology* 90(3): 701–709.

Alkire, M. T., J. R. McReynolds, E. L. Hahn, and A. N. Trivedi. 2007. Thalamic microinjection of nicotine reverses sevoflurane-induced loss of righting reflex in the rat. *Anesthesiology* 107(2):264–272.

Andrade, J., D. J. Sapsford, D. Jeevaratnum, A. J. Pickworth, and J. G. Jones. 1996. The coherent frequency in the electroencephalogram as an objective measure of cognitive function during propofol sedation. *Anesth Analg* 83(6):1279–1284.

Angel, A. 1991. The G. L. Brown lecture. Adventures in anaesthesia. *Exp Physiol* 76(1):1–38.

Antognini, J. F., and E. Carstens. 2002. *In vivo* characterization of clinical anaesthesia and its components. *Br J Anaesth* 89(1):156–166.

Antognini, J. F., L. Barter, and E. Carstens. 2005. Overview movement as an index of anesthetic depth in humans and experimental animals. *Comp Med* 55(5):413–418.

Antognini, J. F., E. Carstens, and V. Buzin. 1999. Isoflurane depresses motoneuron excitability by a direct spinal action: an F-wave study. *Anesth Analg* 88(3):681–685.

Baars, B. J. 2005. Global workspace theory of consciousness: toward a cognitive neuroscience of human experience. *Prog Brain Res* 150:45–53.

Baars, B. J., W. P. Banks, and J. B. Newman. 2003. *Essential sources in the scientific study of consciousness*. Cambridge, MA: MIT Press.

Badgaiyan, R. D. 2005. Conscious awareness of retrieval: an exploration of the cortical connectivity. *Int J Psychophysiol* 55(2):257–262.

Banoub, M., J. E. Tetzlaff, and A. Schubert. 2003. Pharmacologic and physiologic influences affecting sensory evoked potentials: implications for perioperative monitoring. *Anesthesiology* 99(3):716–737.

Bassett, D. S., and E. Bullmore. 2006. Small-world brain networks. *Neuroscientist* 12(6):512–523.

Bassett, D. S., A. Meyer-Lindenberg, S. Achard, T. Duke, and E. Bullmore. 2006. Adaptive reconfiguration of fractal small-world human brain functional networks. *Proc Natl Acad Sci USA* 103(51):19518–19523.

Berg-Johnsen, J., and I. A. Langmoen. 1986. The effect of isoflurane on unmyelinated and myelinated fibres in the rat brain. *Acta Physiol Scand* 127(1):87–93.

Berg-Johnsen, J., and I. A. Langmoen. 1990. Mechanisms concerned in the direct effect of isoflurane on rat hippocampal and human neocortical neurons. *Brain Res* 507(1):28–34.

Biswal, B., F. Z. Yetkin, V. M. Haughton, and J. S. Hyde. 1995. Functional connectivity in the motor cortex of resting human brain using echo-planar MRI. *Magn Reson Med* 34(4):537–641.

Biswal, B., A. G. Hudetz, F. Z. Yetkin, V. M. Haughton, and J. S. Hyde. 1997. Hypercapnia reversibly suppresses low-frequency fluctuations in the human motor cortex during rest using echo-planar MRI. *J Cereb Blood Flow Metab* 17(3):301–308.

Block, N. 2005. Two neural correlates of consciousness. *Trends Cogn Sci* 9(2):46–52.

Bonhomme, V., G. Plourde, P. Meuret, P. Fiset, and S. B. Backman. 2000. Auditory steady-state response and bispectral index for assessing level of consciousness during propofol sedation and hypnosis. *Anesth Analg* 91(6):1398–1403.

Bouyer, J. J., M. F. Montaron, A. Rougeul, and P. Buser. 1980. Parietal electrocortical rhythms in the cat: their relation to a behavior of focused attention and possible mesencephalic control through a dopaminergic pathway. *C R Seances Acad Sci D* 291(9):779–783.

Bressler, S. L., R. Coppola, and R. Nakamura. 1993. Episodic multiregional cortical coherence at multiple frequencies during visual task performance. *Nature* 366(6451):153–156.

Buckner, R. L., J. Sepulcre, T. Talukdar, F. M. Krienen, H. Liu, T. Hedden, J. R. Andrews-Hanna, R. A. Sperling, and K. A. Johnson. 2009. Cortical hubs revealed by intrinsic functional connectivity: mapping, assessment of stability, and relation to Alzheimer's disease. *J Neurosci* 29(6):1860–1873.

Bullmore, E., and O. Sporns. 2009. Complex brain networks: graph theoretical analysis of structural and functional systems. *Nat Rev Neurosci* 10(3):186–198.

Buzsaki, G., L. W. Leung, and C. H. Vanderwolf. 1983. Cellular bases of hippocampal EEG in the behaving rat. *Brain Res* 287(2):139–171.

Cammarota, J. P., Jr., and B. Onaral. 1998. State transitions in physiologic systems: a complexity model for loss of consciousness. *IEEE Trans Biomed Eng* 45(8):1017–1023.

Cauller, L. 1995. Layer I of primary sensory neocortex: where top-down converges upon bottom-up. *Behav Brain Res* 71(1–2):163–170.

Cauller, L. J., and A. T. Kulics. 1991. The neural basis of the behaviorally relevant N1 component of the somatosensory-evoked potential in SI cortex of awake monkeys: evidence that backward cortical projections signal conscious touch sensation. *Exp Brain Res* 84(3):607–619.

Chalmers, D. J. 1996. *The conscious mind: in search of a fundamental theory, philosophy of mind series.* New York: Oxford University Press.

Chapin, J. K., B. D. Waterhouse, and D. J. Woodward. 1981. Differences in cutaneous sensory response properties of single somatosensory cortical neurons in awake and halothane anesthetized rats. *Brain Res Bull* 6(1):63–70.

Coogan, T. A., and A. Burkhalter. 1990. Conserved patterns of cortico-cortical connections define areal hierarchy in rat visual cortex. *Exp Brain Res* 80(1):49–53.

Crick, F., and C. Koch. 2003. A framework for consciousness. *Nat Neurosci* 6(2):119–126.

Del Cul, A., S. Baillet, and S. Dehaene. 2007. Brain dynamics underlying the nonlinear threshold for access to consciousness. *PLoS Biol* 5(10):e260.

Desmedt, J. E., and C. Tomberg. 1994. Transient phase-locking of 40 Hz electrical oscillations in prefrontal and parietal human cortex reflects the process of conscious somatic perception. *Neurosci Lett* 168(1–2):126–129.

Detsch, O., C. Vahle-Hinz, E. Kochs, M. Siemers, and B. Bromm. 1999. Isoflurane induces dose-dependent changes of thalamic somatosensory information transfer. *Brain Res* 829(1–2):77–89.

Devor, M., and V. Zalkind. 2001. Reversible analgesia, atonia, and loss of consciousness on bilateral intracerebral microinjection of pentobarbital. *Pain* 94(1):101–112.

Dong, H. L., S. Fukuda, E. Murata, and T. Higuchi. 2006. Excitatory and inhibitory actions of isoflurane on the cholinergic ascending arousal system of the rat. *Anesthesiology* 104(1): 122–133.

Donoghue, J. P., J. N. Sanes, N. G. Hatsopoulos, and G. Gaal. 1998. Neural discharge and local field potential oscillations in primate motor cortex during voluntary movements. *J Neurophysiol* 79(1):159–173.

Dyhrfjeld-Johnsen, J., V. Santhakumar, R. J. Morgan, R. Huerta, L. Tsimring, and I. Soltesz. 2007. Topological determinants of epileptogenesis in large-scale structural and functional models of the dentate gyrus derived from experimental data. *J Neurophysiol* 97(2): 1566–1587.

Eger, E. I., II, and J. M. Sonner. 2006. Anaesthesia defined (gentlemen, this is no humbug). *Best Pract Res Clin Anaesthesiol* 20(1):23–29.

Engel, A. K., P. Konig, A. K. Kreiter, and W. Singer. 1991. Interhemispheric synchronization of oscillatory neuronal responses in cat visual cortex. *Science* 252(5010):1177–1179.

Erchova, I. A., M. A. Lebedev, and M. E. Diamond. 2002. Somatosensory cortical neuronal population activity across states of anaesthesia. *Eur J Neurosci* 15(4):744–752.

Erdelyi, M. H. 2004. Subliminal perception and its cognates: theory, indeterminacy, and time. *Conscious Cogn* 13(1):73–91.

Fell, J., G. Widman, B. Rehberg, C. E. Elger, and G. Fernandez. 2005. Human mediotemporal EEG characteristics during propofol anesthesia. *Biol Cybern* 92(2):92–100.

Fiset, P., T. Paus, T. Daloze, G. Plourde, P. Meuret, V. Bonhomme, N. Hajj-Ali, S. B. Backman, and A. C. Evans. 1999. Brain mechanisms of propofol-induced loss of consciousness in humans: a positron emission tomographic study. *J Neurosci* 19(13):5506–5513.

Franks, N. P. 2008. General anaesthesia: from molecular targets to neuronal pathways of sleep and arousal. *Nat Rev Neurosci* 9(5):370–386.

Fries, P., J. H. Reynolds, A. E. Rorie, and R. Desimone. 2001. Modulation of oscillatory neuronal synchronization by selective visual attention. *Science* 291(5508):1560–1563.

Garrido, M. I., J. M. Kilner, S. J. Kiebel, and K. J. Friston. 2007. Evoked brain responses are generated by feedback loops. *Proc Natl Acad Sci USA* 104(52):20961–20966.

Gong, G., Y. He, L. Concha, C. Lebel, D. W. Gross, A. C. Evans, and C. Beaulieu. 2009. Mapping anatomical connectivity patterns of human cerebral cortex using *in vivo* diffusion tensor imaging tractography. *Cereb Cortex* 19(3):524–536.

Gusnard, D. A., and M. E. Raichle. 2001. Searching for a baseline: functional imaging and the resting human brain. *Nat Rev Neurosci* 2(10):685–694.

Hagmann, P., M. Kurant, X. Gigandet, P. Thiran, V. J. Wedeen, R. Meuli, and J. P. Thiran. 2007. Mapping human whole-brain structural networks with diffusion MRI. *PLoS ONE* 2(7):e597.

Hagmann, P., L. Cammoun, X. Gigandet, R. Meuli, C. J. Honey, V. J. Wedeen, and O. Sporns. 2008. Mapping the structural core of human cerebral cortex. *PLoS Biol* 6(7):e159.

Hameroff, S. R. 2006. The entwined mysteries of anesthesia and consciousness: is there a common underlying mechanism? *Anesthesiology* 105(2):400–412.

Hameroff, S. R., A. W. Kaszniak, and A. Scott. 1996. *Toward a science of consciousness: the first Tucson discussions and debates, Complex adaptive systems.* Cambridge, MA: MIT Press.

Hameroff, S. R., A. W. Kaszniak, and A. Scott. 1998. *Toward a science of consciousness II: the second Tucson discussions and debates, complex adaptive systems.* Cambridge, MA: MIT Press.

Hartikainen, K., and M. G. Rorarius. 1999. Cortical responses to auditory stimuli during isoflurane burst suppression anaesthesia. *Anaesthesia* 54(3):210–214.

Hartikainen, K., M. Rorarius, K. Makela, J. Perakyla, E. Varila, and V. Jantti. 1995. Visually evoked bursts during isoflurane anaesthesia. *Br J Anaesth* 74(6):681–685.

Hartikainen, K. M., M. Rorarius, J. J. Perakyla, P. J. Laippala, and V. Jantti. 1995. Cortical reactivity during isoflurane burst-suppression anesthesia. *Anesth Analg* 81(6):1223–1228.

He, Y., Z. Chen, G. Gong, and A. Evans. 2009. Neuronal networks in Alzheimer's disease. *The Neuroscientist* 15(4):333–350.

He, Y., J. Wang, L. Wang, Z. J. Chen, C. Yan, H. Yang, H. Tang, C. Zhu, Q. Gong, Y. Zang, and A. C. Evans. 2009. Uncovering intrinsic modular organization of spontaneous brain activity in humans. *PLoS ONE* 4(4):e5226.

Heinke, W., and S. Koelsch. 2005. The effects of anesthetics on brain activity and cognitive function. *Curr Opin Anaesthesiol* 18(6):625–631.

Heinke, W., C. J. Fiebach, C. Schwarzbauer, M. Meyer, D. Olthoff, and K. Alter. 2004. Sequential effects of propofol on functional brain activation induced by auditory language processing: an event-related functional magnetic resonance imaging study. *Br J Anaesth* 92(5):641–650.

Hentschke, H., C. Schwarz, and B. Antkowiak. 2005. Neocortex is the major target of sedative concentrations of volatile anaesthetics: strong depression of firing rates and increase of GABAA receptor-mediated inhibition. *Eur J Neurosci* 21(1):93–102.

Hermer-Vazquez, R., L. Hermer-Vazquez, and S. Srinivasan. 2009. A putatively novel form of spontaneous coordination in neural activity. *Brain Res Bull* 79(1):6–14.

Hobson, J. A., and E. F. Pace-Schott. 2002. The cognitive neuroscience of sleep: neuronal systems, consciousness and learning. *Nat Rev Neurosci* 3(9):679–693.

Hofbauer, R. K., P. Fiset, G. Plourde, S. B. Backman, and M. C. Bushnell. 2004. Dose-dependent effects of propofol on the central processing of thermal pain. *Anesthesiology* 100(2):386–394.

Honey, C. J., and O. Sporns. 2008. Dynamical consequences of lesions in cortical networks. *Hum Brain Mapp* 29(7):802–809.

Honey, C. J., O. Sporns, L. Cammoun, X. Gigandet, J. P. Thiran, R. Meuli, and P. Hagmann. 2009. Predicting human resting-state functional connectivity from structural connectivity. *Proc Natl Acad Sci USA* 106(6):2035–2040.

Hudetz, A. G. 2006. Suppressing Consciousness: mechanisms of general anesthesia. *Semin Anesthesia Perioperative Med Pain* 25(4):196–204.

Hudetz, A. G. 2008. Are we unconscious during general anesthesia? *Int Anesthesiol Clin* 46(3): 25–42.

Hudetz, A. G., and O. A. Imas. 2007. Burst activation of the cerebral cortex by flash stimuli during isoflurane anesthesia in rats. *Anesthesiology* 107(6):983–991.

Hudetz, A. G., J. D. Wood, and J. P. Kampine. 2003. Cholinergic reversal of isoflurane anesthesia in rats as measured by cross-approximate entropy of the electroencephalogram. *Anesthesiology* 99(5):1125–1131.

Ikeda, H., and M. J. Wright. 1974a. Effect of halothane-nitrous oxide anaesthesia on the behaviour of 'sustained' and 'transient' visual cortical neurones. *J Physiol* 237(2):20P–21P.

Ikeda, H., and M. J. Wright. 1974b. Sensitivity of neurones in visual cortex (area 17) under different levels of anaesthesia. *Exp Brain Res* 20(5):471–484.

Imas, O. A., K. M. Ropella, J. D. Wood, and A. G. Hudetz. 2004. Halothane augments event-related gamma oscillations in rat visual cortex. *Neuroscience* 123(1):269–278.

Imas, O. A., K. M. Ropella, B. D. Ward, J. D. Wood, and A. G. Hudetz. 2005a. Volatile anesthetics disrupt frontal-posterior recurrent information transfer at gamma frequencies in rat. *Neurosci Lett* 387(3):145–150.

Imas, O. A., K. M. Ropella, B. D. Ward, J. D. Wood, and A. G. Hudetz . 2005b. Volatile anesthetics enhance flash-induced gamma oscillations in rat visual cortex. *Anesthesiology* 102(5):937–947.

Imas, O. A., K. M. Ropella, J. D. Wood, and A. G. Hudetz. 2006. Isoflurane disrupts anterio-posterior phase synchronization of flash-induced field potentials in the rat. *Neurosci Lett* 402(3):216–221.

Jackson, M. E., and L. J. Cauller. 1998. Neural activity in SII modifies sensory evoked potentials in SI in awake rats. *Neuroreport* 9(15):3379–3382.

James, W. 1890. *The principles of psychology.* New York: H. Holt and Co.

John, E. R. 2002. The neurophysics of consciousness. *Brain Res Brain Res Rev* 39(1):1–28.

John, E. R., L. S. Prichep, W. Kox, P. Valdes-Sosa, J. Bosch-Bayard, E. Aubert, M. Tom, F. diMichele, and L. D. Gugino. 2001. Invariant reversible QEEG effects of anesthetics. *Conscious Cogn* 10(2):165–183.

Jugovac, I., O. Imas, and A. G. Hudetz. 2006. Supraspinal anesthesia: behavioral and electroencephalographic effects of intracerebroventricularly infused pentobarbital, propofol, fentanyl and midazolam. *Anesthesiology* 105(4):1–15.

Keil, A., M. M. Muller, W. J. Ray, T. Gruber, and T. Elbert. 1999. Human gamma band activity and perception of a gestalt. *J Neurosci* 19(16):7152–7161.

Kerssens, C., S. Hamann, S. Peltier, X. P. Hu, M. G. Byas-Smith, and P. S. Sebel. 2005. Attenuated brain response to auditory word stimulation with sevoflurane: a functional magnetic resonance imaging study in humans. *Anesthesiology* 103(1):11–19.

Knight, R. T. 2007. Neuroscience. Neural networks debunk phrenology. *Science* 316(5831): 1578–1579.

Koenig, T., L. Prichep, D. Lehmann, P. V. Sosa, E. Braeker, H. Kleinlogel, R. Isenhart, and E. R. John. 2002. Millisecond by millisecond, year by year: normative EEG microstates and developmental stages. *Neuroimage* 16(1):41–48.

Kral, A., J. Tillein, R. Hartmann, and R. Klinke. 1999. Monitoring of anaesthesia in neurophysiological experiments. *Neuroreport* 10(4):781–787.

Kroeger, D., and F. Amzica. 2007. Hypersensitivity of the anesthesia-induced comatose brain. *J Neurosci* 27(39):10597–10607.

Kulli, J., and C. Koch. 1991. Does anesthesia cause loss of consciousness? *Trends Neurosci* 14(1):6–10.

Lamme, V. A., and P. R. Roelfsema. 2000. The distinct modes of vision offered by feedforward and recurrent processing. *Trends Neurosci* 23(11):571–579.

Laureys, S. 2005. The neural correlate of (un)awareness: lessons from the vegetative state. *Trends Cogn Sci* 9(12):556–559.

Lee, U., S. Kim, G. J. Noh, B. M. Choi, E. Hwang, and G. A. Mashour. 2009. The directionality and functional organization of frontoparietal connectivity during consciousness and anesthesia in humans. *Conscious Cogn*. In press. [Epub ahead of print (May 12)].

Lee, U., G. A. Mashour, S. Kim, G. J. Noh, and B. M. Choi. 2009. Propofol induction reduces the capacity for neural information integration: implications for the mechanism of consciousness and general anesthesia. *Conscious Cogn* 18(1):56–64.

Libet, B. 1982. Brain stimulation in the study of neuronal functions for conscious sensory experiences. *Hum Neurobiol* 1(4):235–242.

Libet, B., D. K. Pearl, D. E. Morledge, C. A. Gleason, Y. Hosobuchi, and N. M. Barbaro. 1991. Control of the transition from sensory detection to sensory awareness in man by the duration of a thalamic stimulus. The cerebral 'time-on' factor. *Brain* 114(Pt 4):1731–1757.

Lipschutz, J. H., R. M. Pascuzzi, J. Bognanno, and T. Putty. 1991. Bilateral anterior cerebral artery infarction resulting from explosion-type injury to the head and neck. *Stroke* 22(6): 813–815.

Lydic, R. 1996. Reticular modulation of breathing during sleep and anesthesia. *Curr Opin Pulm Med* 2(6):474–481.

Lydic, R., and J. F. Biebuyck. 1994. Sleep neurobiology: relevance for mechanistic studies of anaesthesia. *Br J Anaesth* 72(5):506–508.

Ma, J., B. Shen, L. S. Stewart, I. A. Herrick, and L. S. Leung. 2002. The septohippocampal system participates in general anesthesia. *J Neurosci* 22(2):RC200.

Maloney, K. J., E. G. Cape, J. Gotman, and B. E. Jones. 1997. High-frequency gamma electroencephalogram activity in association with sleep-wake states and spontaneous behaviors in the rat. *Neuroscience* 76(2):541–555.

Mashour, G. A. 2004. Consciousness unbound: toward a paradigm of general anesthesia. *Anesthesiology* 100(2):428–433.

Mashour, G. A., S. A. Forman, and J. A. Campagna. 2005. Mechanisms of general anesthesia: from molecules to mind. *Best Pract Res Clin Anaesthesiol* 19(3):349–364.

Melloni, L., C. Molina, M. Pena, D. Torres, W. Singer, and E. Rodriguez. 2007. Synchronization of neural activity across cortical areas correlates with conscious perception. *J Neurosci* 27(11): 2858–2865.

Metzinger, T. 2000. *Neural correlates of consciousness : empirical and conceptual questions.* Cambridge, MA: MIT Press.

Meunier, D., S. Achard, A. Morcom, and E. Bullmore. 2009. Age-related changes in modular organization of human brain functional networks. *Neuroimage* 44(3):715–723.

Meuret, P., S. B. Backman, V. Bonhomme, G. Plourde, and P. Fiset. 2000. Physostigmine reverses propofol-induced unconsciousness and attenuation of the auditory steady state response and bispectral index in human volunteers. *Anesthesiology* 93(3):708–717.

Miller, J. W., C. M. Hall, K. D. Holland, and J. A. Ferrendelli. 1989. Identification of a median thalamic system regulating seizures and arousal. *Epilepsia* 30(4):493–500.

Muller, M. M., T. Gruber, and A. Keil. 2000. Modulation of induced gamma band activity in the human EEG by attention and visual information processing. *Int J Psychophysiol* 38(3): 283–299.

Munglani, R., J. Andrade, D. J. Sapsford, A. Baddeley, and J. G. Jones. 1993. A measure of consciousness and memory during isoflurane administration: the coherent frequency. *Br J Anaesth* 71(5):633–641.

Munk, M. H., P. R. Roelfsema, P. Konig, A. K. Engel, and W. Singer. 1996. Role of reticular activation in the modulation of intracortical synchronization. *Science* 272(5259):271–274.

Naghavi, H. R., and L. Nyberg. 2005. Common fronto-parietal activity in attention, memory, and consciousness: shared demands on integration? *Conscious Cogn* 14(2):390–425.

Nelson, L. E., T. Z. Guo, J. Lu, C. B. Saper, N. P. Franks, and M. Maze. 2002. The sedative component of anesthesia is mediated by GABA(A) receptors in an endogenous sleep pathway. *Nat Neurosci* 5(10):979–984.

Ogawa, T., K. Shingu, M. Shibata, M. Osawa, and K. Mori. 1992. The divergent actions of volatile anaesthetics on background neuronal activity and reactive capability in the central nervous system in cats. *Can J Anaesth* 39(8):862–872.

Osa, M., M. Ando, and E. Adachi-Usami. 1989. Human flash visually evoked cortical potentials under different levels of halothane anesthesia. *Nippon Ganka Gakkai Zasshi* 93(2):265–270.

Pascual-Leone, A., and V. Walsh. 2001. Fast backprojections from the motion to the primary visual area necessary for visual awareness. *Science* 292(5516):510–512.

Pawela, C. P., B. B. Biswal, Y. R. Cho, D. S. Kao, R. Li, S. R. Jones, M. L. Schulte, H. S. Matloub, A. G. Hudetz, and J. S. Hyde. 2008. Resting-state functional connectivity of the rat brain. *Magn Reson Med* 59(5):1021–1029.

Pawela, C. P., A. G. Hudetz, B. D. Ward, M. L. Schulte, R. Li, D. S. Kao, M. C. Mauck, Y. R. Cho, J. Neitz, and J. S. Hyde. 2008. Modeling of region-specific fMRI BOLD neurovascular response functions in rat brain reveals residual differences that correlate with the differences in regional evoked potentials. *Neuroimage* 41(2):525–534.

Peltier, S. J., C. Kerssens, S. B. Hamann, P. S. Sebel, M. Byas-Smith, and X. Hu. 2005. Functional connectivity changes with concentration of sevoflurane anesthesia. *Neuroreport* 16(3): 285–288.

Pfurtscheller, G., D. Flotzinger, and C. Neuper. 1994. Differentiation between finger, toe and tongue movement in man based on 40 Hz EEG. *Electroencephalogr Clin Neurophysiol* 90(6):456–460.

Plourde, G., and C. Villemure. 1996. Comparison of the effects of enflurane/N$_2$O on the 40-Hz auditory steady- state response versus the auditory middle-latency response. *Anesth Analg* 82(1):75–83.

Plourde, G., J. Baribeau, and V. Bonhomme. 1997. Ketamine increases the amplitude of the 40-Hz auditory steady-state response in humans. *Br J Anaesth* 78(5):524–529.

Plourde, G., P. Belin, D. Chartrand, P. Fiset, S. B. Backman, G. Xie, and R. J. Zatorre. 2006. Cortical processing of complex auditory stimuli during alterations of consciousness with the general anesthetic propofol. *Anesthesiology* 104(3):448–457.

Plourde, G., C. Villemure, P. Fiset, V. Bonhomme, and S. B. Backman. 1998. Effect of isoflurane on the auditory steady-state response and on consciousness in human volunteers. *Anesthesiology* 89(4):844–851.

Rabe, L. S., L. Moreno, B. M. Rigor, and N. Dafny. 1980. Effects of halothane on evoked field potentials recorded from cortical and subcortical nuclei. *Neuropharmacology* 19(9): 813–825.

Raichle, M. E., A. M. MacLeod, A. Z. Snyder, W. J. Powers, D. A. Gusnard, and G. L. Shulman. 2001. A default mode of brain function. *Proc Natl Acad Sci USA* 98(2):676–682.

Rampil, I. J., and B. S. King. 1996. Volatile anesthetics depress spinal motor neurons. *Anesthesiology* 85(1):129–134.

Rampil, I. J., and M. J. Laster. 1992. No correlation between quantitative electroencephalographic measurements and movement response to noxious stimuli during isoflurane anesthesia in rats. *Anesthesiology* 77(5):920–925.

Rampil, I. J., P. Mason, and H. Singh. 1993. Anesthetic potency (MAC) is independent of forebrain structures in the rat. *Anesthesiology* 78(4):707–712.

Rees, G., G. Kreiman, and C. Koch. 2002. Neural correlates of consciousness in humans. *Nat Rev Neurosci* 3(4):261–270.

Riehle, A., F. Grammont, M. Diesmann, and S. Grun. 2000. Dynamical changes and temporal precision of synchronized spiking activity in monkey motor cortex during movement preparation. *J Physiol Paris* 94(5–6):569–582.

Ries, C. R., and E. Puil. 1999a. Ionic mechanism of isoflurane's actions on thalamocortical neurons. *J Neurophysiol* 81(4):1802–1809.

Ries, C. R., and E. Puil. 1999b. Mechanism of anesthesia revealed by shunting actions of isoflurane on thalamocortical neurons. *J Neurophysiol* 81(4):1795–1801.

Robbins, S. E. 2004. On time, memory and dynamic form. *Conscious Cogn* 13(4):762–788.

Robson, J. G. 1967. The effects of anesthetic drugs on cortical units. *Anesthesiology* 28(1): 144–154.

Rodriguez, E., N. George, J. P. Lachaux, J. Martinerie, B. Renault, and F. J. Varela. 1999. Perception's shadow: long-distance synchronization of human brain activity. *Nature* 397(6718):430–433.

Roelfsema, P. R., A. K. Engel, P. Konig, and W. Singer. 1997. Visuomotor integration is associated with zero time-lag synchronization among cortical areas. *Nature* 385(6612):157–161.

Rojas, M. J., J. A. Navas, S. A. Greene, and D. M. Rector. 2008. Discrimination of auditory stimuli during isoflurane anesthesia. *Comp Med* 58(5):454–457.

Rozenfeld, H. D., and D. Ben-Avraham. 2007. Percolation in hierarchical scale-free nets. *Phys Rev E Stat Nonlin Soft Matter Phys* 75(6 Pt 1):061102.

Salvador, R., J. Suckling, M. R. Coleman, J. D. Pickard, D. Menon, and E. Bullmore. 2005. Neurophysiological architecture of functional magnetic resonance images of human brain. *Cereb Cortex* 15(9):1332–1342.

Sarnthein, J., H. Petsche, P. Rappelsberger, G. L. Shaw, and A. von Stein. 1998. Synchronization between prefrontal and posterior association cortex during human working memory. *Proc Natl Acad Sci USA* 95(12):7092–7096.

Schiff, N. D., and F. Plum. 2000. The role of arousal and "gating" systems in the neurology of impaired consciousness. *J Clin Neurophysiol* 17(5):438–452.

Schwender, D., M. Daunderer, S. Mulzer, S. Klasing, U. Finsterer, and K. Peter. 1997. Midlatency auditory evoked potentials predict movements during anesthesia with isoflurane or propofol. *Anesth Analg* 85(1):164–173.

Schwender, D., M. Daunderer, N. Schnatmann, S. Klasing, U. Finsterer, and K. Peter. 1997. Midlatency auditory evoked potentials and motor signs of wakefulness during anaesthesia with midazolam. *Br J Anaesth* 79(1):53–58.

Schwender, D., S. Klasing, P. Conzen, U. Finsterer, E. Poppel, and K. Peter. 1996. Midlatency auditory evoked potentials during anaesthesia with increasing endexpiratory concentrations of desflurane. *Acta Anaesthesiol Scand* 40(2):171–176.

Schwender, D., E. Weninger, M. Daunderer, S. Klasing, E. Poppel, and K. Peter. 1995. Anesthesia with increasing doses of sufentanil and midlatency auditory evoked potentials in humans. *Anesth Analg* 80(3):499–505.

Shao, Z., and A. Burkhalter. 1996. Different balance of excitation and inhibition in forward and feedback circuits of rat visual cortex. *J Neurosci* 16(22):7353–7365.

Shulman, R. G., F. Hyder, and D. L. Rothman. 2003. Cerebral metabolism and consciousness. *C R Biol* 326(3):253–273.

Srinivasan, R., D. P. Russell, G. M. Edelman, and G. Tononi. 1999. Increased synchronization of neuromagnetic responses during conscious perception. *J Neurosci* 19(13):5435–5448.

Steinmetz, P. N., A. Roy, P. J. Fitzgerald, S. S. Hsiao, K. O. Johnson, and E. Niebur. 2000. Attention modulates synchronized neuronal firing in primate somatosensory cortex. *Nature* 404(6774):187–190.

Steriade, M. 1998. Corticothalamic networks, oscillations, and plasticity. *Adv Neurol* 77: 105–134.

Steyn-Ross, M. L., D. A. Steyn-Ross, J. W. Sleigh, and L. C. Wilcocks. 2001. Toward a theory of the general-anesthetic-induced phase transition of the cerebral cortex. I. A thermodynamics analogy. *Phys Rev E Stat Phys Plasmas Fluids Relat Interdiscip Topics* 64(1–1):011917.

Steyn-Ross, D. A., M. L. Steyn-Ross, L. C. Wilcocks, and J. W. Sleigh. 2001. Toward a theory of the general-anesthetic-induced phase transition of the cerebral cortex. II. Numerical simulations, spectral entropy, and correlation times. *Phys Rev E Stat Nonlin Soft Matter Phys* 64(1 Pt 1):011918.

Stins, J. F. 2009. Establishing consciousness in non-communicative patients: a modern-day version of the Turing test. *Conscious Cogn* 18(1):187–192.

Stuss, D. T., T. W. Picton, and M. P. Alexander. 2001. Consciousness, self awareness, and the frontal lobes. In *The frontal lobes and neuropsychiatric illness*. Washington, DC: American Psychiatric Pub.

Super, H., H. Spekreijse, and V. A. Lamme. 2001. Two distinct modes of sensory processing observed in monkey primary visual cortex (V1). *Nat Neurosci* 4(3):304–310.

Swindale, N. V. 2003. Neural synchrony, axonal path lengths, and general anesthesia: a hypothesis. *Neuroscientist* 9(6):440–445.

Tallon-Baudry, C., and O. Bertrand. 1999. Oscillatory gamma activity in humans and its role in object representation. *Trends Cogn Sci* 3(4):151–162.

Tallon-Baudry, C., A. Kreiter, and O. Bertrand. 1999. Sustained and transient oscillatory responses in the gamma and beta bands in a visual short-term memory task in humans. *Vis Neurosci* 16(3):449–459.

Tigwell, D. A., and J. Sauter. 1992. On the use of isoflurane as an anaesthetic for visual neurophysiology. *Exp Brain Res* 88(1):224–228.

Tiitinen, H., J. Sinkkonen, K. Reinikainen, K. Alho, J. Lavikainen, and R. Naatanen. 1993. Selective attention enhances the auditory 40-Hz transient response in humans. *Nature* 364(6432):59–60.

Tononi, G. 2004. An information integration theory of consciousness. *BMC Neurosci* 5(1):42.

Treisman, A. 1996. The binding problem. *Curr Opin Neurobiol* 6(2):171–178.

Vanderwolf, C. H. 2000. Are neocortical gamma waves related to consciousness? *Brain Res* 855(2):217–224.

Velly, L. J., M. F. Rey, N. J. Bruder, F. A. Gouvitsos, T. Witjas, J. M. Regis, J. C. Peragut, and F. M. Gouin. 2007. Differential dynamic of action on cortical and subcortical structures of anesthetic agents during induction of anesthesia. *Anesthesiology* 107(2):202–212.

Velmans, M., and S. Schneider. 2007. *The Blackwell companion to consciousness*. Malden, MA; Oxford: Blackwell Pub.

Veselis, R. A., V. A. Feshchenko, R. A. Reinsel, B. Beattie, and T. J. Akhurst. 2005. Propofol and thiopental do not interfere with regional cerebral blood flow response at sedative concentrations. *Anesthesiology* 102(1):26–34.

Villeneuve, M. Y., and C. Casanova. 2003. On the use of isoflurane versus halothane in the study of visual response properties of single cells in the primary visual cortex. *J Neurosci Methods* 129(1):19–31.

Vincent, J. L., G. H. Patel, M. D. Fox, A. Z. Snyder, J. T. Baker, D. C. Van Essen, J. M. Zempel, L. H. Snyder, M. Corbetta, and M. E. Raichle. 2007. Intrinsic functional architecture in the anaesthetized monkey brain. *Nature* 447(7140):83–86.

Vogt, B. A., and S. Laureys. 2005. Posterior cingulate, precuneal and retrosplenial cortices: cytology and components of the neural network correlates of consciousness. *Prog Brain Res* 150:205–217.

von Stein, A., C. Chiang, and P. Konig. 2000. Top-down processing mediated by interareal synchronization. *Proc Natl Acad Sci USA* 97(26):14748–14753.

von Stein, A., P. Rappelsberger, J. Sarnthein, and H. Petsche. 1999. Synchronization between temporal and parietal cortex during multimodal object processing in man. *Cereb Cortex* 9(2):137–150.

Wang, J., L. Wang, Y. Zang, H. Yang, H. Tang, Q. Gong, Z. Chen, C. Zhu, and Y. He. 2009. Parcellation-dependent small-world brain functional networks: a resting-state fMRI study. *Hum Brain Mapp* 30(5):1511–1523.

Weiskrantz, L. 1990. The Ferrier lecture, 1989. Outlooks for blindsight: explicit methodologies for implicit processes. *Proc R Soc Lond B Biol Sci* 239(1296):247–278.

Wheeler, M. A., and D. T. Stuss. 2003. Remembering and knowing in patients with frontal lobe injuries. *Cortex* 39(4–5):827–846.

Woods, J. W. 1964. Behavior of chronic decerebrate rats. *J Neurophysiol* 27:635–644.

Young, G. B., and S. E. Pigott. 1999. Neurobiological basis of consciousness. *Arch Neurol* 56(2):153–157.

Zeman, A. 2002. *Consciousness: a user's guide*. New Haven, CT: Yale University Press.

# Chapter 6
# Anesthesia and the Thalamocortical System

Michael T. Alkire

**Abstract** Neuroimaging the effect of anesthesia in the human brain reveals that anesthesia suppresses the functioning of the brain both in a global- and regional-specific manner. The suppression of activity generally has a global component to it (i.e., the entire brain seems to shut down), but on top of that a few select regions appear to be more suppressed than the rest of the brain (i.e., some areas appear to be more sensitive to the suppression effects of anesthesia). These regionally sensitive areas include the parietal and frontal cortical regions along with a consistently observed effect that shows a suppression of thalamic activity. It remains unknown as to which of these effects caused by anesthesia are the most important for producing a loss of consciousness. This chapter will discuss some recent findings from anesthesia research that serve to implicate a role for the thalamocortical system in mediating consciousness.

**Keywords** Anesthesia · arousal · awareness · central medial thalamus · desflurane · sevoflurane · thalamocortical · unconsciousness

It is now well established, as a general rule, that brain metabolism decreases in a fairly uniform manner as the dose of an anesthetic agent increases (Alkire et al. 1995; Alkire, Haier, Gianzero et al. 1997; Alkire, Haier, Shah et al. 1997; Alkire 1998; Michenfelder 1988; Drummond and Patel 2000). However, every rule has its exception, and for anesthesia the exception is provided by the dissociative anesthetic agent ketamine. Brain metabolism is actually slightly increased during ketamine anesthesia (Langsjo et al. 2004). Nevertheless, with most anesthetics the point at which consciousness is lost seems to occur in the range of a 30–60% reduction in overall brain metabolic rate (Alkire 1998). This is similar to the level of decreased brain metabolism found with deep sleep stages (Buchsbaum et al. 1989).

---

M.T. Alkire (✉)
UCIMC, Department of Anesthesiology and Perioperative Care, 101 The City Dr. South, Bldg 53, Route 81-A, Orange, CA, 92868, USA
e-mail: malkire@uci.edu

A. Hudetz, R. Pearce (eds.), *Suppressing the Mind*, Contemporary Clinical Neuroscience, DOI 10.1007/978-1-60761-462-3_6,
© Humana Press, a part of Springer Science+Business Media, LLC 2010

The global suppression effect of anesthesia might imply that some minimal amount of energy utilization is required to maintain consciousness or that consciousness itself is a fairly global phenomenon that occurs throughout much of the brain. As such, it may depend upon a certain amount of ongoing network interactions that will be occurring among any number of various brain regions (Mashour 2004; Hudetz 2006; Alkire, Hudetz, and Tononi 2008). It might also mean that anesthesia works at a molecular site in the brain that is fairly uniformly represented and yet widely dispersed throughout, such as the inhibitory gamma-aminobutyric acid (GABA) channel (Hentschke, Schwarz, and Antkowiak 2005; Franks 2008). Because many anesthetic drugs directly enhance the functioning of GABA channels, one might easily propose the hypothesis that the more GABA channels there are in a particular brain area, the more that particular area should be metabolically suppressed during anesthesia with an agent that is thought to work through actions on the GABA channel. Evidence in support of this idea was found with human brain imaging data that examined the intravenous general anesthetic agent propofol, a presumed GABA agonist. It appears that the brain "turns off" with propofol in a regional manner that corresponds to the number of GABA receptors found in a particular region of interest (Alkire and Haier 2001). Yet, a similar examination on the regional suppressive effects of the inhalational anesthetic agent isoflurane did not show any regional correspondence with the underlying GABA receptor density (Alkire and Haier 2001). This suggests that inhalational agents might not suppress regional brain activity by their actions on local GABA channels or that they do more than just act at the GABA channels. The latter idea is supported by a number of recent studies that now show that a few key channels may be the relevant molecular targets for the effects of anesthetics on consciousness (Franks 2008; Alkire, Hudetz, and Tononi 2008; Alkire et al. 2009).

Despite the fairly large reduction in brain metabolism associated with the loss of consciousness endpoint during anesthesia, there are a couple of fairly consistent regional findings that emerge from human brain imaging data investigating the regional effects of various anesthetic agents (Alkire and Miller 2005; Alkire, Hudetz, and Tononi 2008). In comparing a number of different agents, we find that the brainstem reticular formation and the thalamus are generally the most suppressed subcortical areas identified during anesthetic-induced unconsciousness studies. Additionally, a cortical network of parietal to frontal brain areas is often suppressed. This parietal-frontal cortical network is often identified in studies looking at resting state functional connectivity (Vincent et al. 2007), and it seems to play an important role in attention (Fan et al. 2005), intelligence (Jung and Haier 2007), movement initiation (Desmurget et al. 2009), and perhaps even consciousness itself (Alkire, Hudetz, and Tononi 2008).

The pattern of regional brain metabolic activity suppression found with many anesthetics is also found in humans during normal deep sleep (Nofzinger et al. 2002). In essence, the brain "turns off" during sleep to a level that is comparable to what happens during a light plane of general anesthesia. This is an interesting discovery because it seems to suggest that anesthesia may be acting in part by simply suppressing the brain's normal arousal mechanisms or possibly by even hijacking the normal mechanisms of sleep induction (Alkire, Haier, and Fallon 2000; Lydic and Baghdoyan 2005; Franks 2008). Indeed, the electrophysiology of light general anesthesia is somewhat similar to that found with deep sleep. In both conditions, recording the electroencephalogram (EEG) from the scalp reveals that the electrical activity of the brain is in a similar state; both show high voltage waves of activity that have a periodic slow rhythm to their oscillations. Perhaps more importantly, at least one anesthetic agent, halothane can induce a state of "anesthetic sleep" in which animals will show EEG spindles that are morphologically identical to those found with natural sleep (Keifer et al. 1994). It is hard to understand how this could be possible if anesthesia was not somehow interacting with normal sleep pathways.

Yet, anesthesia is not sleep. One can easily be aroused from sleep, but one is not easily aroused from anesthesia. This simple observation suggests that even if anesthesia is hijacking various components of sleep neurocircuitry and causing the brain to enter a sleep-like state, some additional effect must be occurring to inhibit the systems in the brain that mediate arousal. Indeed, a specific and dissociable effect on arousal mediated through the orexinergic system has been shown in mice with mutant orexin receptors (Kelz et al. 2008). The idea that anesthesia causes unconsciousness by suppressing arousal has been around for decades. Indeed, shortly after the discovery of an ascending reticular activating system (ARAS) it was proposed that anesthetics might work by suppressing the ARAS, thus causing a sleep-like state (French, Verzeano, and Magoun 1953). Our understanding of arousal systems is now much more advanced than when the original ARAS mechanism of anesthesia was first proposed, yet the concept still applies. If anesthesia suppresses activity in the brain's network nodes necessary for arousal, then arousal itself should go away, leaving one with an unconscious brain. A number of presumed arousal nodes could then be selected as important targets for further study. A nonexclusive list would include any and all areas of the brainstem thought to be important for arousal, including the mesopontine tegmental anesthesia area (MPTA), which seems to be particularly important in the induction of the anesthetic state, at least with GABAergic agents (Abulafia, Zalkind, and Devor 2009). The basal forebrain (Dringenberg and Olmstead 2003; Jones 2003; Dong et al. 2006), the so-called intralaminar thalamus (Bogen 1997; Jones 2003), and the amygdala should all be considered (Dringenberg, Saber, and Cahill 2001). Indeed, even the cortex itself will have arousing properties as it feeds back down onto the lower brainstem structures that mediated arousal (French, Livingston, and Hernandez-Peon 1953). It is important to note that older nomenclature considered the thalamus as part of the higher brainstem (Penfield 1958). This is a source of much confusion in the historical literature, as nowadays we generally talk about the thalamus as the thalamus and consider it to be greatly removed from the brainstem.

What is it that arousal centers or nodes do? Why are they arousal centers? Generally, they provide an excitatory input (or suppress an inhibitory influence) onto another group of cells. This, in turn, will increase their chance of firing an action potential because they will be depolarized closer to their action potential firing threshold (Llinas and Steriade 2006). In other words, arousal nodes contribute to or control the membrane potential of other cells that need to be "turned on" in order for consciousness to occur. Does this mean that the cells that are being turned on are the "consciousness neurons?" Not necessarily, as they might simply be some cells that act as a relay for even more cells. Ultimately, though, a point must be reached where the arousal of one group of cells leads to activating the cells that do contain "a" or "the" neural correlates of consciousness (NCC). Thus, depending on what changes occur when one makes the transition from a state of unconsciousness to a state of consciousness, one must eventually find the cells that contain "the" NCC or at least "a" NCC. So, to find consciousness, we search for cells that seem to change their activity with a change in the state of consciousness. Once localized, we have found cells that are either involved with switching the brain to an aroused awake state or that represent an NCC. Work in patients recovering from a comatose state suggests that an important area for having an awake and aware brain may be the posterior medial parietal cortex (Laureys 2005; Vogt and Laureys 2005).

Our studies with brain imaging of anesthesia revealed that the thalamus and its interactions with the cortex form an important area of interaction related to changes in the state of consciousness. We termed this interaction the "thalamic conscious-ness switch" (Alkire, Haier, and Fallon 2000) to emphasize the network change it implied and the idea that something in the functional activity of the thalamus changes with the change of state. This did not mean, however, that the thalamus is necessarily the site of action (Alkire, Hudetz, and Tononi 2008). In fact, the entire thalamocortical system tends to change state together, so a switch in thalamic firing implies a concurrent switch in the thalamocortical system (Thalamocortical, retic-ulothalamic, corticothalamic, cortico-reticulo-thalamic, and cortico-cortical cells). Ultimately, the thalamocortical system tends to fire either in a bursting pattern (asso-ciated with sleep and anesthesia) or in a tonic pattern (associated with wakefulness and no anesthesia) (Steriade 1994; Steriade, Timofeev, and Grenier 2001). From this idea it can be hypothesized that anesthetics might act to eliminate consciousness by changing the firing pattern of cells in the thalamocortical system. What determines the pattern of firing is the level of the cells' membrane potential. Hyperpolarization (taking the cells' membrane potential away from the action potential firing thresh-old) is associated with bursting activity and sleep states, whereas depolarization (moving the cells membrane potential closer to or over the firing threshold) is associ-ated with tonic firing and wakefulness. The change in thalamocortical firing patterns is associated with the state of vigilance for all creatures that have a thalamocortical-like system – from fish to frogs, to turtles, to birds, to rats, to cats, and to primates and humans (Llinas and Steriade 2006).

The localized change in thalamic activity is now a commonly reported finding seen with numerous anesthetic agents and deep sleep (Alkire and Miller 2005). Whether this represents a direct effect of the various agents on the thalamus itself, or whether it represents a suppression of cortical thalamic feedback, is still unknown (Antkowiak 1999; Velly et al. 2007; Alkire, Hudetz, and Tononi 2008). It may be a direct effect as general anesthetics hyperpolarize neurons in the vertebrate central nervous system in a manner that correlates with their potency as anesthetics (Nicoll and Madison 1982). The more potent they are as anesthetics, the more they tend to hyperpolarize neurons. Importantly, long ago this was noted to be due to a change in a potassium conductance (Nicoll and Madison 1982). Additionally, Sugiyama, Muteki, and Shimoji (1992) found that the switch from tonic to burst firing occurred in thalamocortical cells at the reversal potential of potassium. At about the same time, Ries and Puil (1999a,19996) found that the shunting of current in thalamo-cortical cells caused by isoflurane could be attributed to a change in a potassium conductance. Yet, these direct effects do not rule out the possibility of additional feedback effects when thalamocortical cells are examined within intact networks. Angel's work with evoked potentials reveals that the block in sensory throughput that seems to occur in the thalamus with various anesthetics is removed when the overlying cortex is inactivated (Angel 1991). Also, other in vitro work strongly sug-gests that it is the glutamatergic corticothalamic feedback that controls most of the level of membrane hyperpolarization that is evident during anesthesia (Detsch et al. 1999; Vahle-Hinz et al. 2001, 2007).

We proposed that all of these observations are correct and must be incorporated into a singular theory of why anesthesia causes a loss of consciousness. We accom-plish this by noting that the common ground for both direct cellular effects and system-wide neuroanatomic influences would converge on the thalamocortical sys-tem to change its state of firing (Alkire, Haier, and Fallon 2000). We did not limit our understanding of anesthetic-induced unconsciousness to a simple singular receptor or ion channel. Rather, we allowed all of them to be included according to some rel-evant proportion that had yet to be determined. Years later, we were delighted to see that many subsequent reviews embraced our conceptualization (Arhem, Klement, and Nilsson 2003; John and Prichep 2005; Franks 2008).

To investigate further what the suppression of thalamic activity means, we used functional connectivity analysis to try and get some idea of where the change in activity is coming from. We found that the thalamus produced a reduced effect on the premotor and motor cortex during anesthetic-induced unconsciousness in our published report (White and Alkire 2003), but we also found that the influence from the cortex back down to the thalamus was diminished with anesthesia (White and Alkire 2002). We tried to communicate the disconnection of thalamus and cortex in our paper, but it seems that this was often read as anesthesia setting up some kind of blockade to information transfer through the thalamus. This is not really the correct interpretation. Rather, it is that the effective influence of the thalamus on the cortex and of the cortex on the thalamus is greatly reduced during anesthetic-induced unconsciousness.

Subsequent to this focus on thalamocortical system changes with anesthesia, we asked the question, "Which area turns off first?" Is it the thalamus that is decreasing its activity and disconnecting from the cortex, or is it the cortex that is decreasing activity in key regions and ultimately causing a change in the thalamus, which is then seen with the brain scanning technique? To try and get at this issue we reasoned that imaging the lowest dose of an inhaled agent, yet imaged using the highest resolution PET scanner available, might reveal where in the brain the anesthesia was "first" working. Of course, we suspected that it would really be working everywhere, but nevertheless, perhaps some region would emerge as being particularly sensitive to a very low dose of anesthesia.

As luck would have it, we chose to investigate this question initially with sevoflurane because of sevoflurane's effects on emotional memory (Alkire et al. 2008). We anticipated from our animal work that we might find that sevoflurane would have a particular ability to simply "turn off" the amygdala (Alkire and Nathan 2005). To test this, we gave volunteers a very low dose of sevoflurane anesthesia at 0.25%. To put this dose of anesthesia in perspective, this is about 1/8–1/10th the dose a person would need for surgery. This is in fact such a low dose that it is just barely past where the on/off switch of the vaporizer turns on at. Indeed, the first number printed on the side of the vaporizer dial is 0.3%. Usually, just to knock someone out with this drug the dial needs to be turned past the 1.0% mark (Kodaka, Johansen, and Sebel 2005). Subjects breathing this dose of sevoflurane could smell the gas, and a number of them said that it made them feel sleepy. Many spontaneously reported that it felt like they were somewhat drunk, and we soon tried to quantify the subjective effects by asking the subjects to rate their state of "drunkenness" in the approximate equivalent number of "beers." Most rated it as being similar to having consumed one to two beers, though one felt it was equivalent to having had as many as 11 or 12. Upon examining the results of the PET scans, we were somewhat surprised to see that this low dose of anesthesia was still associated with a rather large reduction in global brain metabolism of around 17%. An example of this global suppression effect is shown in Fig. 6.1A.

We then used statistical parametric mapping to discover where sevoflurane might have a regionally specific effect in our group of subjects. This revealed a localized effect on the thalamus, with some effects also evident on the cerebellum and occipital cortex. Other work has suggested that the subjective sensation of sleepiness during exposure to drugs might be related to the regional metabolic suppression of the thalamus (Volkow et al. 1995; Schreckenberger et al. 2004). Generally, this drug-induced suppression is induced with the use of a benzodiazepine, a GABA agonist. However, the effect of sevoflurane was occurring at a dose that might be a bit too low for causing significant effects at GABA channels (Eger et al. 2001). Other work in anesthesia channel research suggests that a more sensitive channel, which might be affected by this low dose of sevoflurane, could be the neuronal nicotinic acetylcholine channel (Flood, Ramirez-Latorre, and Role 1997; Violet et al. 1997). We wondered whether a connection could be made between the sensitivity of nicotinic channels to inhaled anesthetics and the regional findings evident in our PET scan results. We found some interesting work with human brain imaging that visualized the distribution of regional nicotinic channel density. This work revealed that nicotinic channels are densely expressed in the living human thalamus (Gallezot et al. 2005).

This seemed to represent an interesting correlation between the actions of the sevoflurane anesthetic at the cellular level and its actions for influencing regional brain activity. To be sure, there are other anesthetic-sensitive channels that are also densely expressed in the thalamus and could account for sevoflurane's regional effect. Yet, these now apparently obvious connections suggested to us that we needed to at least explore the possibility that there may be a cause and effect relationship occurring here between sevoflurane's actions on nicotinic channels and its

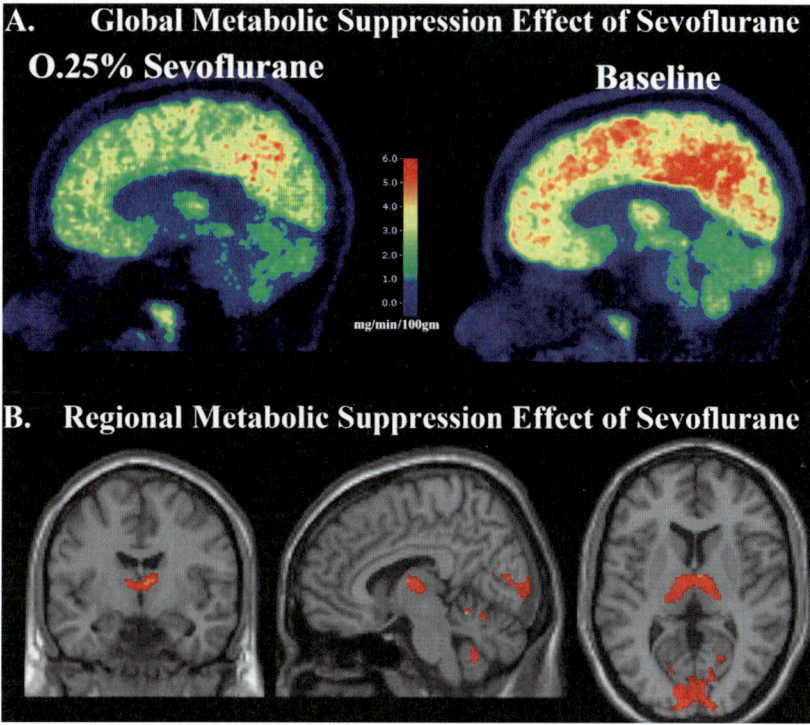

**Fig. 6.1** **A** The global metabolic suppression effect of low-dose sevoflurane anesthesia is shown in sagittal brain slices from one representative subject who was scanned on two separate occasions. The scans were obtained on a high-resolution positron emission tomography (PET) scanner. The scans represent regional cerebral glucose utilization according to the scale shown. The figure reflects that global brain metabolism decreased 17% from baseline even with the low dose of sevoflurane used. **B** The regional metabolic suppression effect of low-dose sevoflurane anesthesia is shown in three separate views that cut through the thalamus. The highlighted areas (*red*) show the brain regions that are significantly more suppressed with sevoflurane than the rest of the brain (shown at a threshold value of $p < 0.005$) in the group of subjects that were studied. These areas are the most suppressed by sevoflurane anesthesia and thus show up in this subtraction image between the drug and no drug conditions. The comparison of this image with the findings from studies investigating the regional density of nicotine binding sites reveals a potential explanation for this regional effect (Gallezot et al. 2005). Sevoflurane's effect on regional metabolism may be caused in part by its actions on localized nicotinic receptors

effects on causing a state of drowsiness by suppressing the regional activity of the thalamus.

To test this, we anesthetized rats in a box full of sevoflurane and then injected a microscopic amount of nicotine directly into their thalamus. We anticipated that if sevoflurane was knocking them out by disturbing the functioning of the nicotinic channels in the thalamus, then we might be able to overcome this regional effect by adding more nicotine to the thalamus directly. We found that we were able to reverse the unconsciousness of sevoflurane anesthesia with this maneuver (Alkire et al. 2007).

Does this mean that the site and mechanism mediating the unconsciousness of anesthesia has been found? Perhaps, it does, but probably not. There are many other components to the story of anesthetic-induced unconsciousness, and all we may really have shown is that one of the important nodes to be considered in any complete understanding of anesthetic-induced unconsciousness happens to reside in the thalamus. Indeed, some evidence suggests that the spot we found in the thalamus is important for regulating arousal and seizures (Miller et al. 1989) and that certain drugs injected into this region can also lead to a suppression of arousal (Miller and Ferrendelli 1990). Yet, other brain regions such as the MPTA can also be targeted and shown to be important specific sites that suppress arousal when microinjected with anesthetic agents (Abulafia, Zalkind, and Devor 2009).

Moreover, the dose of nicotine we used was at a relatively large concentration, and this means that it may have had effects on channels other than just the nicotinic channels. Indeed, it is known that nicotine can have a two-pronged attack on arousal. It seems to directly enhance arousal by enhancing actions at the nicotinic channel, but it also seems to turn off the suppression of arousal by blocking actions at voltage-gated potassium channels (Wang et al. 2000). The link to voltage-gated potassium channels is intriguing because recent work suggests a molecular link between sleep and voltage-gated potassium channels. One potential molecular mechanism of overlap between sleep and anesthesia might be related to effects on voltage-gated potassium channels. Cirelli and coworkers screened a number of mutant fruit fly lines in order to discover those that did not sleep very much. They then genetically determined what was different in the flies that they had created that hardly slept and found that their mutation was in the gene that encoded a voltage-gated potassium channel of the Kv1 or Shaker family (Cirelli et al. 2005).

This is an interesting link to anesthesia because the name "Shaker" came from early studies on fruit flies, where it was noticed that such flies anesthetized with ether would shake their legs vigorously (Trout and Kaplan 1970). Years ago, Tinklenberg et al. (1991) investigated the link between anesthesia and Shaker channels by quantifying how much anesthesia various Shaker mutants needed to become anesthetized (Tinklenberg et al. 1991). They then compared the amount of increased anesthetic requirements needed for the different mutants with the degree of dysfunction occurring in the various versions of mutated voltage-gated potassium channels. They found that higher anesthetic requirements were needed for those mutants with the greatest degree of dysfunction in the Shaker channel. This suggests that anesthesia might work in part by hijacking the proper functioning of the Shaker channels in vivo. A confirmation of this logic was recently seen by Weber and colleagues who investigated the anesthetic requirements of the fruit flies created by the Cirelli group (Weber et al. 2009). In line with the Tinklenberg work, they also found that the flies that do not sleep very much have an increased requirement for isoflurane anesthesia. Taken together, these facts strongly focus attention on what role might be played by the voltage-gated potassium channels in the mechanisms of anesthetic-induced unconsciousness.

We recently investigated the link between thalamic actions of anesthetics and the possibility that our nicotinic arousal effects might have been due to actions on voltage-gated potassium channels in the thalamus. We borrowed some antibody from Chiara Cirelli, M.D., Ph.D. and Giulio Tononi, M.D., Ph.D. that they had made specifically in order to block the currents mediated by the KV1.2 potassium channel. When this antibody was microinjected into the thalamus of rats rendered unconscious with either sevoflurane or desflurane anesthesia, the animals woke up and started to move around in the anesthesia chamber still filled with anesthetic gas (Alkire et al. 2009). This observation taken together with the previous findings strongly points our attention towards the thalamus as a key brain site involved with mediating the effects of anesthetics on consciousness.

# References

Abulafia, R., V. Zalkind, and M. Devor. 2009. Cerebral activity during the anesthesia-like state induced by mesopontine microinjection of pentobarbital. *J Neurosci* 29(11):7053–7064.

Alkire, M. T. 1998. Quantitative EEG correlations with brain glucose metabolic rate during anesthesia in volunteers. *Anesthesiology* 89(2):323–333.

Alkire, M. T., and R. J. Haier. 2001. Correlating *in vivo* anaesthetic effects with *ex vivo* receptor density data supports a GABAergic mechanism of action for propofol, but not for isoflurane. *Br J Anaesth* 86(5):618–626.

Alkire, M. T., and J. Miller. 2005. General anesthesia and the neural correlates of consciousness. *Prog Brain Res* 150:229–244.

Alkire, M. T., and S. V. Nathan. 2005. Does the amygdala mediate anesthetic-induced amnesia? Basolateral amygdala lesions block sevoflurane-induced amnesia. *Anesthesiology* 102(4):754–760.

Alkire, M. T., R. J. Haier, and J. H. Fallon. 2000. Toward a unified theory of narcosis: brain imaging evidence for a thalamocortical switch as the neurophysiologic basis of anesthetic-induced unconsciousness. *Conscious Cogn* 9(3):370–386.

Alkire, M. T., A. G. Hudetz, and G. Tononi. 2008. Consciousness and anesthesia. *Science* 322(5903):876–880.

Alkire, M. T., R. J. Haier, S. J. Barker, N. K. Shah, J. C. Wu, and Y. J. Kao. 1995. Cerebral metabolism during propofol anesthesia in humans studied with positron emission tomography. *Anesthesiology* 82(2):393–403.

Alkire, M. T., R. J. Haier, M. V. Gianzero, C. M. Chan, L. M. Smalling, B. P. Jacobsen, and C. T. Anderson. 1997. A positron emission tomography study of human cerebral metabolism during halothane anesthesia. *Anesthesiology* 87(3):A175.

Alkire, M. T., R. J. Haier, N. K. Shah, and C. T. Anderson. 1997. Positron emission tomography study of regional cerebral metabolism in humans during isoflurane anesthesia. *Anesthesiology* 86(3):549–557.

Alkire, M. T., J. R. McReynolds, E. L. Hahn, and A. N. Trivedi. 2007. Thalamic microinjection of nicotine reverses sevoflurane-induced loss of righting reflex in the rat *Anesthesiology* 107:264–272.

Alkire, M. T., R. Gruver, J. Miller, J. R. McReynolds, E. L. Hahn, and L. Cahill. 2008. Neuroimaging analysis of an anesthetic gas that blocks human emotional memory. *Proc Natl Acad Sci USA* 105(5):1722–1727.

Alkire, M. T., C. D. Asher, A. M. Franciscus, and E. L. Hahn. 2009. Thalamic microinfusion of antibody to a voltage-gated potassium channel restores consciousness during anesthesia. *Anesthesiology* 110(4):766–773.

Angel, A. 1991. The G. L. Brown lecture. Adventures in anaesthesia. *Exp Physiol* 76(1):1–38.

Antkowiak, B. 1999. Different actions of general anesthetics on the firing patterns of neocortical neurons mediated by the GABA(A) receptor. *Anesthesiology* 91(2):500–511.

Arhem, P., G. Klement, and J. Nilsson. 2003. Mechanisms of anesthesia: towards integrating network, cellular, and molecular level modeling. *Neuropsychopharmacology* 28(Suppl 1):S40–S47.

Bogen, J. E. 1997. Some neurophysiologic aspects of consciousness. *Semin Neurol* 17(2):95–103.

Buchsbaum, M. S., J. C. Gillin, J. Wu, E. Hazlett, N. Sicotte, R. M. Dupont, and W. E. Bunney, Jr. 1989. Regional cerebral glucose metabolic rate in human sleep assessed by positron emission tomography. *Life Sci* 45(15):1349–1356.

Cirelli, C., D. Bushey, S. Hill, R. Huber, R. Kreber, B. Ganetzky, and G. Tononi. 2005. Reduced sleep in Drosophila Shaker mutants. *Nature* 434(7037):1087–1092.

Desmurget, M., K. T. Reilly, N. Richard, A. Szathmari, C. Mottolese, and A. Sirigu. 2009. Movement intention after parietal cortex stimulation in humans. *Science* 324(5928):811–813.

Detsch, O., C. Vahle-Hinz, E. Kochs, M. Siemers, and B. Bromm. 1999. Isoflurane induces dose-dependent changes of thalamic somatosensory information transfer. *Brain Res* 829(1–2):77-89.

Dong, H. L., S. Fukuda, E. Murata, and T. Higuchi. 2006. Excitatory and inhibitory actions of isoflurane on the cholinergic ascending arousal system of the rat. *Anesthesiology* 104(1): 122–133.

Dringenberg, H. C., and M. C. Olmstead. 2003. Integrated contributions of basal forebrain and thalamus to neocortical activation elicited by pedunculopontine tegmental stimulation in urethane-anesthetized rats. *Neuroscience* 119(3):839–853.

Dringenberg, H. C., A. J. Saber, and L. Cahill. 2001. Enhanced frontal cortex activation in rats by convergent amygdaloid and noxious sensory signals. *Neuroreport* 12(11):2395–2398.

Drummond, J. C., and P. Patel. 2000. Cerebral blood flow and metabolism. In *Anesthesia*, edited by R. D. Miller. New York: Churchill-Livingstone.

Eger, E. I., II, D. M. Fisher, J. P. Dilger, J. M. Sonner, A. Evers, N. P. Franks, R. A. Harris, J. J. Kendig, W. R. Lieb, and T. Yamakura. 2001. Relevant concentrations of inhaled anesthetics for in vitro studies of anesthetic mechanisms. *Anesthesiology* 94(5):915–921.

Fan, J., B. D. McCandliss, J. Fossella, J. I. Flombaum, and M. I. Posner. 2005. The activation of attentional networks. *Neuroimage* 26(2):471–479.

Flood, P., J. Ramirez-Latorre, and L. Role. 1997. Alpha 4 beta 2 neuronal nicotinic acetylcholine receptors in the central nervous system are inhibited by isoflurane and propofol, but alpha 7-type nicotinic acetylcholine receptors are unaffected. *Anesthesiology* 86(4):859–865.

Franks, N. P. 2008. General anaesthesia: from molecular targets to neuronal pathways of sleep and arousal. *Nat Rev Neurosci* 9(5):370–386.

French, J. D., R. B. Livingston, and R. Hernandez-Peon. 1953. Cortical influences upon the arousal mechanism. *Trans Am Neurol Assoc* 3(78th Meeting):57–58.

French, J. D., M. Verzeano, and H. W. Magoun. 1953. A neural basis of the anesthetic state. *Arch Neurol Psychiatry* 69(4):519–529.

Gallezot, J. D., M. Bottlaender, M. C. Gregoire, D. Roumenov, J. R. Deverre, C. Coulon, M. Ottaviani, F. Dolle, A. Syrota, and H. Valette. 2005. In vivo imaging of human cerebral nicotinic acetylcholine receptors with 2-18F-fluoro-A-85380 and PET. *J Nucl Med* 46(2): 240–247.

Hentschke, H., C. Schwarz, and B. Antkowiak. 2005. Neocortex is the major target of sedative concentrations of volatile anaesthetics: strong depression of firing rates and increase of GABAA receptor-mediated inhibition. *Eur J Neurosci* 21(1):93–102.

Hudetz, A.G. 2006. Suppressing consciousness: mechanisms of general anesthesia. *Semin Anesthesia Perioperative Med Pain* 25:196–204.

John, E. R., and L. S. Prichep. 2005. The anesthetic cascade: a theory of how anesthesia suppresses consciousness. *Anesthesiology* 102(2):447–471.

Jones, B. E. 2003. Arousal systems. *Front Biosci* 8:s438–s451.

Jung, R. E., and R. J. Haier. 2007. The Parieto-Frontal Integration Theory (P-FIT) of intelligence: converging neuroimaging evidence. *Behav Brain Sci* 30(2):135–154.

Keifer, J. C., H. A. Baghdoyan, L. Becker, and R. Lydic. 1994. Halothane decreases pontine acetylcholine release and increases EEG spindles. *Neuroreport* 5(5):577–580.

Kelz, M. B., Y. Sun, J. Chen, Q. Cheng Meng, J. T. Moore, S. C. Veasey, S. Dixon, M. Thornton, H. Funato, and M. Yanagisawa. 2008. An essential role for orexins in emergence from general anesthesia. *Proc Natl Acad Sci USA* 105(4):1309–1314.

Kodaka, M., J. W. Johansen, and P. S. Sebel. 2005. The influence of gender on loss of consciousness with sevoflurane or propofol. *Anesth Analg* 101(2):377–381.

Langsjo, J. W., E. Salmi, K. K. Kaisti, S. Aalto, S. Hinkka, R. Aantaa, V. Oikonen, T. Viljanen, T. Kurki, M. Silvanto, and H. Scheinin. 2004. Effects of subanesthetic ketamine on regional cerebral glucose metabolism in humans. *Anesthesiology* 100(5):1065–1071.

Laureys, S. 2005. The neural correlate of (un)awareness: lessons from the vegetative state. *Trends Cogn Sci* 9(12):556–559.

Llinas, R. R., and M. Steriade. 2006. Bursting of thalamic neurons and states of vigilance. *J Neurophysiol* 95(6):3297–3308.

Lydic, R., and H. A. Baghdoyan. 2005. Sleep, anesthesiology, and the neurobiology of arousal state control. *Anesthesiology* 103(6):1268–1295.

Mashour, G. A. 2004. Consciousness unbound: toward a paradigm of general anesthesia. *Anesthesiology* 100(2):428–433.

Michenfelder, J.D. 1988. *Anesthesia and the brain*. New York: Churchill-Livingstone.

Miller, J. W., and J. A. Ferrendelli. 1990. Characterization of GABAergic seizure regulation in the midline thalamus. *Neuropharmacology* 29(7):649–655.

Miller, J. W., C. M. Hall, K. D. Holland, and J. A. Ferrendelli. 1989. Identification of a median thalamic system regulating seizures and arousal. *Epilepsia* 30(4):493–500.

Nicoll, R. A., and D. V. Madison. 1982. General anesthetics hyperpolarize neurons in the vertebrate central nervous system. *Science* 217(4564):1055–1057.

Nofzinger, E. A., D. J. Buysse, J. M. Miewald, C. C. Meltzer, J. C. Price, R. C. Sembrat, H. Ombao, C. F. Reynolds, T. H. Monk, M. Hall, D. J. Kupfer, and R. Y. Moore. 2002. Human regional cerebral glucose metabolism during non-rapid eye movement sleep in relation to waking. *Brain* 125(Pt 5):1105–1115.

Penfield, W. 1958. Centrencephalic integrating system. *Brain* 81(2):231–234.

Ries, C. R., and E. Puil. 1999a. Ionic mechanism of isoflurane's actions on thalamocortical neurons. *J Neurophysiol* 81(4):1802–1809.

Ries, C. R., and E. Puil. 1999b. Mechanism of anesthesia revealed by shunting actions of isoflurane on thalamocortical neurons. *J Neurophysiol* 81(4):1795–1801.

Schreckenberger, M., C. Lange-Asschenfeld, M. Lochmann, K. Mann, T. Siessmeier, H. G. Buchholz, P. Bartenstein, and G. Grunder. 2004. The thalamus as the generator and modulator of EEG alpha rhythm: a combined PET/EEG study with lorazepam challenge in humans. *Neuroimage* 22(2):637–644.

Steriade, M. 1994. Sleep oscillations and their blockage by activating systems. *J Psychiatry Neurosci* 19(5):354–358.

Steriade, M., I. Timofeev, and F. Grenier. 2001. Natural waking and sleep states: a view from inside neocortical neurons. *J Neurophysiol* 85(5):1969–1985.

Sugiyama, K., T. Muteki, and K. Shimoji. 1992. Halothane-induced hyperpolarization and depression of postsynaptic potentials of guinea pig thalamic neurons in vitro. *Brain Res* 576(1):97–103.

Tinklenberg, J. A., I. S. Segal, T. Z. Guo, and M. Maze. 1991. Analysis of anesthetic action on the potassium channels of the Shaker mutant of Drosophila. *Ann N Y Acad Sci* 625:532–539.

Trout, W. E., and W. D. Kaplan. 1970. A relation between longevity, metabolic rate, and activity in shaker mutants of Drosophila melanogaster. *Exp Gerontol* 5(1):83–92.

Vahle-Hinz, C., O. Detsch, M. Siemers, E. Kochs, and B. Bromm. 2001. Local GABA(A) receptor blockade reverses isoflurane's suppressive effects on thalamic neurons in vivo.

Vahle-Hinz, C., O. Detsch, M. Siemers, and E. Kochs. 2007. Contributions of GABAergic and glutamatergic mechanisms to isoflurane-induced suppression of thalamic somatosensory information transfer. *Exp Brain Res* 176(1):159–172.

Velly, L. J., M. F. Rey, N. J. Bruder, F. A. Gouvitsos, T. Witjas, J. M. Regis, J. C. Peragut, and F. M. Gouin. 2007. Differential dynamic of action on cortical and subcortical structures of anesthetic agents during induction of anesthesia. *Anesthesiology* 107(2):202–212.

Vincent, J. L., G. H. Patel, M. D. Fox, A. Z. Snyder, J. T. Baker, D. C. Van Essen, J. M. Zempel, L. H. Snyder, M. Corbetta, and M. E. Raichle. 2007. Intrinsic functional architecture in the anaesthetized monkey brain. *Nature* 447(7140):83–86.

Violet, J. M., D. L. Downie, R. C. Nakisa, W. R. Lieb, and N. P. Franks. 1997. Differential sensitivities of mammalian neuronal and muscle nicotinic acetylcholine receptors to general anesthetics. *Anesthesiology* 86(4):866–874.

Vogt, B. A., and S. Laureys. 2005. Posterior cingulate, precuneal and retrosplenial cortices: cytology and components of the neural network correlates of consciousness. *Prog Brain Res* 150:205–217.

Volkow, N. D., G. J. Wang, R. Hitzemann, J. S. Fowler, N. Pappas, P. Lowrimore, G. Burr, K. Pascani, J. Overall, and A. P. Wolf. 1995. Depression of thalamic metabolism by lorazepam is associated with sleepiness. *Neuropsychopharmacology* 12(2):123–132.

Wang, H., H. Shi, L. Zhang, M. Pourrier, B. Yang, S. Nattel, and Z. Wang. 2000. Nicotine is a potent blocker of the cardiac A-type K(+) channels. Effects on cloned Kv4.3 channels and native transient outward current. *Circulation* 102(10):1165–1171.

Weber, B., C. Schaper, D. Bushey, M. Rohlfs, M. Steinfath, G. Tononi, C. Cirelli, J. Scholz, and B. Bein. 2009. Increased volatile anesthetic requirement in short-sleeping Drosophila mutants. *Anesthesiology* 110(2):313–316.

White, N.S., and M.T. Alkire. 2002. Network activity changes during general anesthesia provide support for neurobiological theories of conscious awareness. *American Society of Anesthesiologists annual meeting*, Orlando, Fl, A-796.

White, N. S., and M. T. Alkire. 2003. Impaired thalamocortical connectivity in humans during general-anesthetic-induced unconsciousness. *Neuroimage* 19(2 Pt 1):402–411.

# Chapter 7
# Anesthesia-Induced State Transitions in Neuronal Populations

Jamie Sleigh, Moira Steyn-Ross, Alistair Steyn-Ross,
Logan Voss, and Marcus Wilson

**Abstract** It is a simple observation that the function of the central nervous system changes abruptly at certain critical brain concentrations of the anesthetic drug. This can be viewed as analogous to "state transitions" in systems of interacting particles, which have been extensively studied in the physical sciences. Theoretical models of the electroencephalogram (EEG) are in semi-quantitative agreement with experimental data and show some features suggestive of anesthetic-induced state transitions (critical slowing, the biphasic effect, increase in spatial correlations, and entropy changes). However, the EEG is only an indirect marker of the (unknown) networks of cortical connectivity that are required for the formation of consciousness. It is plausible that anesthetic-induced alteration in cortical and corticothalamic network topology prevents the formation of a giant component in the neuronal population network, and hence induces unconsciousness.

**Keywords** General anesthesia, state transitions, consciousness, entropy, networks, continuum models

## Introduction

The content of this chapter arises from an absurdly simple observation that occurs every day in every anesthesiologist's practice – namely that patients commonly wake up *suddenly* after general anesthesia. The following recent clinical vignette is an example of this phenomenon. Mrs X had received a 2-h propofol total intravenous anesthetic for breast cancer surgery. She had also received muscle relaxant, 6 mg morphine, and local anesthetic infiltration in the wound for postoperative analgesia. At 11:13 am, she was completely unresponsive to verbal command and sternal pressure. At 11:15 am, she awoke to verbal command and her trachea was extubated. By 11:16 am, she demonstrated a high level of cognitive function as demonstrated by the conversation below.

J. Sleigh (✉)
Waikato Clinical School, University of Auckland, Hamilton, New Zealand
e-mail: sleighj@waikatodhb.govt.nz

A. Hudetz, R. Pearce (eds.), *Suppressing the Mind*, Contemporary Clinical
Neuroscience, DOI 10.1007/978-1-60761-462-3_7,
© Humana Press, a part of Springer Science+Business Media, LLC 2010

Anesthesiologist: "Mrs X, the operation is finished, it has all gone well. Do you know where you are?"

Mrs X: "On a beach in Fiji – just kidding, I am in the Waikato Hospital."

Anesthesiologist: "The armpit node was normal."

Mrs X: "Good, so I didn't need the clearance of the nodes from the armpit."

Mrs X was clearly able to joke about a Fijian holiday and was able to understand that the sentinel node had come back from frozen section as showing no cancer; the wide local excision was thought sufficient to clear the cancer, and she had not needed an additional axillary clearance.

It is possible to estimate the propofol concentrations in the brain effect site using various pharmacokinetic and pharmacodynamic modeling techniques. Although the absolute values may be questionable because they are based on population parameters, the trend in these values is reliable. At 11:13 am (when Mrs X was quite unresponsive), the effect-site concentration of propofol was calculated to be 1.3 mg/L. At 11:16 am (when Mrs X was alert and demonstrating impressive reasoning ability), the effect-site concentration of propofol was 1.2 mg/L. Clearly there had been a huge change in Mrs X's cognitive function that occurred in response to a very small change in brain propofol concentration. This is the essence of the idea about state transitions. As humans we are naturally biased toward the assumption of linear cause and effect relationships – i.e., a big change in output implies that there was big change in the causative input. This is obviously not the case in the transitions into and out of general anesthesia, where a very small change in brain propofol concentration results in a big change in cognitive function.

As an aside, there is debate as to whether entering or losing the state of consciousness is a continuous process or a binary process. We take the pragmatic approach that consciousness is primarily measured by assessing the degree of complexity of responsiveness to external stimuli. There is the problem that the occurrence of the internal *state* (such as the presence or absence of cognition) is conflated with the strength of the arousal *stimulus*. There are a lot of different possible responses to noxious stimuli – e.g., the classical Glasgow Coma Scale measures (motor withdrawal, moaning, or eye opening). The highest response is memory and cognition – which would involve perception, thought, and self-awareness. There is clearly a continuous gradation in magnitude of possible noxious stimuli, but the actual response at any particular level of stimuli is approximately binary – e.g., the absence or presence of flexion withdrawal or the absence or presence of cognition. Thus, we suggest that if we separate out the state of consciousness from the arousal stimulus, then consciousness is more of a categorical phenomenon than a continuous one. As described later (see Fig. 7.1), the dynamics of state transitions naturally capture these sudden transitions in behavior.

In this chapter, we will initially explore changes in state of the electrical field of the brain that occur as the brain concentration of the anesthetic drugs change. We will then move to the more controversial idea that the changes in state of the electrical field might be proxy indicators of a more abstract change in state of effective neuronal communication – that itself may form the neurophysiological

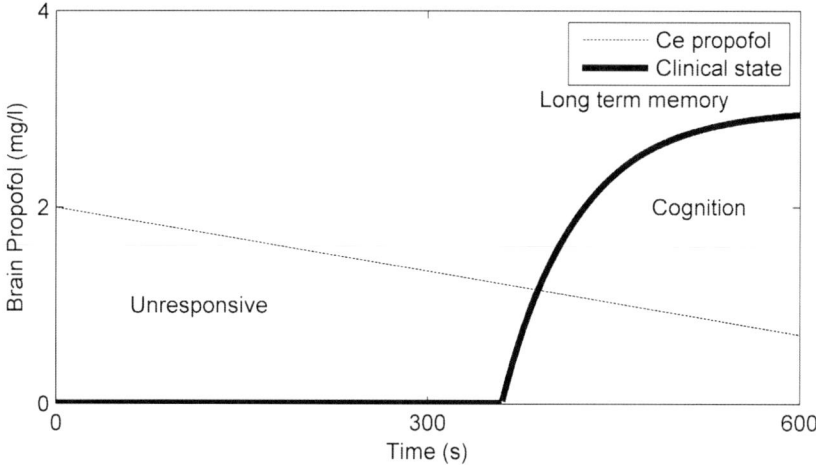

**Fig. 7.1** A diagram of changes in clinical responsiveness to differing brain concentrations of propofol. (Ce = effect-site propofol concentration, Clinical state = arbitrary measure of responsiveness, ranging from 0 = no response to painful stimulus to maximum = the alert and fully amnestic state)

underpinning of the phenomenon of conscious awareness (i.e., general anesthesia causes a "connectivity" state transition within networks of connecting neuronal populations).

## What Do We Mean by the Term "State Transition?"

Typically, state transitions happen without much warning – something suddenly appears out of nothing. During general anesthesia, we may think of the central nervous system as a sort of complicated mathematical function. The inputs into the system might be the brain concentration of propofol and degree of surgical stimulus, with the output of the system being the level of cognitive function/responsiveness. A diagram of this relationship is shown in Fig. 7.1. This diagram illustrates that there is a narrow range of drug concentrations over which the target nervous system dramatically changes from anesthetized to alert states. We may therefore talk about general anesthesia as inducing a "state transition." Since the input hasn't changed very much over the minute during the wake-up period (e.g., from 360 to 420 s), we can assume that it must be the internal dynamics of the system that have undergone dramatic alterations in interactions that allow a new mode of function by the system as a whole.

Typically, state transitions are characteristically marked by: (1) a *qualitative* transformation in behavior as well as (2) *quantitative* changes in activity – in response to very small changes in the controlling input variable. In this chapter, we will concentrate on transitions in state of consciousness that have been induced by changes in concentration of general anesthetic drugs. However, it must be noted that there are myriad transitions that occur in almost every area of physiology and pathology and at all scales of size (e.g., transitions in ion channel configuration, the presence or absence of a neuronal action potential, immobility in general anesthesia, slow-wave sleep to REM sleep, apnea to regular breathing, and sinus rhythm to atrial fibrillation). Usually, these involve an abrupt change in the absolute level of activity or changing from a steady-state to some form of oscillation. Mathematically, these changes are termed "bifurcations." In statistical physics, there is a large body of work that is devoted to the explanation and description of abrupt changes in the state of physical substances, such as the transition from liquid water to solid ice, sols to gels, and ferromagnetism to diamagnetism. It may be possible to borrow many of the methods of statistical physics to understand how the anesthetic-induced changes in interactions between neuronal populations could cause the ablation of consciousness. However – as will be expanded upon later – translation of methods derived in one scientific discipline (physics) to a different discipline (biology) is appealing but often beset with significant difficulties. The main problem is that we are probably measuring the wrong thing. In physics, the object of interest that is undergoing the transition is the same as the quantity that is being measured. For example, in describing the steam to liquid water transition, we might measure how physical density changes with physical temperature. From the theory of phase transitions, we could explain these variations by changes in molecular interactions and kinetic energy – because temperature is (without exception) directly correlated with molecular kinetic energy. In physiology, it is different. Although we might (with difficulty) accurately measure changes in states of neurons' electrical activity between the anesthetized and awake states, we are using the neuronal electrical activity as a *proxy* for the real measurement of interest – namely, cognitive activity. If we assume that the state of consciousness is dependent on sufficient complexity of effective interneuronal communication, the enterprise will therefore only be successful in so far as effective neuronal connectivity (probably unmeasurable at the present time) correlates with (measurable) neuronal electrical activity. Most of the time this is the case – but as will be described later, not always.

Because transitions often involve a change from solid to liquid or gas in physics, they have been usually called "phase transitions," meaning a transformation from a solid "phase" to a liquid "phase" of matter. However, the term "phase" is often also used as part of Fourier analysis to describe a point in an oscillatory cycle. To avoid confusion, we will use the term "state transition" (rather than "phase transition") to describe the changes in state of the nervous system.

## Types of Transition

State transitions can be technically classified according to the mathematical behavior of their equations. If these equations show a nonanalytic discontinuity in the first-order differential, it is called a "first-order" transition. If the discontinuity lies in the second (or greater) order, they are known as "continuous" transitions.

First-order transitions show characteristic features of an *instantaneous* change in state as some control parameter is varied, and they also show hysteresis. For example, if the anesthesia-responsiveness transition was a first-order transition, we would expect that the change in state would occur at a higher concentration of anesthetic if the direction of the transition was from consciousness to unconsciousness and a lower concentration for the transition from unconsciousness to consciousness. Because of the difficulties in actually measuring the effect-site concentrations of anesthetic drugs, it is unknown if even this simple observation is true or false. Continuous transitions are not instantaneous and they show symmetry breaking. As previously mentioned, state transitions have two interlinked components: the qualitative change in *output* from the system and the causative quantitative changes that occur within and between the *internal components* of the system. We have identified the output from our system – namely, the clinical states of responsiveness (consciousness) or unresponsiveness (anesthesia). The question arises – what are the quantitative changes in the internal workings of the central nervous system that cause the qualitative output to change?

## Theoretical Models of the Electrical Activity of the Cerebral Cortex

To answer this question, it is necessary to be able to quantitatively describe how each neuron responds to its inputs, how it is connected to other neurons, and how general anesthetic drugs influence these responses. This approach naturally links the molecular and neuronal level of actions of general anesthetic drugs, with the changes in function of the whole animal. However, the obvious problem is that we would need enormous knowledge of neuronal function and enormous computing power to replicate the $10^{11}$ neurons and $10^{15}$ connections in the human brain. One solution to this problem – at a very abstract level – would be to represent the anesthetic effects on the brain as a cellular automaton (Sleigh and Galletly 1997), but it seemed that more specific neurophysiological–neuropharmacological details are necessary for proper understanding of anesthetic action. Over the years, a number of "mean field" models have been proposed. Early examples were published by Wilson and Cowan (1972), Freeman (1972, 1994), and Wright and Liley (1995). There have been a number of models that specifically describe anesthesia and seizures (Steyn-Ross et al. 1999; Steyn-Ross, Steyn-Ross, Wilcocks et al. 2001; Bojak and Liley 2005; Liley and Bojak 2005; Breakspear et al. 2006). These models include differing degrees of neurobiological reality, but their essence is that individual neuron activity can be replaced by the average activity of a group of neurons. They are therefore also known as continuum models. Since the seminal work by Ramon y Cajal a century ago, we have been accustomed to understanding the nervous system as consisting of the interactions of a collection of discrete units (neurons and their attendant glia). However, there is a good case to be made that the units that make up the nervous system are much less discrete than would seem at the first glance – the so-called neuropil. There is extensive diffusion of neuroactive substances both extracellularly and intracellularly (via gap junctions), and there are significant ephaptic effects that are the result of the electrical fields that are generated by neuronal activity. Adjacent neurons are highly connected and tend not to act independently (Bullock et al. 2005). Indeed, it has been argued that encoding information using populations of neurons – rather than individual neurons – is one form of error correction in the central nervous system. For these reasons, the use of a mean-field model may be justified.

## Detailed Description of Our Theoretical Mean-Field Model

In our work, we have used modifications of mean-field models by Robinson et al. (2003) and Wright and Liley (1995). We have used models that were formulated to have changes in mean neuronal soma potential (not rate of production of action potentials) as their output. This enables us to directly compare the output of the model with changes in local field potential [and even the scalp electroencephalogram (EEG)], which can be easily measured from real experimental preparations. The model consists of a set of equations that describe how populations of inhibitory neurons and excitatory neurons interact. We have used the term "macrocolumn" to describe the population of neurons of interest. It is defined as the population of neurons (between 20,000 and 100,000) that lie more or less within the bounds of the dendritic arbor of a typical cortical inhibitory neuron. The equations are shown below in Box 1, and typical parameters for the human cortex are shown in Box 2. The inputs consist of rates of action potentials from the local (excitatory and inhibitory) and distant (only excitatory) cortical neurons and subcortical neuronal input (in the form of white noise). The time courses of the inputs are not instantaneous, but follow the time course of typical inhibitory/excitatory postsynaptic potentials (IPSP, EPSP). The amplitude of the IPSP/EPSPs is also modified by the ionic reversal potentials. In later forms of the model, we have included the effects of neuronal gap junctions (modeled as Ohmic diffusivity; Steyn-Ross et al. 2007) and intrinsic voltage-sensitive ion channels (to explain some of the patterns of slow-wave sleep; Wilson et al. 2006).

## Box 1. Mathematical description of the mean-field model.

Excitatory pyramidal neurons are denoted with subscript $e$ and inhibitory interneurons with subscript $i$. We have used the convention of $a \rightarrow b$, meaning the direction of transmission in the synaptic connections is from the presynaptic nerve $a$ to postsynaptic nerve $b$. The superscript "$sc$" indicates random subcortical input that is independent of the cortical membrane potential. The time evolutions of the mean neuronal soma membrane potential ($V_a$) in each population of neurons in response to synaptic input ($\rho_a \, \psi_{ab} \Phi_{ab}$) are given by the following set of equations:

$$\tau_e \frac{\partial V_e}{\partial t} = V_e^{\text{rest}} - V_e + \Delta V_e^{\text{rest}} + \lambda \rho_e \psi_{ee} \Phi_{ee} + \rho_i \psi_{ie} \Phi_{ie}, \tag{1}$$

$$\tau_i \frac{\partial V_i}{\partial t} = V_i^{\text{rest}} - V_i + \lambda \rho_e \psi_{ei} \Phi_{ei} + \rho_i \psi_{ii} \Phi_{ii}, \tag{2}$$

where $\tau_a$ are the time constants of the neurons, $\rho_a$ are the strength of the postsynaptic potentials, and $\psi_{ab}$ are the weighting functions that allow for the effects of reversal potentials and are described by Eq. (3):

$$\psi_{ab} = \frac{V_a^{\text{rev}} - V_b}{V_a^{\text{rev}} - V_a^{\text{rest}}}, \tag{3}$$

$\Phi_{ab}$ are the synaptic input spike-rate densities, which are described by the following equations:

$$\left( \frac{\partial^2}{\partial t^2} + 2\gamma_{ee} \frac{\partial}{\partial t} + \gamma_{ee}^2 \right) \Phi_{ee} = \gamma_{ee}^2 \left( N_{ee}^{\alpha} \phi_{ee} + N_{ee}^{\beta} Q_e + \phi_{ee}^{sc} \right), \tag{4}$$

$$\left( \frac{\partial^2}{\partial t^2} + 2\gamma_{ei} \frac{\partial}{\partial t} + \gamma_{ei}^2 \right) \Phi_{ei} = \gamma_{ei}^2 \left( N_{ei}^{\alpha} \phi_{ei} + N_{ei}^{\beta} Q_e + \phi_{ei}^{sc} \right), \tag{5}$$

$$\left( \frac{\partial^2}{\partial t^2} + 2\gamma_{ie} \frac{\partial}{\partial t} + \gamma_{ie}^2 \right) \Phi_{ee} = \gamma_{ie}^2 \left( N_{ie}^{\beta} Q_i + \phi_{ie}^{sc} \right), \tag{6}$$

$$\left( \frac{\partial^2}{\partial t^2} + 2\gamma_{ii} \frac{\partial}{\partial t} + \gamma_{ii}^2 \right) \Phi_{ii} = \gamma_{ii}^2 \left( N_{ii}^{\beta} Q_i + \phi_{ii}^{sc} \right), \tag{7}$$

where $\gamma_{ab}$ are the synaptic rate constants, $N^{\alpha}$ are the number of long-range connections, and $N^{\beta}$ are the number of local, intramacrocolumn connections. The mean axonal velocity is given by $v$ and the characteristic length (the length at which the connectivity approximately decays to $1/e$) is $1/\Lambda_{ea}$. The spatial interactions among and within the macrocolumns are described by the two equations:

$$\left( \frac{\partial}{\partial t^2} + 2v\Lambda_{ee} \frac{\partial}{\partial t} + v^2 \Lambda_{ee}^2 - v^2 \nabla^2 \right) \phi_{ee} = v^2 \Lambda_{ee}^2 Q_e, \tag{8}$$

$$\left( \frac{\partial}{\partial t^2} + 2v\Lambda_{ei} \frac{\partial}{\partial t} + v^2 \Lambda_{ei}^2 - v^2 \nabla^2 \right) \phi_{ei} = v^2 \Lambda_{ei}^2 Q_e, \tag{9}$$

The population firing rate of the neurons is related to their mean soma potential by sigmoidal functions:

$$Q_e(V_e) - \frac{Q_e^{max}}{1 + \exp[-\pi(V_e - \theta_e)/\sqrt{3}\sigma_e]}, \tag{10}$$

$$Q_i(V_i) - \frac{Q_i^{max}}{1 + \exp[-\pi(V_i - \theta_i)/\sqrt{3}\sigma_i]}, \tag{11}$$

where $\theta_a$ describes the inflexion point voltage and $\sigma_a$ describes the standard deviation of the threshold potential. The parameters and ranges used in our simulations are shown below ($e$ and $i$ refer to values assigned to excitatory and inhibitory cell populations).

## Box 2. Parameters for the cortex model.

| Symbol | Description | Value |
|--------|-------------|-------|
| $\tau_{e,i}$ | Membrane time constant | 20, 20 ms |
| $Q_{e,i}$ | Maximum firing rates | 30, 60 s$^{-1}$ |
| $\theta_{e,i}$ | Sigmoid thresholds | –58, –58 mV |
| $\sigma_{e,i}$ | Standard deviation of thresholds | 3, 4 mV |
| $\rho_{e,i}$ | Gain per synapse at resting voltage | 0.001, –0.001 mVs |
| $V_{e,i\ i}^{rev}$ | Cell reversal potential | 0, –70 mV |
| $V_{e,i}^{rest}$ | Cell resting potential | –64, –64 mV |
| $N_{ea}^{\alpha}$ | Long-range $e$ to $e$ or $i$ connectivity | 1500 |
| $N_{ea}^{\beta}$ | Short-range $e$ to $e$ or $i$ connectivity | 1000 |
| $N_{ia}^{\alpha}$ | Short-range $i$ to $e$ or $i$ connectivity | 500, 250 |
| $\langle\phi_{ea}^{sc}\rangle$ | Mean $e$ to $e$ or $i$ subcortical flux | 50 s$^{-1}$ |
| $\gamma_{ea}$ | Excitatory synaptic rate constant | 60 s$^{-1}$ |
| $\gamma_{ia}$ | Inhibitory synaptic rate constant | 40 s$^{-1}$ |
| $L_{x,y}$ | Spatial length of cortex | 25 cm |
| $a_{mac}$ | Area of macrocolumn | 1.0 mm$^2$ |
| $\Lambda_{ea}$ | Inverse length connection scale | 14 cm$^{-1}$ |
| $v$ | Mean axonal conduction speed | 140 cm s$^{-1}$ |

The model was implemented using Matlab (Mathworks, Natick, MA, USA) in a simulation of 25×25 cm cortex. This is approximately the area of the human cortex. The primary output was the time course of the mean soma potential of a cortical macrocolumn.

# Entering and Leaving the State of General Anesthesia: Theoretical Results

General anesthesia is a phenomenon that is fairly easy to study. The effects on the central nervous system are not subtle, usually reversible, and extraordinarily reproducible in all circumstances and in all animals. How does the theoretical model (as described above) behave if it is given a general anesthetic drug? To do this, we need to change some of the parameters in the model in a fashion that would realistically simulate the known neuronal effects of anesthetic drugs – and then match the "voltage" output of the model (the AC fluctuations in mean soma potential) with that observed in real experiments. It is not known for certain how anesthetic drugs work; however, they are well-studied drugs, and we are able to quantify many of their effects. The whole process is iterative. If these effects easily produce anesthesia-like changes in the model, then the model itself acts to indirectly validate these molecular-scale observations as being plausibly causative in the production of general anesthesia. Many anesthetic drugs act (directly and indirectly) to increase the effects of the inhibitory neurotransmitter gamma-aminobutyric acid (GABA; Wakamori, Ikemoto, and Akaike 1991; Mihic et al. 1994; Zimmerman, Jones, and Harrison 1994; Antkowiak 1999; Campagna, Miller, and Forman 2003; Grasshoff et al. 2006). There is good evidence that GABAergic effects are very important in explaining the actions of propofol and etomidate (Jurd et al. 2003; Rudolph and Antkowiak 2004). Therefore, the first (and simplest) idea might be to model the effects of general anesthesia as an increase in the IPSP. The total electrical charge that is generated in the postsynaptic membrane during an IPSP event is measured by the area under the IPSP voltage-time curve. In Fig. 7.2, we show an example of the changes in the mean soma potential (the "pseudoEEG") of the model – and the changes in frequency spectrum – as we slowly increase the IPSP area from its starting value of –1.0 to a final value of –2.0. This would be the equivalent of increasing the brain propofol concentrations from 0 to about 6 µg/ml. The pseudoEEG is computed as the fluctuations (after subtraction of the DC component) in soma potential averaged over the grid position and four nearest neighbors. We have also filtered out frequencies above 45 Hz. We can see that the pseudoEEG suddenly changes from a low-amplitude high-frequency pattern to a pattern in which the spectral power is concentrated in the low frequencies. We also note that there is a critical point where this change in behavior occurs (relative IPSP area $\approx$ –1.4).

Figure 7.2 shows the results of a simulation run, but it is known from formal theoretical analysis of the set of equations that at this critical point the system undergoes a first-order state transition if applied to a single macrocolumn (Steyn-Ross et al. 1999). It must be remembered that in order to simulate an EEG obtained from a real cerebral cortex, we are applying the anesthetic drug to a large two-dimensional *grid of macrocolumns*, which do not all transit between states simultaneously due to the vagaries of numerical simulation and the white noise that drives the system. This is the reason why the transition in the simulation is not as pure as the theory might suggest.

If a state transition in the electrical activity of the cerebral cortex is truly occurring as a patient becomes anesthetized, then a few other important questions and

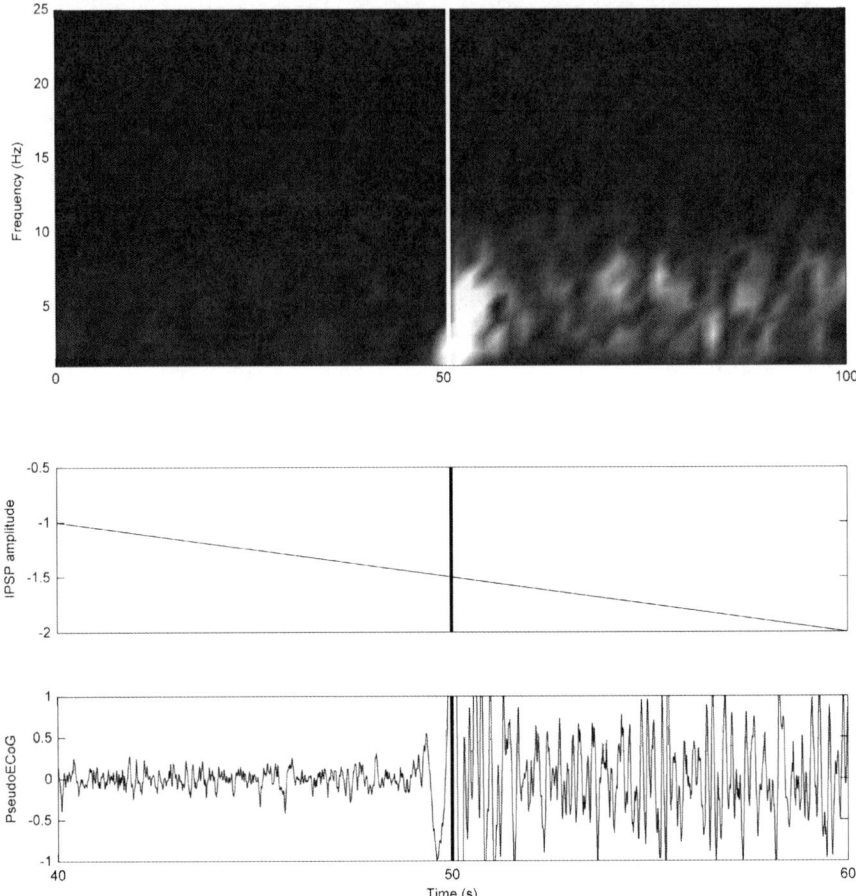

**Fig. 7.2** An example of changes in the model mean soma potential with increasing concentration of the anesthetic drug. The effect of the anesthetic drug was modeled as increasing negative area under the IPSP (*second graph*). The *bottom graph* shows the pseudoECoG output from the model, and the *top graph* shows the spectrogram of the pseudoECoG. The spectral power at each frequency and time point is shown by tone. White is high power and black is low power. The *thick vertical lines* show the critical point of transition from a high-firing to a low-firing state. It is assumed that the high-firing state is representative of the activity of the cortex in the alert state and that the low-firing mode is analogous to the anesthetized state. Note the sudden transition causes some transient filter artifact

predictions arise from the theoretical model. As is described below, many (but not all) of these are observed in real experiments.

## Entering and Leaving the State of General Anesthesia: Experimental Observations

There are numerous studies that have examined in detail the transition from the alert state to the unresponsive/anesthetized state. We may compare the changes observed

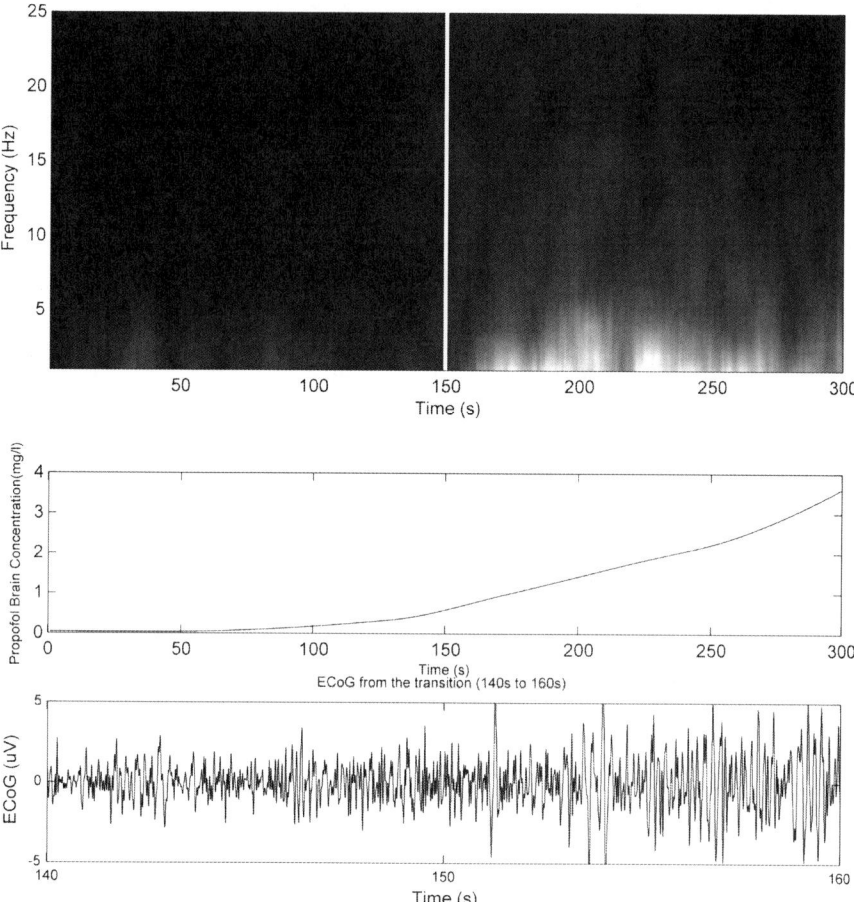

**Fig. 7.3** Changes in real ECoG recordings from a sheep undergoing a slow induction of anesthesia with a continuous propofol infusion. The *top graph* is the spectrogram. The spectral power at each frequency and time point is shown by tone. *White* is high power and *black* is low power. The *middle graph* shows the brain concentrations of propofol (Voss et al. 2007). The lower graph is a 20-s sample of the raw ECoG around the point of unresponsiveness

in the theoretical model with those obtained from real experiments. In Fig. 7.3, we show this process occurring slowly in a sheep that is receiving an intravenous infusion of propofol [see Voss et al. (2007) for the details of the experiments]. We have chosen this as an example for two reasons. First, we are able to measure the brain concentration of propofol directly, and we can also use the electrocorticogram (ECoG) as a more accurate indicator of brain activity than the scalp EEG – and hence minimize the confounding problems of skull attenuation of the brain signal and electromyographic interference. Second, the slowness of the induction serves to exaggerate the abruptness of the changes in the ECoG that can be clearly seen at 150 s. These changes consist of an increase in amplitude and a shift to low frequencies in the ECoG. These changes are semi-quantitatively equivalent to those obtained from the theoretical model (see Fig. 7.2 for comparison).

Do proxy indicators of change or impending change exist? From the clinical point of view, it would be useful to be able to predict the state transition. Are there signs that the patient is getting close to waking up? Perhaps, this would warn the anesthesiologist to increase the dose of the anesthetic drug. Unfortunately, one of the properties of systems that show state transitions is that any changes before and after the transition are much smaller than those observed at the transition point itself. However, there are a number of subtle phenomena that would be predicted to occur close to the critical point. These are the features that – if seen in a real anesthetic-consciousness transition – would be strongly indicative that the real cortex is undergoing a state transition similar to that seen in the theoretical model.

## Critical Slowing Near the Transition Point

As the system approaches the critical point, it is less vigorously attracted to its steady-state values. Therefore, if the system is perturbed by some random input, it may take longer to settle back down to its previous activity. This is known as "critical slowing" (see Fig. 7.2 lower graph just before transition). The system may also transiently jump to its other state (known as intermittency) in response to noisy input. It is conceivable (but unproven) that many of the exaggerated movements and responses that occur during the induction of (or recovery from) general anesthesia are manifestations of a dynamic system that is close to a state transition. We may imagine the central nervous system as being acutely sensitive to internal and external fluctuations because it is "not really sure" what steady state it wants to enter.

## Increased Amplitude – The Biphasic Effect

Another feature of proximity to the critical point is increased amplitude of the fluctuations (see Fig. 7.2 lower graph around transition). This phenomenon is related to the critical slowing as described above and may provide an explanation of the so-called "biphasic effect" of many general anesthetics on the EEG. It has been known for decades that the amplitude of the EEG increases initially and then decreases with progressively deeper anesthesia (Kuizenga, Kalkman, and Hennis 1998; Kuizenga, Wierda, and Kalkman 2001). The modeling of the biphasic effect on general anesthesia has been nicely reviewed by Foster et al (Foster, Bojak, and Liley 2008).

## Long-Distance Spatial Correlations

The theory also predicts the appearance of long-distance spatial correlations when the system is near the critical point (coincident with the critical slowing). The practical measurement of EEG correlations is fraught with difficulty. However, the EEG of the anesthetized state demonstrates increased temporal and spatial coherence, as compared to the EEG of the awake state. In a large study of patients undergoing general anesthesia, Gugino et al. (2001) have shown a sudden increase in coherence in the gamma band (30–45 Hz) of the EEG between the hemispheres that is maximal around the point of loss of consciousness. This is somewhat paradoxical. In general anesthesia (and in slow-wave sleep), there is widespread correlation of fluctuations of neuronal voltage as compared to those found in the EEG of the awake state (Greenberg, Houweling, and Kerr 2008). These correlations in EEG voltage are *increasing* during anesthesia when the actual effective connectivity between neurons is almost certainly *decreasing*, as measured by various evoked potential studies (Angel and Arnott 1999; Massimini et al. 2005). This is clear evidence of the problem (as stated earlier and discussed later) that there is not a one-to-one correspondence between patterns of EEG and patterns of neuronal connectivity. In fact, it could be argued that anesthetics cause a majority of neurons to fire synchronously, hence inhibit the ability to form synchronous neuronal assemblies that are specific to each cognitive frame (see section on network theory and general anesthesia and the brain).

## Changes in Entropy

In physical state transitions, the most obvious change occurs in the thermodynamic entropy of the system. Some parallels can be drawn with systems of neuronal interactions – and this has been explored in some detail in a previous publication (Sleigh et al. 2004). The configurational entropy contained within the EEG signal can be estimated using a variety of measures [e.g., the approximate entropy (Bruhn et al. 2003), the spectral entropy (Steyn-Ross, Steyn-Ross, Wilcocks et al. 2001; Steyn-Ross, Steyn-Ross, Sleigh et al. 2001; Ellerkmann et al. 2004; Vakkuri et al. 2004), or the permutation entropy (Olofsen, Sleigh, and Dahan 2008)]. The spectral entropy can be envisaged as being an indicator of the "size" of the phase space of the system. If the spectral entropy is high, it indicates that the system is actively exploring a phase space of high dimension and many different modes of behavior are available to the system. If the spectral entropy is low, it implies that the system is working in a constrained phase space and has a limited repertoire of responses. With appropriate preprocessing, all the measures of EEG configurational entropy may be used effectively in clinical anesthesia as measures of depth of anesthesia. More precisely, they have a markedly nonlinear relationship with anesthetic drug effect, showing a precipitous decrease around the point of the transition to anesthesia and an increase when the patient awakens. Examples of the changes in spectral entropy for the experiment and the theory are shown in Fig. 7.4. It can be seen that the abrupt state change is nicely mirrored in the abrupt change in spectral entropy – although for various technical reasons, the absolute values in the anesthetized state differ.

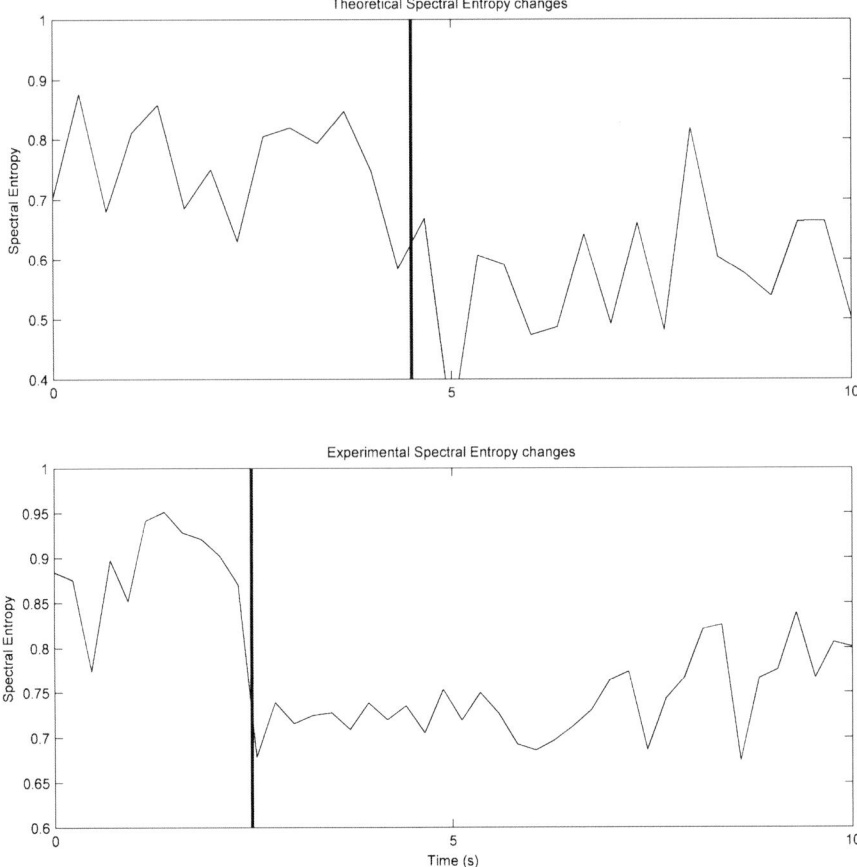

**Fig. 7.4** Examples of changes in spectral entropy with state transition. These are the changes in spectral entropy that are calculated from the pseudoECoG and real ECoG time series as shown in Figs. 2 and 3. The *vertical thick black line* in the *upper diagram* marks the point of the state transition in the theoretical model and in the *lower diagram* marks the point of loss of responsiveness in the experiment

## The Problem of General Anesthesia and the Active Cortex

There is one glaring problem with the aforementioned theory of unconsciousness. Namely, that although the decrease in spectral entropy (and activity) of the EEG signal often coincides with loss of responsiveness, this is not always the case. There are quite a number of situations where the patient becomes unresponsive even when there is much electrical activity in the cortex. The patient undergoes a *clinical* state change (i.e., becomes unresponsive) well before the apparent state change in the *EEG signal*. The classical example of this is the loss of responsiveness when the patient has been given ketamine or nitrous oxide (Sleigh and Barnard 2004), but it also occurs in 20–50% of patients undergoing anesthesia with traditional

GABAergic anesthetics (Baker, Sleigh, and Smith 2000; Strachan and Edwards 2000; Gunawardane, Murphy, and Sleigh 2002; Muncaster, Sleigh, and Williams 2003; Vanluchene et al. 2004; McKay et al. 2006). In our model, we assume that the number of connections between neurons decays approximately exponentially with increasing spatial separation. We assume that nearby neurons are active more or less synchronously with their neighbors. Taken as a long-term (ergodic) average over space and time, this appears to be correct (Bullock et al. 2005; Song et al. 2005). However, this probably not true for the effective connectivity matrix for each discrete $\sim 100$ ms "frame" of consciousness (Quiroga et al. 2005). Within each macrocolumn there is a continually changing subpopulation between 1 and 5% of the neurons that undergo a transition to the high-firing state for each cognitive frame (Kerr, Greenberg, and Helmchen 2005; Mainy et al. 2008). The other >95% of the neurons in each macrocolumn are quiescent. These data suggest that for each of the fundamental building blocks of the stream of consciousness, the effective connectivity matrix is different from the spatially homogenous assumptions of the mean-field model. This again highlights the issue that measured or modeled changes in the electrical field activity of the brain may not always accurately map one-to-one onto the effective interneuronal communication.

## Network Theory and General Anesthesia and the Brain

Twenty years ago, Baars introduced the hypothesis that the neural basis of the phenomenon of consciousness could be explained in terms of a concept he called the "global workspace" (Baars and Franklin 2003, 2007). These ideas have been refined with increasing detail and complexity, but in essence he proposed that:

(1) The brain consists of a collection of interacting but somewhat separate (and spatially widely distributed) subnetworks – each having a particular specialized function.
(2) Each of these networks is, by itself, subconscious.
(3) These networks can compete and cooperate with each other.
(4) At any particular point in time, these networks can be fleetingly integrated together in the "global workspace" – and this gives rise to the phenomenon of perceptual awareness.

As is further discussed in this volume, it is only possible for this process to arise if there is sufficient complexity of neuronal information processing and adequate integration of these subnetworks (Tononi, Sporns, and Edelman 1994; Sporns, Tononi, and Edelman 2002; Tononi and Sporns 2003). General anesthesia is not subtle. Unlike almost all other drug treatments, it works every time. Therefore, general anesthesia is a unique tool for investigating the neural machinery of consciousness. If the global workspace hypothesis is true in some sense, then general anesthesia should act by destroying the global workspace. Any theory of consciousness that does not explain the phenomenon of general anesthesia is immediately falsified. It is possible to use network theory to further develop these ideas. A network is generally defined as a set of nodes (or vertices) and a set of links between these nodes (faces). In the case of a nervous system, we can envisage that the neurons (or perhaps small populations of neurons) are effectively the nodes, and their axons and synaptic connections make up the links. Because the "strength" of the synapses (links) and the responsiveness of the neurons (nodes) change with time, the use of a particular type of network called a "fitness model" would seem to be appropriate. In these models, there is a parameter associated with each node (called its "fitness"). It is more likely that an effective link will form between two nodes if they have high fitness than if they have low fitness. Fitness is therefore similar to "synaptic weight" in Hebbian models. In real neurons, a high fitness would occur if (1) the internal state of the neuron favored effective connection – e.g., if the neuron was in a depolarized state; and/or (2) the synaptic connection was particularly strong (i.e., some sort of long-term potentiation effect at that synapse, such as large post-synaptic potentials caused by a large glutamate release or amplification by NMDA activation). It can be envisaged that the neuronal-synaptic fitness would be a very evanescent and context-sensitive parameter.

Using fMRI data, it also seems that the topology of connections in the brain follows the so-called "small-world" pattern (Salvador et al. 2005, 2007; Achard et al. 2006; He, Chen, and Evans 2007; Chen et al. 2008). This pattern implies that there is greater local connectivity than a random network, combined with a shorter path length between a given pair of distant nodes, than would be expected in a regular network. Functionally, these types of networks would allow balance between local processing and global integration, would facilitate rapid long-distance transfer of information, and would be resistant to damage. Parameters that are used to define the mesoscale topology of connectivity have been estimated for the awake human brain using fMRI (Achard et al. 2006). The clustering coefficient is the ratio of the number of nearest neighbors of a node divided by the total number of possible connections to the nearest neighbors; it can take on values lying between 0 and 1. The minimum path length is the mean number of links that lie in the shortest route between any two nodes in the network. Over the whole brain, the mean path length was found to be relatively low (~2.8 links) and the clustering coefficient relatively high (~0.55). There have been some compelling studies that show that changes in cortical network topology are important in the development of seizures (Netoff et al. 2004; Ponten, Bartolomei, and Stam 2007; Reijneveld et al. 2007), but to date no one has shown how these parameters change with GABAergic anesthesia.

All this would be of only passing interest to us, except that networks are the archetypal system for displaying state transitions (Goltsev, Dorogovtsev, and Mendes 2003). In fact, the shape of the graph of "state of cognition" in Fig. 7.1 is derived from a mathematical function that describes just such a phase transition (Wallace 2005) This abstract model consists of a simple network of M nodes, connected by $(1/2) \times a \times M$ edges (links). The parameter "$a$" serves to control the connectivity of the network. When "$a$" is <1, the number of links between the nodes is less than half the number of nodes, and the network consists of a loose collection of many small disconnected tree-like structures. If the brain was in this state of dynamics, it would consist of a number of separate modules of activity – lacking a global workspace. Once the value of the parameter "$a$" increases to above the critical point (>1 in this simple case), there is an abrupt change in the topology of the network so that it is dominated by a single large cluster that connects across the whole network – this is termed the "giant component."

The size of the so-called "giant component" ($g(a)$) is given by the equation:

$$g(a) = 1 + W(-a \times \exp(-a))/a,$$

where $W$ is the Lambert-$W$ function.

It is tempting to make the analogy with the brain and postulate that the formation of Baars' "global workspace" is in fact the appearance of a giant component in the brain. The problem of understanding how general anesthetics work to impair consciousness may therefore be framed in a slightly more specific manner as: "How could anesthetic drugs act on network dynamics to prevent the formation of the giant component?"

There are few published experimental data that could help answer this question. If the theoretical network topology framework is valid, the obvious inference is that general anesthetic drugs must act to force the network activity to the left in Fig. 7.1 – below the consciousness-state transition. Using results from network topology, what are some of the possible mechanisms by which this could be achieved?

(1) The most obvious manipulation would be to decrease the global number and strength of connections between modules by:

   (a) decreasing excitatory or increasing inhibitory, synaptic strength
   (b) changing the intrinsic excitability of the neurons in the modules. This is determined principally by their neuromodulatory environment. For example, a brainstem cholinergic projection to the neocortex will tend to depolarize the resting membrane potential of the neurons, making any excitatory synaptic input more likely to initiate an action potential in the downstream neuron.

(2) A less obvious manipulation would be to alter local clustering topology.

As previously described, the clustering coefficient is a measure of local nonuniformity of connections. Interestingly, a strongly clustered network requires a higher number of effective edges for the state transition to occur (Serrano and Boguna 2006) and a giant component to form. Conversely, a less clustered network is able to form a giant component even when many of its links have been removed. Thus, one of the ways that general anesthesia could destroy consciousness is by increasing local clustering. Perhaps that is how ketamine works? As an aside, various memory processes (such as long-term potentiation of synapses) will act to increase local clustering. We may speculate that this is the mechanism by which we become fatigued, with the local clustering being reversed by natural sleep.

Experimental evidence for either of these mechanisms is sparse, but somewhat supportive. Massimini et al. (2005) attempted to investigate changes in cortical connectivity in natural slow-wave sleep – not anesthesia – versus the awake state. They stimulated the cerebral cortex directly with transcranial magnetic stimulation. They found that in the awake state, the stimulus resulted in a series of waves that spread widely across the cortex and lasted hundreds of milliseconds. In contrast, when the subject was in non-REM (slow wave) sleep, the initial local wave was larger, but the spreading wave dramatically died away in space and time. They concluded that the loss of consciousness that occurred in slow-wave sleep was associated with a decrease in effective connectivity in the brain. In particular, it would seem that the topology of connectivity changes from the small-world pattern. There is selective impairment of long-range connections, but local connectivity may even be increased (Imas et al. 2004), which would have the effect of increasing the clustering coefficient, thus invoking mechanism #2 above. Sufficient long-range connectivity (mechanism #1) is clearly essential for the formation of a giant component. There is slowly accumulating evidence for the idea that general anesthetics act also to impair that long-range connectivity (Hudetz, Wood, and Kampine 2003; Ramani and Wardhan 2008). Using fMRI to measure changes in connectivity during sevoflurane anesthesia, Peltier estimated that functional connectivity decreased by 78% at 1.0% and 98% at 2.0% sevoflurane concentrations (Peltier et al. 2005).

## Conclusions

(1) The transition from the state of alert cognition to general anesthesia shows many features of a state transition and explains the observed changes in spectral entropy, spatial correlations, and biphasic effects.
(2) The physical transition in the electrical neuronal activity is usually – but not always – an indicator of a transition in the underlying connectivity.
(3) It is plausible that anesthetic-induced alteration in cortical and corticothalamic network topology prevents the formation of a giant component, hence inducing unconsciousness.

## References

Achard, S., R. Salvador, B. Whitcher, J. Suckling, and E. Bullmore. 2006. A resilient, low-frequency, small-world human brain functional network with highly connected association cortical hubs. J Neurosci 26(1):63–72.

Angel, A., and R. H. Arnott. 1999. The effect of etomidate on sensory transmission in the dorsal column pathway in the urethane-anaesthetized rat. Eur J Neurosci 11(7): 2497–2505.

Antkowiak, B. 1999. Different actions of general anesthetics on the firing patterns of neocortical neurons mediated by the GABA(A) receptor. Anesthesiology 91(2):500–511.

Baars, B. J., and S. Franklin. 2003. How conscious experience and working memory interact. Trends Cogn Sci 7(4):166–172.

Baars, B.J., and Franklin, S. 2007. An architectural model of conscious and unconscious brain functions: Global Workspace Theory and IDA. Neural Netw 20(9):955–961.

Baker, G.W., J. W. Sleigh, and P. Smith. 2000. Electroencephalographic indices related to hypnosis and amnesia during propofol anaesthesia for cardioversion. Anaesthesia Intens Care 28: 386–391.

Bojak, I., and D. T. Liley. 2005. Modeling the effects of anesthesia on the electroencephalogram. Phys Rev E Stat Nonlin Soft Matter Phys 71(4 Pt 1):041902.

Breakspear, M., J. A. Roberts, J. R. Terry, S. Rodrigues, N. Mahant, and P. A. Robinson. 2006. A unifying explanation of primary generalized seizures through nonlinear brain modeling and bifurcation analysis. Cereb Cortex 16(9):1296–1313.

Bruhn, J., T. W. Bouillon, L. Radulescu, A. Hoeft, E. Bertaccini, and S. L. Shafer. 2003. Correlation of approximate entropy, bispectral index, and spectral edge frequency 95 (SEF95) with clinical signs of "anesthetic depth" during coadministration of propofol and remifentanil. Anesthesiology 98(3):621–627.

Bullock, T. H., M. V. Bennett, D. Johnston, R. Josephson, E. Marder, and R. D. Fields. 2005. Neuroscience. The neuron doctrine, redux. Science 310(5749):791–793.

Campagna, J. A., K. W. Miller, and S. A. Forman. 2003. Mechanisms of actions of inhaled anesthetics. N Engl J Med 348(21):2110–2124.

Chen, Z. J., Y. He, P. Rosa-Neto, J. Germann, and A. C. Evans. 2008. Revealing modular architecture of human brain structural networks by using cortical thickness from MRI. Cereb Cortex 18(10):2374–2381.

Ellerkmann, R. K., V. M. Liermann, T. M. Alves, I. Wenningmann, S. Kreuer, W. Wilhelm, H. Roepcke, A. Hoeft, and J. Bruhn. 2004. Spectral entropy and bispectral index as measures of the electroencephalographic effects of sevoflurane. Anesthesiology 101(6): 1275–1282.

Foster, B. L., I. Bojak, and D. T. Liley. 2008. Population based models of cortical drug response: insights from anaesthesia. Cogn Neurodyn 2(4):283–296.

Freeman, W. J. 1972. Linear analysis of the dynamics of neural masses. Annu Rev Biophys Bioeng 1:225–256.

Freeman, W.J. 1994. Characterization of state transitions in spatially distributed, chaotic, nonlinear, dynamical systems in cerebral cortex. Integr Physiol Behav Sci 29(3):294–306.

Goltsev, A. V., S. N. Dorogovtsev, and J. F. Mendes. 2003. Critical phenomena in networks. Phys Rev E Stat Nonlin Soft Matter Phys 67(2 Pt 2):026123.

Grasshoff, C., B. Drexler, U. Rudolph, and B. Antkowiak. 2006. Anaesthetic drugs: linking molecular actions to clinical effects. Curr Pharm Des 12(28):3665–3679.

Greenberg, D. S., A. R. Houweling, and J. N. Kerr. 2008. Population imaging of ongoing neuronal activity in the visual cortex of awake rats. Nat Neurosci 11(7):749–751.

Gugino, L. D., R. J. Chabot, L. S. Prichep, E. R. John, V. Formanek, and L. S. Aglio. 2001. Quantitative EEG changes associated with loss and return of consciousness in healthy adult volunteers anaesthetized with propofol or sevoflurane. Br J Anaesth 87(3):421–428.

Gunawardane, P, Murphy P, Sleigh J. 2002. Bispectral index monitoring during electroconvulsive therapy under propofol anaesthesia. Br J Anaesthesia 88:184–187.

He, Y., Z. J. Chen, and A. C. Evans. 2007. Small-world anatomical networks in the human brain revealed by cortical thickness from MRI. Cereb Cortex 17(10):2407–2419.

Hudetz, A. G., J. D. Wood, and J. P. Kampine. 2003. Cholinergic reversal of isoflurane anesthesia in rats as measured by cross-approximate entropy of the electroencephalogram. Anesthesiology 99(5):1125–1131.

Imas, O. A., K. M. Ropella, J. D. Wood, and A. G. Hudetz. 2004. Halothane augments event-related gamma oscillations in rat visual cortex. Neuroscience 123(1):269–278.

Jurd, R., M. Arras, S. Lambert, B. Drexler, R. Siegwart, F. Crestani, M. Zaugg, K. E. Vogt, B. Ledermann, B. Antkowiak, and U. Rudolph. 2003. General anesthetic actions in vivo strongly attenuated by a point mutation in the GABA(A) receptor beta3 subunit. FASEB J 17(2): 250–252.

Kerr, J. N., D. Greenberg, and F. Helmchen. 2005. Imaging input and output of neocortical networks in vivo. Proc Natl Acad Sci USA 102(39):14063–14068.

Kuizenga, K., C. J. Kalkman, and P. J. Hennis. 1998. Quantitative electroencephalographic analysis of the biphasic concentration-effect relationship of propofol in surgical patients during extradural analgesia. Br J Anaesth 80(6):725–732.

Kuizenga, K., J. M. Wierda, and C. J. Kalkman. 2001. Biphasic EEG changes in relation to loss of consciousness during induction with thiopental, propofol, etomidate, midazolam or sevoflurane. Br J Anaesth 86(3):354–360.

Liley, D. T., and I. Bojak. 2005. Understanding the transition to seizure by modeling the epileptiform activity of general anesthetic agents. J Clin Neurophysiol 22(5):300–313.

Mainy, N., J. Jung, M. Baciu, P. Kahane, B. Schoendorff, L. Minotti, D. Hoffmann, O. Bertrand, and J. P. Lachaux. 2008. Cortical dynamics of word recognition. Hum Brain Mapp 29(11):1215–1230.

Massimini, M., F. Ferrarelli, R. Huber, S. K. Esser, H. Singh, and G. Tononi. 2005. Breakdown of cortical effective connectivity during sleep. Science 309(5744):2228–2232.

McKay, I. D. H., L. J. Voss, J. W. Sleigh, J. P. Barnard, and E. K. Johannsen. 2006. Pharmacokinetic-pharmacodynamics modeling the hypnotic effect of sevoflurane using the spectral entropy of the electroencephalogram. Anesthesia Analgesia 102:91–97.

Mihic, S. J., S. J. McQuilkin, E. I. Eger, II, P. Ionescu, and R. A. Harris. 1994. Potentiation of gamma-aminobutyric acid type A receptor-mediated chloride currents by novel halogenated compounds correlates with their abilities to induce general anesthesia. Mol Pharmacol 46(5):851–857.

Muncaster, A, Sleigh J, Williams M. 2003. Changes in consciousness, conceptual memory, and quantitative electroencephalographical measures during recovery from sevoflurane- and remifentanil-based anesthesia. Anesthesia Analgesia 96:720–725.

Netoff, T. I., R. Clewley, S. Arno, T. Keck, and J. A. White. 2004. Epilepsy in small-world networks. J Neurosci 24(37):8075–8083.

Olofsen, E., J. W. Sleigh, and A. Dahan. 2008. Permutation entropy of the electroencephalogram: a measure of anaesthetic drug effect. Br J Anaesth 101(6):810–821.

Peltier, S. J., C. Kerssens, S. B. Hamann, P. S. Sebel, M. Byas-Smith, and X. Hu. 2005. Functional connectivity changes with concentration of sevoflurane anesthesia. Neuroreport 16(3):285–288.

Ponten, S. C., F. Bartolomei, and C. J. Stam. 2007. Small-world networks and epilepsy: graph theoretical analysis of intracerebrally recorded mesial temporal lobe seizures. Clin Neurophysiol 118(4):918–927.

Quiroga, R. Q., L. Reddy, G. Kreiman, C. Koch, and I. Fried. 2005. Invariant visual representation by single neurons in the human brain. Nature 435(7045):1102–1107.

Ramani, R., and R. Wardhan. 2008. Understanding anesthesia through functional imaging. Curr Opin Anaesthesiol 21(5):530–536.

Reijneveld, J. C., S. C. Ponten, H. W. Berendse, and C. J. Stam. 2007. The application of graph theoretical analysis to complex networks in the brain. Clin Neurophysiol 118(11): 2317–2331.

Robinson, P. A., C. J. Rennie, D. L. Rowe, S. C. O'Connor, J. J. Wright, E. Gordon, and R. W. Whitehouse. 2003. Neurophysical modeling of brain dynamics. Neuropsychopharmacology 28(Suppl 1):S74–S79.

Rudolph, U., and B. Antkowiak. 2004. Molecular and neuronal substrates for general anaesthetics. Nat Rev Neurosci 5(9):709–720.

Salvador, R., J. Suckling, M. R. Coleman, J. D. Pickard, D. Menon, and E. Bullmore. 2005. Neurophysiological architecture of functional magnetic resonance images of human brain. Cereb Cortex 15(9):1332–1342.

Salvador, R., A. Martinez, E. Pomarol-Clotet, S. Sarro, J. Suckling, and E. Bullmore. 2007. Frequency based mutual information measures between clusters of brain regions in functional magnetic resonance imaging. Neuroimage 35(1):83–88.

Serrano, M. A., and M. Boguna. 2006. Clustering in complex networks. II. Percolation properties. Phys Rev E Stat Nonlin Soft Matter Phys 74(5 Pt 2):056115.

Sleigh, J. W., and J. P. Barnard. 2004. Entropy is blind to nitrous oxide. Can we see why? Br J Anaesth 92(2):159–161.

Sleigh, J. W., and D. C. Galletly. 1997. A model of the electrocortical effects of general anaesthesia. Br J Anaesth 78(3):260–263.

Sleigh, J. W., D. A. Steyn-Ross, M. L. Steyn-Ross, C. Grant, and G. Ludbrook. 2004. Cortical entropy changes with general anaesthesia: theory and experiment. Physiol Meas 25(4): 921–934.

Song, S., P. J. Sjostrom, M. Reigl, S. Nelson, and D. B. Chklovskii. 2005. Highly nonrandom features of synaptic connectivity in local cortical circuits. PLoS Biol 3(3):e68.

Sporns, O., G. Tononi, and G. M. Edelman. 2002. Theoretical neuroanatomy and the connectivity of the cerebral cortex. Behav Brain Res 135(1–2):69–74.

Steyn-Ross, M. L., D. A. Steyn-Ross, J. W. Sleigh, and D. T. Liley. 1999. Theoretical electroencephalogram stationary spectrum for a white-noise-driven cortex: evidence for a general anesthetic-induced phase transition. Phys Rev E Stat Phys Plasmas Fluids Relat Interdiscip Topics 60(6 Pt B):7299–7311.

Steyn-Ross, M. L., D. A. Steyn-Ross, J. W. Sleigh, and L. C. Wilcocks. 2001. Toward a theory of the general-anesthetic-induced phase transition of the cerebral cortex. I. A thermodynamics analogy. Phys Rev E Stat Nonlin Soft Matter Phys 64(1 Pt 1):011917.

Steyn-Ross, D. A., M. L. Steyn-Ross, L. C. Wilcocks, and J. W. Sleigh. 2001. Toward a theory of the general-anesthetic-induced phase transition of the cerebral cortex. II. Numerical simulations, spectral entropy, and correlation times. Phys Rev E Stat Nonlin Soft Matter Phys 64(1 Pt 1):011918.

Steyn-Ross, M. L., D. A. Steyn-Ross, M. T. Wilson, and J. W. Sleigh. 2007. Gap junctions mediate large-scale Turing structures in a mean-field cortex driven by subcortical noise. Phys Rev E Stat Nonlin Soft Matter Phys 76(1 Pt 1):011916.

Strachan, A, Edwards N. 2000. Randomizes placebo-controlled trial to assess the effect of remifentanil and propofol on bispectral index and sedation. Br J Anaesthesia 84(4):489–490.

Tononi, G., and O. Sporns. 2003. Measuring information integration. BMC Neurosci 4:31.

Tononi, G., O. Sporns, and G. M. Edelman. 1994. A measure for brain complexity: relating functional segregation and integration in the nervous system. Proc Natl Acad Sci USA 91(11):5033–5037.

Vakkuri, A., A. Yli-Hankala, P. Talja, S. Mustola, H. Tolvanen-Laakso, T. Sampson, and H. Viertio-Oja. 2004. Time-frequency balanced spectral entropy as a measure of anesthetic drug effect in central nervous system during sevoflurane, propofol, and thiopental anesthesia. Acta Anaesthesiol Scand 48(2):145–153.

Vanluchene, A, Struys M, Heyse B, Mortier E. 2004. Spectral entropy measurement of patient responsiveness during propofol and remifentanil. A comparison with bispectral index. Br J Anaesthesia 93(5):645–654.

Voss, L. J., G. Ludbrook, C. Grant, R. Upton, and J. W. Sleigh. 2007. A comparison of pharmacokinetic/pharmacodynamic versus mass-balance measurement of brain concentrations of intravenous anesthetics in sheep. Anesth Analg 104(6):1440–1446, table of contents.

Wakamori, M., Y. Ikemoto, and N. Akaike. 1991. Effects of two volatile anesthetics and a volatile convulsant on the excitatory and inhibitory amino acid responses in dissociated CNS neurons of the rat. J Neurophysiol 66(6):2014–2021.

Wallace, R. 2005. A Global Workspace perspective on mental disorders. Theor Biol Med Model 2:49.

Wilson, H. R., and J. D. Cowan. 1972. Excitatory and inhibitory interactions in localized populations of model neurons. Biophys J 12(1):1–24.

Wilson, M. T., D. A. Steyn-Ross, J. W. Sleigh, M. L. Steyn-Ross, L. C. Wilcocks, and I. P. Gillies. 2006. The K-complex and slow oscillation in terms of a mean-field cortical model. J Comput Neurosci 21(3):243–257.

Wright, J. J., and D. T. Liley. 1995. Simulation of electrocortical waves. Biol Cybern 72(4): 347–356.

Zimmerman, S. A., M. V. Jones, and N. L. Harrison. 1994. Potentiation of gamma-aminobutyric acidA receptor Cl-current correlates with in vivo anesthetic potency. J Pharmacol Exp Ther 270(3):987–991.

# Chapter 8
# Anesthesia Awareness: When the Mind Is Not Suppressed

George A. Mashour

**Abstract** In approximately 1–2 general anesthetics/1000, adequate hypnosis and amnesia are not achieved, leading to intraoperative awareness and subsequent explicit recall. "Anesthesia awareness," as it is sometimes called, is one of the most feared complications of surgery by both patients and physicians alike. A significant proportion of these patients develops long-term psychological sequelae, including posttraumatic stress disorder. In this chapter I discuss the incidence, risk factors, prevention, and postoperative consequences of anesthesia awareness.

**Keywords** Awareness during general anesthesia, awareness with explicit recall, intraoperative awareness, anesthesia awareness, anesthetic depth, depth of anesthesia monitor

## Introduction

"Anesthesia awareness" is a problem that is receiving increased attention by anesthesia providers, patients, and the general public. Before proceeding, it is important to define the problem, which is variably known as "anesthesia awareness," "intraoperative awareness," or "awareness during general anesthesia." Confusion arises from this terminology because "awareness" in the context of cognitive science denotes *only* the experiential or phenomenal aspect of consciousness. In the setting of clinical anesthesiology, however, "awareness" denotes *both* the experience of intraoperative events *and* the subsequent explicit recall of these events. The dissociation of these two cognitive processes has been demonstrated in the clinical context with the isolated forearm technique (Kerssens, Klein, and Bonke 2003).

G.A. Mashour (✉)
Assistant Professor of Anesthesiology and Neurosurgery, University of Michigan Medical School; Director, Division of Neuroanesthesiology, University of Michigan Health Systems, 1H247 UH, SPC 5048, 1500 East Medical Center Drive, 48109-5048, Ann Arbor, MI, USA
e-mail: gmashour@umich.edu

A. Hudetz, R. Pearce (eds.), *Suppressing the Mind*, Contemporary Clinical Neuroscience, DOI 10.1007/978-1-60761-462-3_8,
© Humana Press, a part of Springer Science+Business Media, LLC 2010

In a study using a tourniquet to exclude the hand from the effects of neuromuscular blockers, approximately 66% of patients were able to give at least one unequivocal response to multiple commands during general anesthesia. Of those that responded, only 25% actually remembered the event. Thus, a higher proportion of individuals experienced intraoperative consciousness compared to postoperative explicit recall – these data illustrate that anesthesia awareness is not simply a problem of consciousness, but also of memory.

Although the first report of anesthesia awareness dates to the first public demonstration of surgery under ether in 1846, this perioperative complication did not become a focus of investigation until a century later. Winterbottom (1950) reported the first case of awareness associated with neuromuscular blockade. Hutchinson et al. (1961) reported the first study on the incidence of awareness, but the first academic symposium on the subject did not take place until 1989. JCAHO (2004) issued a sentinel event alert that helped bring the problem to the forefront.

There are numerous reasons why anesthesia awareness is an important subject of academic inquiry. First and foremost, it is a source of distress and psychological sequelae for our patients. It is important to note that this distress is not limited to patients who have actually experienced awareness – many patients presenting for surgery are terrified at the very prospect of "waking up" (Fitzgerald and Elder 2008). Second, our ability or failure to suppress consciousness is central to the public perception of our field. This is evidenced culturally in the realm of film: while the movie *Coma* (1978) reflected the collective fear of never waking up from anesthesia, *Awake* (2007) reflected the collective fear of never going to sleep in the first place. It is the subject of awareness that generates articles in the lay press (Healy 2008) rather than, for example, our ability to manage the airway or manipulate the cardiovascular system. Third, there are medicolegal consequences to awareness (Kent and Domino 2007), which may become more significant in light of recent attention to the subject. Finally, understanding how to detect and prevent awareness in a patient who is otherwise behaviorally suppressed may provide insight into the problem of consciousness (Mashour and LaRock 2008).

## The Incidence of Anesthesia Awareness

There have been large, prospective, multicenter studies on the incidence of awareness in both the United States and Europe. Sandin et al. (2000) found 19 awareness events in a total of 11,785 cases (0.16%), while Sebel et al. (2004) found 25 awareness events in a total of 19,575 cases (0.13%). Taken together, these studies suggest that the incidence of anesthesia awareness is approximately 1–2 cases/1000 in the general population, with high-risk cases 10 times more common (1 case/100). The qualitative description of awareness events ranges from isolated auditory perceptions to the experience of being awake, paralyzed, and in pain. Severe emotional distress often accompanies such experiences. Although the incidence of approximately 0.15% can legitimately be regarded as rare, there could still be as many as 45,000 patients each year in the United States who experience awareness (assuming 30 million general anesthetics were delivered).

The established incidence of awareness came into question after the report of Pollard et al. ( 2007) In a multicenter study of a regional health-care system, only six awareness events in 87,381 cases were identified, for an incidence of 0.0068%. If the risk were indeed this small, there would be implications for both the study of awareness and the necessity of standardized preventive strategies such as electroencephalographic monitoring. This leads to the question of how we adjudicate between disparate reports of awareness incidence in the literature. When evaluating an article, it is helpful to consider aspects of "how?", "when?", and "upon whom?" the study was conducted.

The "how" question bears on methodology, including the instrument to detect awareness, as well as whether the study was prospective or retrospective. For example, the Pollard study employed a form of the Brice Interview (Brice, Hetherington, and Utting 1970) that was distinct from the studies of Sandin and Sebel (see Table 8.1).

**Table 8.1**  Two forms of the Brice interview

| Pollard et al. (2007) | Sandin et al. (2000), Sebel et al. (2004, 2007) |
| --- | --- |
| 1. What was the last thing you remember before surgery? | 1. What is the last thing you remember before anesthesia? |
| 2. What is the first thing you remember once you woke up? | 2. What is the first thing you recall after waking up? |
| 3. Did you have any dreams when you were asleep for surgery? | 3. Do you recall anything in between? |
| 4. Were you put to sleep gently? | 4. Did you have any dreams during surgery? |
| 5. Did you have any problems going to sleep? | 5. What was the worst thing about your surgery and anesthesia? |

Note that question #3 differs between the studies. It has been argued (Sebel et al. 2007) that the lack of an explicit question assessing recall is a contributor to the lower incidence in the Pollard study. Although counterintuitive, it has been shown that patients do not report awareness events unless specifically asked (Moerman, Bonke, and Oosting 1993). Furthermore, the timing of the interview is important, with greater capture of events at later interview times. The reluctance to speak about awareness during earlier interviews may reflect the difficulty processing the trauma. It also raises the possibility of false memory formation after repeated interviews.

The Pollard study was retrospective, using a quality improvement database. Mashour et al. (2009) conducted a follow-up study in which they showed a 3-fold higher incidence of awareness compared to Pollard study using a retrospective database that did not even include a structured interview. Further bringing the data into question, they also found that the incidence of awareness complaints in patients receiving general anesthesia was not statistically different from those who received regional anesthesia or monitored anesthesia care. Based on these findings, it was suggested that retrospective methodologies may have too low a resolution to study the incidence of awareness. That being said, even prospective studies using the standard Brice interview have yielded disparate results. Errando et al. (2008) found an incidence approaching 1% for the general population, an incidence almost an order of magnitude higher than that reported by Sandin et al. and Sebel et al.

As clinicians become ever more vigilant about anesthesia awareness, it is possible that the incidence is changing over time. Thus, asking the "when" question about a study may be increasingly important. Recent studies of obstetric and pediatric populations – classically considered high risk – have demonstrated an incidence of awareness that is more consistent with the general population. Paech et al. (2008) found approximately 2/1000 emergent Cesarean sections, considerably lower than the 1% incidence typically quoted. Similarly, Davidson et al. (2008) recently showed an incidence of approximately 1/500 in a pediatric population, compared to 0.8% in his previous study (Davidson et al. 2005). Changes in anesthetic protocol over time may account for the changing incidence of awareness. Indeed, one might also argue that the decreased incidence in the Pollard study resulted from their standardized anesthetic protocol and rare use of total intravenous anesthesia, rather than methodological factors.

Finally, one must clearly understand "on whom" the study was conducted when evaluating reports of incidence in the literature, as different populations have different associated risk factors. In the next section, we will consider patients at high risk for anesthesia awareness.

## Risk Factors for Anesthesia Awareness

Not surprisingly, insufficient anesthesia is the major cause of awareness. This may result from both the side of the anesthesia provider and that of the patient. Situations in which end-organ perfusion is critical are associated with increased risk; certain cardiac surgeries, hypovolemic trauma, and emergent Cesarean delivery (in which the fetus is the "end organ") are classically high-risk cases (see Table 8.2). This is insufficient anesthesia by design, in which a choice is made to minimize anesthesia in order to avoid life-threatening physiologic sequelae. There are, however, cases in which insufficient anesthesia is not by design. For example, difficult intubations are high-risk cases, as the focus on the airway may lead to prolonged neuromuscular blockade in the absence of sedative-hypnotic redosing. Machine malfunction and lack of vigilance are other important causes. There are pharmacologic and genetic patient factors that may also lead to subtherapeutic dosing. Chronic use of ethanol or opioids may lead to resistance of anesthetic effects (Tammisto and Takki 1973; Tammisto and Tigerstedt 1977; Shafer et al. 1983). One example of genetic influence is the increased minimum alveolar concentration (MAC) for red-haired individuals, which is associated with mutations of the melanocortin-1 receptor (Liem et al. 2004; Xing et al. 2004). There is a suggestion that genetic factors may also play a role in the resistance to the amnestic effects of anesthetics as well. For example, mutations of the $\alpha 5$ GABA$_A$ receptor subunit may confer resistance to the amnestic – but not hypnotic – effects of etomidate (Cheng et al. 2006).

**Table 8.2** High-risk criteria from two major clinical trials

| B-Aware Study, Myles et al, 2004 (Myles et al. 2004)* | B-Unaware Study, Avidan et al, 2008 (Avidan et al. 2008)** |
|---|---|
| *Major Criteria* | *Major Criteria* |
| Chronic benzodiazepine or opioid use | Preoperative long-term use of anticonvulsant agents, opiates, benzodiazepines, or **cocaine** |
| Cardiac ejection fraction < **30%** | Cardiac ejection fraction < **40%** |
| History of anesthesia awareness | History of anesthesia awareness |
| Anticipated difficult intubation | History of difficult intubation or anticipated difficult intubation |
| **Significant cardiovascular impairment, suspected hypotension** | ASA physical status classes 4 or 5 |
| Severe aortic stenosis | Aortic stenosis |
| Severe end-stage lung disease | End-stage lung disease |
| **Hypovolemic trauma** | **Marginal exercise tolerance not resulting from musculoskeletal dysfunction** |
| Pulmonary hypertension | Pulmonary hypertension |
| **Off-pump coronary artery bypass graft surgery** | **Planned open-heart surgery** |
| Heavy alcohol intake | Daily alcohol consumption |
| *Major criteria, continued* | *Minor criteria* |
| **Current protease inhibitor therapy** | **Preoperative beta-blocker use** |
| **Cesarean delivery** | **Chronic obstructive pulmonary disease** |
| | **Moderate exercise tolerance not resulting from musculoskeletal dysfunction** |
| | **Smoking two or more packs of cigarettes per day** |
| | **Obesity, define as body mass index of more than 30** |

Boldface indicates differences in inclusion criteria between the two studies.
* At least one major criterion was necessary for inclusion (no minor criteria in this study).
**At least one major criterion OR two minor criteria were necessary for inclusion.

A comprehensive list of known high-risk situations from the two major trials on awareness prevention is shown in Table 8.2.

The precise risk factors for awareness are as yet unknown. A clearer understanding of risk factors has implications beyond the prevention of awareness in individual patients. Given the relative rarity of awareness, the major clinical trials investigating prevention using electroencephalographic methods (see below) have studied high-risk populations. One critique of the recent B-Unaware study by Avidan and colleagues was that the inclusion of patients satisfying minor criteria led to a lower risk population (Myles, Leslie, and Forbes 2008). This claim brings into question whether the study was appropriately powered to detect a difference in the active comparators, but there is no definitive evidence. Thus, establishing high-risk populations with accuracy will be critical to the future study of awareness prevention.

## Electroencephalographic Monitoring
## and the Prevention of Awareness

The use of electroencephalography (EEG) to prevent intraoperative awareness is a hotly debated issue in the field of anesthesiology. The assessment of anesthetic depth has evolved from stages of anesthesia in 1847 (Snow 1847), to MAC in 1965 (Eger, Saidman, and Brandstater 1965), to the current use of "awareness monitors." However, the sensitivity of EEG to anesthetic agents was demonstrated as early as 1937 by Gibbs, Gibbs, and Lennox (1937). The authors recognized the potential for the intraoperative use of EEG, stating: "A practical application of these observations might be the use of the electroencephalogram as a measure of the depth of anesthesia during surgical operations. The anesthetist and surgeon could have before them on tape or screen a continuous record of the electrical activity of both heart and brain." Of note, this was the same year that Guedel described a refined scheme of stages and planes of anesthesia (Guedel 1937).

Although useful, unprocessed EEG as a gauge of anesthetic depth has several limitations. First, there is no unequivocal electrical signature that is common to all anesthetics. Second, the routine application of EEG electrodes can be time-consuming and the equipment potentially cumbersome. Finally, the analysis of raw EEG data in real time requires a dedicated observer. These limitations have motivated the development of processed EEG monitors for intraoperative use. Such processing typically involves Fourier transformation, which translates raw EEG signals progressing in time to a frequency domain (for review of EEG processing and anesthesia, see Rampil, 1998). This allows for the analysis of the contribution of each frequency band ($\gamma$, $\beta$, $\alpha$, $\theta$, $\delta$) to the overall EEG, providing a "spectrum." Furthermore, different waveforms within a transformed data set can be evaluated for their relationship to one another, providing a "bispectrum." In addition to spectral and bispectral analysis, the presence and ratio of burst suppression is often quantified. An electrical burst pattern followed by quiescent EEG is characteristic of deeper levels of anesthesia. In addition to the various features of spontaneous EEG, evoked potentials (usually of the auditory system) may also be used in the analysis of anesthetic depth (Scheller et al. 2005; Schneider et al. 2005).

There is a variety of depth-of-anesthesia monitors that are commercially available and that employ the EEG features described above. Table 8.3 describes the basic features of some monitors currently in clinical use (Bowdle 2006; Bruhn et al. 2006; Mashour 2006; Voss and Sleigh 2007).

The BIS monitor is the only device that has been formally studied for the prevention of awareness during general anesthesia. Ekman et al. (2004) demonstrated a 0.04% incidence in awareness with BIS monitoring, compared to an incidence of 0.18% in a historical control without BIS monitoring. This represented a significant 5-fold reduction in awareness events and suggested that the BIS monitor may be of value in prevention. Myles et al. (2004) conducted a prospective trial of patients at high risk for awareness (see Table 8.2) and found that the BIS group had significantly less definite awareness events (2 cases, 0.17%) compared to routine anesthetic care (11 cases, 0.91%). These data suggest that the use of a BIS monitor

Table **8.3** EEG-based monitors in clinical use

| Monitor | Company | Features analyzed |
| --- | --- | --- |
| Bispectral index (BIS) | Aspect Medical Systems, Norwood, MA, USA | Analyzes β power, bispectral coherence, burst suppression, and then combines them in a multivariate (and proprietary) algorithm to generate the index |
| Entropy | Datex-Ohmeda/GE Healthcare, Madison, WI, USA | As anesthetic depth increases, the variability of neural function decreases. Based on the concept of information entropy put forth by Shannon (1948), this device analyzes frequency ranges of both EEG (state entropy) and EEG/EMG (response entropy) |
| Narcotrend | Monitortechnik GMBH & Co., Bad Bramstedt & Hanover, Germany | Analyzes stages and substages of anesthesia, a concept based on the EEG changes during sleep. The Narcotrend was based on a similar developmental process as the BIS, with a distinct algorithm |
| Cerebral state | Danmeter, Odense, Denmark | Calculates the $\alpha$ ratio and $\beta$ ratio, and then uses the difference between them to indicate a shift from higher to lower frequencies. Burst suppression also analyzed |
| SEDline | Hospira, Lake Forest, IL, USA | Evolved from the Patient State Analyzer and is based on EEG decoherence that occurs during general anesthesia |
| SNAP II | Stryker, Kalamazoo, MI, USA | Based on spectral analysis of low- and high-frequency components of the EEG |
| A-Line AEP | Danmeter, Odense, Denmark | Stimulus-response technique using auditory-evoked potentials in conjunction with other EEG features |

reduces the incidence of awareness in the high-risk population to that of the general population. In this study, the routine care group did not follow a protocol. Recent data from Avidan et al. (2008) demonstrate that the BIS monitor was no more successful in preventing awareness than a MAC-based protocol. These data suggest that it is an anesthetic protocol, rather than a monitoring modality itself, which is associated with a reduced incidence of awareness. One important difference in these two landmark studies is the use of intravenous anesthetics – the B-Aware trial had a significant number of total intravenous anesthetics, while the B-Unaware trial was focused exclusively on inhalational agents. Since we have no readily available metric for the effects intravenous anesthetics but do for inhalational agents (i.e., MAC), the BIS monitor may have been more efficacious in gauging anesthetic depth and preventing awareness in the B-Aware trial.

One important limitation to the BIS monitor is its relative insensitivity to nitrous oxide, ketamine, and xenon (Dahaba 2005). The common pharmacologic feature of these agents is that they primarily antagonize the glutamatergic NMDA receptor, rather than potentiate the effects of GABA. Nitrous oxide and ketamine are well known to activate the EEG (Yamamura et al. 1981; Sakai et al. 1999) and can therefore increase the BIS value to create a false-positive result. It is of interest to note that recent animal studies demonstrate that in contrast to GABAergic agents, ketamine suppresses activity of the sleep-promoting ventrolateral preoptic nucleus but instead activates wake-promoting centers (Lu et al. 2008). As evidenced by the pathophysiologic conditions of stroke and seizure, unconsciousness can arise from either depressive or excitatory brain states. One might consider the effects of nitrous oxide and ketamine as an excitatory pathway to unconsciousness.

The development of the next generation of awareness monitors should be focused on detecting the neurophysiologic features common to both the depressive and excitatory pathways to general anesthesia. One such feature may be the capacity for information integration (Mashour 2004; Tononi 2004; Alkire, Hudetz, and Tononi 2008). This capacity has been modeled (denoted as $\phi$) and is predicted to decrease during both sleep and seizure activity. Extrapolating to anesthesia, the reduction of information integration may potentially be a common feature of both depressive and excitatory agents. In the first study to measure information integration capacity in the human brain, Lee et al. (2009) have found a significant reduction in information integration after induction with propofol in the $\delta$, $\alpha$, and $\gamma$ bands of the EEG. Using state-space analysis, a spatiotemporal breakdown in the dynamic organization of $\gamma$ wave activity was also identified. The application of information integration theory to the detection of intraoperative awareness represents an exciting new direction in the field.

## Psychological Consequences of Awareness

Fully experiencing a surgical intervention can be considered a major trauma. Indeed, one needs only to consider the description of surgery in the preanesthetic era to appreciate this: "Suffering so great as I underwent cannot be expressed in words... but the blank whirlwind of emotion, the horror of great darkness, and the sense of desertion by God and man, which swept through my mind, and overwhelmed my heart, I can never forget" (Campagna, Miller, and Forman 2003). The experience of intraoperative awareness in the modern era may be yet worse, as patients are now routinely paralyzed – the sense of "desertion by God and man" would likely be exacerbated by the inability to communicate or express what is being experienced. It is therefore not surprising that there are psychological sequelae to awareness events, most notably posttraumatic stress disorder (PTSD). As such, patients reporting awareness should be informed about the possibility of psychological disturbances and should be offered psychiatric counseling if needed.

Prospective studies of PTSD after awareness have found an incidence of 0–44%, whereas retrospective cases series suggest an incidence of 2–56% (reviewed in Mashour, 2009). The diagnosis criteria of PTSD can be summarized as (1) exposure to a traumatic event; (2) re-experiencing of the traumatic event (e.g., intrusive thoughts, nightmares, or flashbacks); (3) avoidance of stimuli associated with the trauma; (4) hyperarousal states; (5) symptoms of >1 month duration; and (6) functional impairment. It is important to note that in patients who have experienced intraoperative awareness, "avoidance of stimuli associated with the trauma" may be tantamount to avoidance of further health care.

In general, known risk factors for PTSD include severity of trauma, female sex, middle age, single status, psychiatric history, low education status, and borderline personality disorder. In patients experiencing awareness, the precise predictors remain unclear. A severe early emotional response was the only predictor found in one major study of PTSD and late psychological symptoms (Samuelsson, Brudin, and Sandin 2007). Treatment modalities for PTSD include cognitive behavioral therapy, exposure or "flooding" therapy, eye movement desensitization reprocessing, and antidepressant medication. A treatment modality to help attenuate the response to health-care settings would be of benefit for patients with awareness-induced PTSD, and who are returning for surgery (Mashour, Jiang, and Osterman 2006). Mashour et al. (2008) have recently reported operating room desensitization as one successful method. Further study is required.

## Dreaming During General Anesthesia

In addition to frank awareness events, there are other subjective states that can occur during general anesthesia. Dreaming during general anesthesia was reported in Henry Bigelow's original report describing the early use of ether in 1846 (Bigelow 1846): "A girl of 16 immediately occupied the chair. After coughing a little, she inhaled during 3 min, and fell asleep, when a molar tooth was extracted, after which she continued to slumber tranquilly during 3 min more. At the moment when force was applied she flinched and frowned, raising her hand to her mouth, but said she had been dreaming a pleasant dream and knew nothing of the operation." As early as 1847, sexual dreaming during anesthesia was reported, with attendant medicolegal implications (Strickland and Butterworth 2007).

Recent studies estimate the incidence of dreaming during anesthesia at 22% (Leslie et al. 2007). Younger individuals, males, those with high dream recall at home, and those receiving propofol were the subjects most likely to dream. Like awareness, the incidence of dreaming depends on the timing of the interview, but in this case higher rates are obtained when querying the patient immediately upon emergence [for review, see Leslie and Skrzypek (2007)]. Pharmacologic factors also play a role in the incidence: Ketamine-based anesthetics have a very high incidence of dream reports (Grace 2003), while scopolamine eliminates dreaming completely when compared to atropine (Toscano, Pancaro, and Peduto 2007).

The nature and mechanism of dream reports after anesthesia has yet to be elucidated. It is as yet unclear if these dreams are generated during emergence, are "near-miss" awareness events, or are intrinsic to the state itself. If intrinsic to the state, it is also unclear what neurobiological mechanism is generating the dreams. The relationship of dreams to intraoperative awareness has been studied in the past decade. Secondary analysis of the patients in the B-Aware trial suggested an association of dreaming with awareness and light anesthesia (as measured by BIS; Leslie et al. 2005). More recent data suggest that there is no association of dreaming with light anesthesia (as measured by BIS), but a 19-fold increased risk of awareness in dreamers (Samuelsson, Brudin, and Sandin 2008a,b). Collectively, these recent data are hard to reconcile, since (1) dreaming confers a 19-fold risk of awareness, (2) dreaming is not related to light anesthesia, but (3) light anesthesia is a major cause of awareness. Again, further study is required.

## Practical Tips for Awareness Prevention and Management

- Be cognizant of the patient's risk factors for awareness. In high-risk cases, consider using a cerebral function monitor.
- If adequate anesthesia cannot be delivered due to physiologic concerns, administer a benzodiazepine or scopolamine for amnesia.
- Complete a thorough machine check.
- Be vigilant regarding anesthetic levels in vaporizers, as well as the function of intravenous lines and pumps infusing anesthetics.
- Minimize the use of neuromuscular blocking agents when possible.
- If a patient reports an awareness event, be supportive and nonjudgmental, with appropriate follow-up.
- If a patient reports an awareness event, discuss the risk of psychological symptoms or PTSD and offer psychiatric counseling.
- In a patient with a history of awareness and PTSD returning for surgery, meet with them in advance to establish trust and a sense of participation in their anesthetic plan. A full general anesthetic with cerebral function monitoring or regional/neuraxial anesthesia with minimal sedation may be preferable over the intermediate state of monitored anesthesia care. Sedative states may be misinterpreted as awareness or may trigger flashbacks (Mashour, Jiang, and Osterman 2006).

## Conclusion

Awareness is a problem of concern for patients, anesthesia providers, and the general public. Although there are many unanswered questions, it is now clear that intraoperative awareness is a real phenomenon that requires serious consideration. Ongoing studies should help elucidate the role of currently used cerebral function monitors, but further work in identifying detectable correlates of intraoperative consciousness and memory formation is required.

**Conflicts** The author has no conflicts of interest to declare. Funding for the author's research on awareness comes from the Foundation for Anesthesia Education & Research and the American Society of Anesthesiologists.

# References

Alkire, M. T., A. G. Hudetz, and G. Tononi. 2008. Consciousness and anesthesia. *Science* 322(5903):876–880.

Avidan, M. S., L. Zhang, B. A. Burnside, K. J. Finkel, A. C. Searleman, J. A. Selvidge, L. Saager, M. S. Turner, S. Rao, M. Bottros, C. Hantler, E. Jacobsohn, and A. S. Evers. 2008. Anesthesia awareness and the bispectral index. *New Engl J Med* 358(11):1097–1108.

Bigelow, HJ. 1846. Insensibility during surgical operations produced by inhalation. *Boston Med Surg J* 35:309–317.

Bowdle, T. A. 2006. Depth of anesthesia monitoring. *Anesthesiol Clin* 24(4):793–822.

Brice, D. D., R. R. Hetherington, and J. E. Utting. 1970. A simple study of awareness and dreaming during anaesthesia. *Br J Anaesth* 42(6):535–542.

Bruhn, J., P. S. Myles, R. Sneyd, and M. M. R. F. Struys. 2006. Depth of anaesthesia monitoring: what's available, what's validated and what's next? *Br J Anaesth* 97(1):85–94.

Campagna, J. A., K. W. Miller, and S. A. Forman. 2003. Mechanisms of actions of inhaled anesthetics. *New Engl J Med* 348(21):2110–2124.

Cheng, V. Y., L. J. Martin, E. M. Elliott, J. H. Kim, H. T. J. Mount, F. A. Taverna, J. C. Roder, J. F. Macdonald, A. Bhambri, N. Collinson, K. A. Wafford, and B. A. Orser. 2006. Alpha5GABAA receptors mediate the amnestic but not sedative-hypnotic effects of the general anesthetic etomidate. *J Neurosci* 26(14):3713–3720.

Dahaba, A. A. 2005. Different conditions that could result in the bispectral index indicating an incorrect hypnotic state. *Anesth Analg* 101(3):765–773.

Davidson, A. J., G. H. Huang, C. Czarnecki, M. A. Gibson, S. A. Stewart, K. Jamsen, and R. Stargatt. 2005. Awareness during anesthesia in children: a prospective cohort study. *Anesth Analg* 100(3):653–661.

Davidson, A. J., S. J. Sheppard, A. L. Engwerda, A. Wong, L. Phelan, C. M. Ironfield, and R. Stargatt. 2008. Detecting awareness in children by using an auditory intervention. *Anesthesiology* 109(4):619–624.

Eger, E. I., II, L. J. Saidman, and B. Brandstater. 1965. Minimum alveolar anesthetic concentration: a standard of anesthetic potency. *Anesthesiology* 26(6):756–763.

Ekman, A., M. L. Lindholm, C. Lennmarken, and R. Sandin. 2004. Reduction in the incidence of awareness using BIS monitoring. *Acta Anaesthesiol Scand* 48(1):20–26.

Errando, C. L., J. C. Sigl, M. Robles, E. Calabuig, J. García, F. Arocas, R. Higueras, E. Del Rosario, D. López, C. M. Peiró, J. L. Soriano, S. Chaves, F. Gil, and R. García-Aguado. 2008. Awareness with recall during general anaesthesia: a prospective observational evaluation of 4001 patients. *Br J Anaesth* 101(2):178–185.

Fitzgerald, B. M., and J. Elder. 2008. Will a 1-page informational handout decrease patients' most common fears of anesthesia and surgery? *J Surg Educ* 65(5):359–363.

Gibbs, F. A., L. E. Gibbs, and W. G. Lennox. 1937. Effect on the electroencephalogram of certain drugs which influence nervous activity. *Arch Intern Med* 60:154–166.

Grace, R. F. 2003. The effect of variable-dose diazepam on dreaming and emergence phenomena in 400 cases of ketamine-fentanyl anaesthesia. *Anaesthesia* 58(9):904–910.

Guedel, A. 1937. *Inhalation anaesthesia: a fundamental guide*. New York, NY: Macmillan.

Healy, B. 2008. Under anesthesia, yet aware. *US News World Rep* 144(9):68.

Hutchinson, R. 1961. Awareness during surgery. A study of its incidence. *Br J Anaesth* 33:463–469.

JCAHO. 2004. Preventing and managing the impact of anesthesia awareness. *Joint Comm Persp* 24(12):10–11.

Kent, C. D., and K. B. Domino. 2007. Awareness: practice, standards, and the law. *Best Pract Res Clin Anaesthesiol* 21(3):369–383.

Kerssens, C., J. Klein, and B. Bonke. 2003. Awareness: monitoring versus remembering what happened. *Anesthesiology* 99(3):570–575.

Lee, U., G. A. Mashour, S. Kim, G.J. Noh, and B.M. Choi. 2009. Propofol Induction Reduces the capacity for neural information integration: implications for the mechanism of consciousness and general anesthesia. *Conscious Cogn* 18(1):56–64.

Leslie, K., and H. Skrzypek. 2007. Dreaming during anaesthesia in adult patients. *Best Pract Res Clin Anaesthesiol* 21(3):403–414.

Leslie, K., P. S. Myles, A. Forbes, M. T. V. Chan, S. K. Swallow, and T. G. Short. 2005. Dreaming during anaesthesia in patients at high risk of awareness. *Anaesthesia* 60(3):239–244.

Leslie, K., H. Skrzypek, M. J. Paech, I. Kurowski, and T. Whybrow. 2007. Dreaming during anesthesia and anesthetic depth in elective surgery patients: a prospective cohort study. *Anesthesiology* 106(1):33–42.

Liem, E. B., C.-M. Lin, M.-I. Suleman, A. G. Doufas, R. G. Gregg, J. M. Veauthier, G. Loyd, and D. I. Sessler. 2004. Anesthetic requirement is increased in redheads. *Anesthesiology* 101(2): 279–283.

Lu, J., L. E. Nelson, N. Franks, M. Maze, N. L. Chamberlin, and C. B. Saper. 2008. Role of endogenous sleep-wake and analgesic systems in anesthesia. *J Comp Neurol* 508(4):648–662.

Mashour, G. A. 2004. Consciousness unbound: toward a paradigm of general anesthesia. *Anesthesiology* 100(2):428–433.

Mashour, G. A. 2006. Monitoring consciousness: EEG-based measures of anesthetic depth. *Semin Anesthesia Perioper Med Pain* 25:205–210.

Mashour, G. A., 2009. Post-traumatic stress disorder after intraoperative awareness and high-risk surgery. *Anesth Analg*, In press.

Mashour, G. A., and E. LaRock. 2008. Inverse zombies, anesthesia awareness, and the hard problem of unconsciousness. *Conscious Cogn* 17(4):1163–1168.

Mashour, G. A., Y. Jiang, and J. Osterman. 2006. Perioperative treatment of patients with a history of intraoperative awareness and post-traumatic stress disorder. *Anesthesiology* 104(4):893–894.

Mashour, G. A., L. Y. J. Wang, R. K. Esaki, and N. N. Naughton. 2008. Operating room desensitization as a novel treatment for post-traumatic stress disorder after intraoperative awareness. *Anesthesiology* 109(5):927–929.

Mashour, G. A., L. Y. J. Wang, C. R. Turner, A. Shanks, J. Vandervest, and K. K. Tremper. 2009. A retrospective study of intraoperative awareness with methodological implications. *Anesth Analg* 108(2):521–526.

Moerman, N., B. Bonke, and J. Oosting. 1993. Awareness and recall during general anesthesia. Facts and feelings. *Anesthesiology* 79(3):454–464.

Myles, P. S., K. Leslie, J. McNeil, A. Forbes, and M. T. Chan. 2004. Bispectral index monitoring to prevent awareness during anaesthesia: the B-Aware randomised controlled trial. *Lancet* 363(9423):1757–1763.

Myles, P. S., K. Leslie, and A. Forbes. 2008. Anesthesia awareness and the bispectral index. *New Engl J Med* 359(4):428–429.

Paech, M. J., K. L. Scott, O. Clavisi, S. Chua, and N. McDonnell. 2008. A prospective study of awareness and recall associated with general anaesthesia for caesarean section. *Intl J Obstetric Anesth* 17(4):298–303.

Pollard, R. J., J. P. Coyle, R. L. Gilbert, and J. E. Beck. 2007. Intraoperative awareness in a regional medical system: a review of 3 years' data. *Anesthesiology* 106(2):269–274.

Rampil, I. J. 1998. A primer for EEG signal processing in anesthesia. *Anesthesiology* 89(4):980–1002.

Sakai, T., H. Singh, W. D. Mi, T. Kudo, and A. Matsuki. 1999. The effect of ketamine on clinical endpoints of hypnosis and EEG variables during propofol infusion. *Acta Anaesthesiol Scand* 43(2):212–216.

Samuelsson, P., L. Brudin, and R. H. Sandin. 2007. Late psychological symptoms after awareness among consecutively included surgical patients. *Anesthesiology* 106 (1):26–32.

Samuelsson, P., L. Brudin, and R. H. Sandin. 2008a. BIS does not predict dreams reported after anaesthesia. *Acta Anaesthesiol Scand* 52(6):810–814.

Samuelsson, P., L. Brudin, and R. H. Sandin. 2008b. Intraoperative dreams reported after general anaesthesia are not early interpretations of delayed awareness. *Acta Anaesthesiol Scand* 52(6):805–809.

Sandin, R. H., G. Enlund, P. Samuelsson, and C. Lennmarken. 2000. Awareness during anaesthesia: a prospective case study. *Lancet* 355(9205):707–711.

Scheller, B., G. Schneider, M. Daunderer, E. F. Kochs, and B. Zwissler. 2005. High-frequency components of auditory evoked potentials are detected in responsive but not in unconscious patients. *Anesthesiology* 103(5):944–950.

Schneider, G., R. Hollweck, M. Ningler, G. Stockmanns, and E. F. Kochs. 2005. Detection of consciousness by electroencephalogram and auditory evoked potentials. *Anesthesiology* 103(5):934–943.

Sebel, P. S., T. Andrew Bowdle, M. M. Ghoneim, I. J. Rampil, R. E. Padilla, T. J. Gan, and K. B. Domino. 2004. The incidence of awareness during anesthesia: a multicenter United States study. *Anesth Analg* 99(3):833–839.

Sebel, P. S., T. Andrew Bowdle, I. J. Rampil, R. E. Padilla, T. J. Gan, M. M. Ghoneim, and K. B. Domino. 2007. Don't ask, don't tell. *Anesthesiology* 107(4):672.

Shafer, A., P. F. White, J. Schüttler, and M. H. Rosenthal. 1983. Use of a fentanyl infusion in the intensive care unit: tolerance to its anesthetic effects? *Anesthesiology* 59(3):245–248.

Snow, J.D. 1847. *On the inhalation of the vapour of ether in surgical operations.* London: John Churchill.

Strickland, R. A., and J. F. Butterworth. 2007. Sexual dreaming during anesthesia: early case histories (1849–1888) of the phenomenon. *Anesthesiology* 106(6):1232–1236.

Tammisto, T, and S. Takki. 1973. Nitrous oxide-oxygen-relaxant anaesthesia in alcoholics: a retrospective study. *Acta Anaesthesiol Scand Supplementum* 53:68–75.

Tammisto, T., and I. Tigerstedt. 1977. The need for fentanyl supplementation of N2O-O2 relaxant anaesthesia in chronic alcoholics. *Acta Anaesthesiol Scand* 21(3):216–221.

Tononi, G. 2004. An information integration theory of consciousness. *BMC Neurosci* 5(1):42.

Toscano, A., C. Pancaro, and V. A. Peduto. 2007. Scopolamine prevents dreams during general anesthesia. *Anesthesiology* 106(5):952–955.

Voss, L., and J. Sleigh. 2007. Monitoring consciousness: the current status of EEG-based depth of anaesthesia monitors. *Best Pract Res Clin Anaesthesiol* 21(3):313–325.

Winterbottom, E. H. 1950. Insufficient anaesthesia. *Br Med J* 1(4647):247.

Xing, Y., J. M. Sonner, E. I. Eger, M. Cascio, and D. I. Sessler. 2004. Mice with a melanocortin 1 receptor mutation have a slightly greater minimum alveolar concentration than control mice. *Anesthesiology* 101(2):544–546.

Yamamura, T., M. Fukuda, H. Takeya, Y. Goto, and K. Furukawa. 1981. Fast oscillatory EEG activity induced by analgesic concentrations of nitrous oxide in man. *Anesth Analg* 60(5): 283–288.

# Chapter 9
# Loss of Recall and the Hippocampal Circuit Effects Produced by Anesthetics

M. Bruce MacIver

**Abstract** In the last decade, there has been a tremendous interest in the effects produced by general anesthetics on learning and memory. This has occurred for several reasons, including the rare but devastating clinical consequences of failing to block recall during surgery, the recognition that amnesia is an important component of anesthesia, a refined understanding of the synaptic basis for information storage, and an interest in using anesthetics as tools to probe learning and memory circuits in the brain. Also during this time, powerful new experimental approaches have been brought into focus to study anesthetic actions on learning. This chapter will review new findings that have emerged from this intersection of combined interest and detailed physiological studies of anesthetic effects on the circuits underlying learning and memory.

**Keywords** Learning and memory · block of recall · consciousness · anesthesia · LTP · LTD · synapse · network · systems · neuronal

It is clear from early studies that anesthetic-induced loss of recall is not the result of a simple "deafferentation" of higher brain regions, which would be expected to occur if anesthetics depressed sensory inputs to cortical areas. This was evident from experiments that demonstrated relatively intact processing of sensory information in both humans (using EEG recordings) and animals (using microelectrode recordings in cortical areas that process various sensory signals, such as visual, auditory, and somatosensory inputs). For example, many experiments have examined the processing of complex visual responses including orientation, movement, form, and color coding through several stages of cortical integration, spanning many visual areas

M.B. MacIver (✉)
Neuropharmacology Lab, Department of Anesthesia, Stanford University School of Medicine, 94305, SUMC S 288 MC 5177, Stanford, CA, USA
e-mail: maciver@stanford.edu

Confidential – submitted to Humana Press for Hudetz and Pearce 2009

A. Hudetz, R. Pearce (eds.), *Suppressing the Mind*, Contemporary Clinical Neuroscience, DOI 10.1007/978-1-60761-462-3_9,
© Humana Press, a part of Springer Science+Business Media, LLC 2010

and involving seven or more serial synaptic connections – all performed in surgically anesthetized animals (Nothdurft, Gallant, and Van Essen 1999; Villeneuve and Casanova 2003). Thus, anesthetics do not simply depress synaptic inputs to higher brain regions, although they may produce some cumulative depression and, more importantly, a distortion of information across a series of synapses. It has become apparent that anesthetics act through a more subtle disruption of synaptic signaling, and these effects will be the focus of this chapter.

## The Hippocampus is an Important Brain Region in Memory

It has long been recognized that the hippocampal formation and perihippocampal circuits in the temporal cortex are key brain components needed for many forms of learning and memory, especially for explicit, declarative experiences. This was first evident in patients who had suffered damage to their hippocampus, secondary to stroke or traumatic injury, and were subsequently found to have markedly diminished learning abilities. The most famous patient in this regard was "HM," who was subjected to extensive psychological tests and found to have profound learning loss, following a bilateral hippocampal injury (Schmolck et al. 2002).

The relationship between memory and hippocampal circuits has been consistently demonstrated in animal models. It is clear that the ability to remember many of the most important kinds of information critically depends on intact hippocampal processing. Not surprisingly, a great deal of attention has been focused on the synaptic physiology of hippocampal circuits to determine the basis for information storage in this brain region.

The hippocampal formation consists of several interconnected cell populations that form a "trisynaptic circuit" (Fig. 9.1). The circuit originates with perforant path inputs from the entorhinal cortex that forms excitatory, glutamate-mediated synapses onto granule neurons located in the dentate gyrus. Granule neurons form excitatory synapses with CA2 to CA3 pyramidal neurons via the mossy fiber pathway, also using glutamate as the neurotransmitter. The Schaffer-collateral pathway from CA3 neurons provides excitatory (glutamate) inputs to CA1 neurons that, in turn, send excitatory outputs to widespread regions of neocortex. All three excitatory fiber pathways also provide glutamate-mediated excitation of local inhibitory

**Fig. 9.1** **A** Diagram of the rat hippocampal formation showing pyramidal neuron regions (CA1 to CA3) and dentate granule neurons (DG). The intrinsic trisynaptic pathway originates with perforant path (pp) inputs to granule neurons, which form synapses with CA2 to CA3 pyramidal neurons via the mossy fiber (mf) pathway. The Schaffer-collateral (sc) pathway from CA3 provides excitatory (glutamate) inputs to CA1 neurons. Schaffer-collateral fibers also provide glutamate-mediated excitation of local inhibitory interneurons (colored cells), which form GABA synapses on granule and pyramidal neurons. **B** A higher resolution diagram of the CA1 region showing the placement of stimulating and recording electrodes used to study synaptic transmission in the Schaffer-collateral pathway

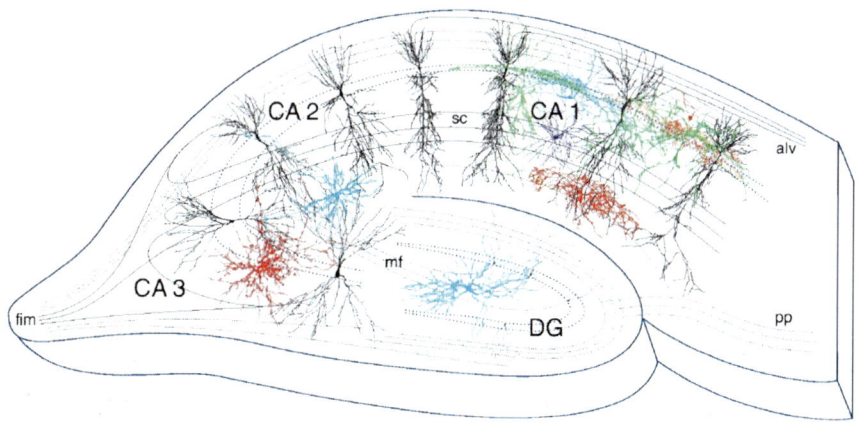

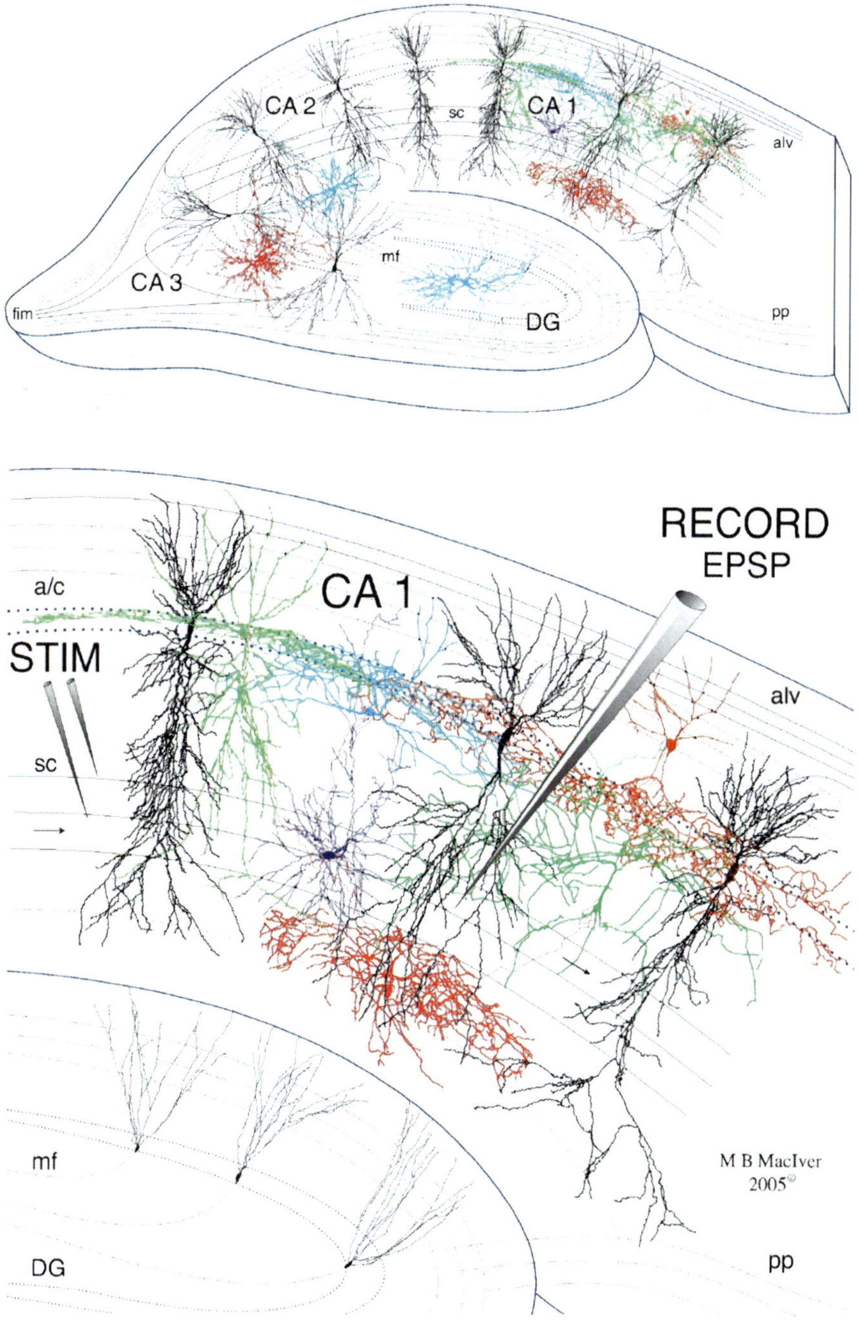

**Fig. 9.1** (continued)

interneurons (colored cells in Fig. 9.1), which form GABA-mediated synapses on all of the neurons in the hippocampal formation (including those that provide inhibition to other inhibitory interneurons). All of the excitatory synapses exhibit an ability to change their connection strength depending on the pattern of preceding action potential activity (Bliss and Lomo 1973; Malinow and Malenka 2002). Higher frequency activity, especially coincident with postsynaptic discharge, increases synaptic strength. Lower frequency activity, in contrast, decreases synaptic strength. This ability to change synaptic strength is termed plasticity, and synaptic plasticity is generally thought to form the basis for storing information in synaptic circuits (Bear and Malenka 1994). The best-studied pathway for plasticity in the brain is the synaptic connection between CA3 and CA1 neurons – the Schaffer-collateral pathway. It should also be noted that inhibitory synapses in the hippocampus can also exhibit plasticity, and even postsynaptic CA1 pyramidal neurons appear to be able to change their intrinsic excitability dependent on preceding action potential activity.

Anesthetics have been shown to depress synaptically evoked neuronal action potential discharge for each of the three main pathways. This appears to result from depressed glutamate excitation and/or enhanced GABA-mediated inhibition at several places in the circuit via multiple mechanisms at each synapse (MacIver and Roth 1988; Pittson, Himmel, and MacIver 2004; Winegar and MacIver 2006).

## Long-Term Potentiation (LTP) as a Synaptic Information Storage Mechanism

As described above, plasticity is thought to provide the cellular basis for storing information in hippocampal circuits. The best characterized form of plasticity is LTP – a persistent increase in synaptic strength is thought to contribute to the earliest stages of long-term memory formation. Several biochemically distinct forms of LTP occur at different types of synapses. Some forms require the activation of NMDA receptors while others do not. Only the most important and well-studied forms of LTP will be considered in this chapter.

## NMDA Receptor-Dependent LTP

LTP exhibits several characteristics that are thought to be important for information storage associated with memory (Bear and Malenka 1994). For example, the long-lasting time course of increased synaptic strength seen with LTP – lasting at least for weeks – is essential for any cellular substrate of memory. Other characteristics of LTP that are thought to be similar to learning and memory include synapse specificity, cooperativity, and associativity. *Synapse specificity* means that only synapses activated specifically by a distinct pattern of activity express LTP. Nearby synapses, even on the same dendritic branch, remain unpotentiated if they do not experience that same activity pattern. *Cooperativity* refers to the fact that a minimum "threshold" number of synaptic inputs must be simultaneously activated to induce LTP. Patterned activity involving too few presynaptic fibers will fail to induce LTP. *Associativity* occurs when a weak pattern of activity in one group of presynaptic inputs occurs together with a stronger input – the weak input is potentiated along with the stronger inputs.

To understand how LTP arises, it is necessary to consider the role played by AMPA and NMDA receptors at synapses. On binding glutamate, AMPA receptors become activated. This allows $Na^+$ and $Ca^{2+}$ to enter the neuron, causing the membrane potential to depolarize. This depolarization relieves a voltage-dependent block of NMDA receptors by $Mg^{2+}$ ions allowing an increased flow of $Ca^{2+}$ current. The requirement for concurrent presynaptic (glutamate release) and postsynaptic (depolarization) activities make the NMDA receptor an ideal *coincidence detector* for LTP induction. How this feature of the receptor translates into specific properties of LTP is clearly evident when considering the physiologic consequences of a high-frequency action potential train needed to induce LTP. High-frequency action potential patterns activate presynaptic fibers, resulting in the release of sufficient amounts of glutamate to ensure effective depolarization of the postsynaptic cell. The extent of the depolarization is directly related to the number of presynaptic fibers activated. If too few fibers are stimulated or the high-frequency train is too short, then the depolarization may be insufficient to fully activate the NMDA receptor. Patterns of activity that enhance the magnitude of postsynaptic depolarization, such as cooperativity or associativity, will enhance NMDA receptor activation and, therefore, are more likely to induce LTP.

An early indication of the importance of NMDA receptor activation in LTP induction was the demonstration that a selective NMDA receptor antagonist, APV, blocks induction of LTP at Schaffer-collateral synapses (Harris, Ganong, and Cotman 1984). The finding that APV blocked the ability to learn new information in rats supports a role for synaptic plasticity in memory (Morris et al. 1986).

## LTP and Second Messengers

The high permeability of NMDA receptor/channels to $Ca^{2+}$ gave a strong indication that an increase in intracellular $Ca^{2+}$ is a primary step in activating the signaling cascades that give rise to LTP. The first demonstration of this came from Lynch et al. (1983), who showed that injecting the calcium chelator EGTA into postsynaptic cells blocks LTP induction. This experiment illustrates that calcium is *necessary* for LTP induction; whether $Ca^{2+}$ is *sufficient* to induce LTP was not, however, clear. To address this question, experiments were conducted using a photolabile caged $Ca^{2+}$ compound. On being exposed to ultraviolet light, the compound liberates $Ca^{2+}$. By preloading CA1 cells with caged $Ca^{2+}$, it was possible to rapidly increase intracellular $Ca^{2+}$ by a brief flash of ultraviolet light. This approach revealed that LTP can be induced by an elevation in intracellular $Ca^{2+}$, indicating that $Ca^{2+}$ is not simply *necessary* but *sufficient* for LTP induction.

The critical role of NMDA receptors and $Ca^{2+}$ in LTP induction at Schaffer collateral-CA1 synapses is now widely accepted. The calcium-activated processes that follow are much less well understood. Many kinase pathways have been implicated, including roles for protein kinase A, protein kinase C, protein kinase G, and tyrosine kinases. Currently, however, one pathway in particular is thought to have a pivotal role. This pathway incorporates the $Ca^{2+}$/calmodulin-dependent protein kinase, CaM kinase II (Nicoll and Malenka 1999). A number of properties of this enzyme make it an attractive candidate as the enzymatic target for the increased intracellular calcium that accompanies LTP induction. First, the enzyme is found at extremely high levels within neurons, especially in the postsynaptic region of excitatory synapses. Second, the enzyme is activated by a calcium-binding protein, calmodulin; therefore, it has the capacity to sense increases in intracellular calcium. Third, the activity of this enzyme is regulated by autophosphorylation, so it remains active after calcium levels return to basal levels. This feature provides a mechanism by which a fleeting signal (the intracellular calcium rise) can be translated into a longer lasting biochemical change, and as such may represent the "molecular switch" for LTP. Extensive experimental studies have implicated CaM kinase II in LTP. For example, the injection of selective inhibitors of CaM kinase II into individual CA1 pyramidal cells abolishes potentiation within 1 h of the high-frequency stimulus being given. Correspondingly, injection of preactivated CaM kinase II slowly potentiates the cells into which it is added. Genetic approaches have also been used to examine the role of CaM kinase II. These experiments have included the generation of "knockout" animals that do not express the alpha isoform of CaM kinase II, the isoform found at high levels in the hippocampus. In these animals, stimuli that would normally induce LTP fail to do so. Finally, the induction of LTP produces an increase in the level of CaM kinase II in the autophosphorylated state (Dai, Hall, and Hell 2009).

## Maintenance of LTP

LTP induction requires NMDA receptor activation, an increase in postsynaptic calcium, and a role for CaM kinase II, which all strongly implicate the postsynaptic cell as the locus of expression for LTP. It is perhaps no surprise, therefore, that a number of models have been proposed that extend this chain of events to produce a mechanism that results in the enhancement of AMPA receptor function and thereby strengthened synaptic transmission. A number of these ideas have become firmly established as critical experimental observations are made. These include the demonstration that CaM kinase II can phosphorylate AMPA receptors and that AMPA receptor-mediated responses are enhanced by CaM kinase II phosphorylation (Vyazovskiy et al. 2008). The simplicity of such models adds to their appeal. There are, however, features of these models that do not adequately address all that is known about LTP. For example, LTP lasts weeks in vivo, whereas a phosphorylation event is generally thought to be sustained only for hours. LTP is also known to be blocked by inhibitors of transcription and translation, results consistent with the long-lasting nature of LTP (Raymond et al. 2000). These observations require one to look beyond posttranslational phosphorylation of AMPA receptors to understand the full complement of mechanisms that underlie LTP.

Enhancement of AMPA receptor function is just one of a number of mechanisms by which synaptic transmission might be enhanced. There could also be presynaptic changes. Using experimental conditions that made it possible to monitor quantal release events, a number of investigations found that the induction of LTP was accompanied by a decrease in the number of failures of transmission. This result is consistent with the idea that the probability of transmitter release increases with LTP. Therefore, LTP may also have a presynaptic expression mechanism (Malenka, Madison, and Nicoll 1986). It has now become evident that both pre- and postsynaptic loci can contribute.

Optical approaches for measuring presynaptic and postsynaptic components of LTP have been used to show that changes occur at both loci. Dyes such as FM1–43 provide a method by which the release of synaptic vesicles from the presynaptic terminal can be monitored both before and after the induction of LTP. These experiments demonstrated that the unloading of the vesicular marker dye FM1–43 was accelerated after induction of LTP (Stanton et al. 2005). This supports the idea that maintenance of LTP has a presynaptic component. An optical approach to monitor postsynaptic responses at single synapses using entry of calcium into dendritic spines demonstrated that induction of LTP is accompanied by both an increase in the probability of transmitter release and an augmentation in the amplitude of the postsynaptic calcium transient. Thus, LTP can have both a presynaptic and a postsynaptic component, and both components offer a number of possible sites of action for anesthetic-induced disruption – hence, possible sites for blocking recall.

## Retrograde Messengers

Evidence supports the idea that LTP has a presynaptic component; yet, since LTP induction is dependent on postsynaptic processes, it is necessary to consider the way in which these two loci communicate. One popular idea is that a molecule is generated by the postsynaptic cell in response to LTP induction. This molecule diffuses in a retrograde manner across the synaptic cleft where it interacts with biochemical pathways that regulate transmitter release. There are now a number of candidate "retrograde messenger" molecules including nitric oxide, arachidonic acid, carbon monoxide, platelet-activating factor, and even the cannabinoid receptor agonist, anandamide, which acts as a retrograde messenger at inhibitory synapses. Despite considerable effort, none of these molecules has been shown convincingly to be a retrograde messenger in LTP.

## Silent Synapses

A new difficulty in interpreting quantal analysis data was illustrated by the finding that NMDA and AMPA receptors are not always colocalized at a synapse. At some synapses within the CNS, NMDA receptors are found in isolation. Such synapses would be functionally inactive or "silent" at the cell resting membrane potential, as the voltage-dependent block by $Mg^{2+}$ of the NMDA receptor would prevent current flow. To reveal the existence of such synapses, the cell membrane potential had to be experimentally manipulated (i.e., depolarized under voltage clamp conditions) to reveal synaptically elicited currents that were not evident at the cell resting membrane potential. These currents were shown to arise from the NMDA receptor, as they are completely blocked by AP5. What has proved to be important about such synapses is that they can undergo LTP and that an "unmasking" process follows the induction of LTP – i.e., AMPA receptors are thought to be inserted into the synapse, transforming it from a "silent" to an active form (Malinow 2003). Evidence supporting activity-dependent trafficking of AMPA receptors came from the demonstration that the GluR1 subunit of the AMPA receptor can migrate to the cell membrane after induction of LTP. These data have significant implications in the debate about the relative importance of presynaptic and postsynaptic mechanisms in establishing LTP: unless it can be shown that a synaptic pathway contains no silent synapses, a decrease in the number of failures of synaptic transmission cannot be assumed to reflect a presynaptic increase in transmitter release.

## Non-NMDA-Dependent LTP

Although the NMDA receptor plays a critical role in the generation of potentiation at many synapses within the CNS, there are some synapses capable of showing long-lasting potentiation in which the NMDA receptor appears to play little or no role. For example, the mossy fiber synapses onto CA3 pyramidal cells show a form of LTP that is NMDA independent. The first clear indication for this came from data obtained by Harris and Cotman (1986), who showed that the NMDA receptor antagonist AP5 did not block potentiation at the mossy fiber pathway. Furthermore, potentiation at these synapses does not require an increase in postsynaptic calcium. Calcium does appear to be the trigger for potentiation because it does not occur if extracellular calcium is removed during inductive stimulation. Details of how the increase in presynaptic calcium serves to augment transmitter release are scant, although a protein kinase A second messenger cascade has been implicated (Pelkey et al. 2008).

## Long-Term Depression

The increase in synaptic strength that accompanies LTP can be *saturated* – i.e., a point is reached where a further increase in synaptic strength cannot occur. During such conditions, the encoding of new information by the brain might be compromised. Behavioral evidence exists to support this view. Normal brain function must, therefore, include a process by which synapses can be selectively weakened. An activity-dependent mechanism able to achieve this has been described at the Schaffer collateral-CA1 cell synapses and is referred to as long-term depression (LTD; Malenka 2003). LTD reduces synaptic efficacy after a sustained period (10–15 min) of low-frequency synaptic stimulation, similar to the 1–3 Hz pattern of activity seen during slow-wave sleep (Huber et al. 2008). Experiments examining the cellular processes that underlie LTD have shown that it shares features in common with LTP: It is dependent on activation of NMDA receptors and requires calcium. The amplitude and temporal characteristics of the calcium increase are, however, different from those required to induce LTP, and it is these differences that are thought to permit the neurons to discriminate between LTP- and LTD-inducing stimuli. The biochemical processes that underlie LTD are somewhat different from those for LTP. For example, it has been shown that calcium-dependent phosphatases (enzymes that dephosphorylate proteins) are essential for LTD induction. In addition, it is apparent that LTD involves a removal of glutamate receptors from synapses, as opposed to the addition seen during LTP.

There remain a number of very basic unresolved issues that remain to be answered about synaptic plasticity: the locus of change that accompanies NMDA receptor-dependent LTP and LTD, the existence and types of the retrograde messenger(s), and the mechanisms by which LTP is maintained for periods extending beyond weeks. Why do different synapses have mechanistically different processes underlying a similar change in efficacy? Perhaps, the most important question of all is whether the synaptic mechanisms described thus far really are processes by which the brain develops memories. Many features of these processes are suggestive of this possibility, and experiments comparing anesthetic effects on LTP and loss of recall in rodent models have also provided results consistent with this possibility – and may provide the definitive link that ultimately addresses this important question.

## Anesthetics Alter Hippocampal Circuit Function In Vivo

If anesthetic-induced loss of recall involves an effect on synaptic plasticity within hippocampal circuits, then we should expect to see an anesthetic-induced change in hippocampal synaptic function at concentrations that block memory formation. There is a limited (but convincing) literature that indicates a marked change in hippocampal synaptic function produced by general anesthetics at concentrations that span the range needed to block memory. The earliest in vivo recordings of hippocampal neurons tested the effects of barbiturates on CA1 neuron synaptic responses and found a marked prolongation of inhibitory postsynaptic potentials (IPSPs; Nicoll et al. 1975). More recently, microEEG recordings of CA1 neuronal activity in awake, behaving rats found a marked increase in amplitude together with

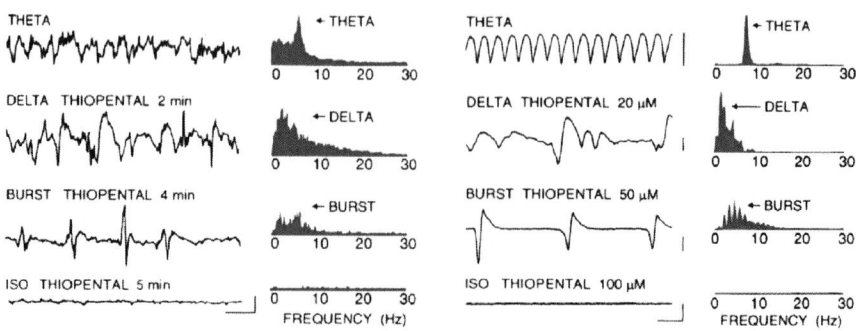

**Fig. 9.2** Thiopental produces a pronounced disruption of EEG signals recorded from awake behaving rats chronically implanted with electrodes (*left*) and from rat brain slices using carbachol-induced EEG (*right*). Note the similarity in EEG patterns seen in vivo and in vitro. Top traces show predrug baseline responses and the lower three traces show effects at increasing concentrations of thiopental. Concentrations associated with loss of recall produce a shift in EEG signals to lower (delta) frequencies. Deeper levels of anesthesia produce burst-suppression patterns and isoelectric activity in both animals and brain slices. Modified from Lukatch and MacIver (1996)

a frequency slowing, produced by concentrations of thiopental that spanned the behavioral effect range from "activation" through loss of righting reflex and deeper planes of anesthetic depth (Fig. 9.2) (MacIver et al. 1996). Thiopental-induced loss of recall is expected to occur somewhere between the activation stage and loss of righting reflex. At these concentrations, a slowing of EEG signals, consistent with a prolongation of GABA-mediated IPSPs, was the most prevalent effect (Lukatch and MacIver 1996). A similar slowing in frequency produced by amnesic concentrations of isoflurane has also been seen in awake mice, with chronically implanted EEG microelectrodes in the CA1 region of hippocampus (Perouansky et al. 2007; Perouansky and Pearce 2009). Thus, anesthetic effects on CA1 neuron electrical activities are clearly consistent with the idea that the loss of recall produced by these agents could come about, at least in part, through effects involving the hippocampus.

## Anesthetic Effects on Ascending Inputs to Hippocampus

Anesthetic effects on inputs to the hippocampal formation have not been systematically studied. Part of the reason for this is that little is known about where these inputs come from, except that a wide range of sensory stimuli are known to drive hippocampal neurons at fairly short latencies. A major input pathway that is known to contribute to memory processing in the hippocampus, and which has been reasonably well characterized, is the ascending cholinergic pathway from the medial septal nucleus. This input pathway shows a marked lack of sensitivity to a number of anesthetics, including barbiturates, ether, urethane, and isoflurane. The major hippocampal response to activating this medial septal input is a pronounced EEG theta rhythm, and this response remains intact even at deep surgical levels of anesthesia (Bland et al. 2003). Thus, it does not seem likely that effects on ascending inputs can fully account for the disruption produced by anesthetics on hippocampal electrical activity. On the other hand, there is also very good evidence that this ascending medial septal input can contribute to anesthetic effects on loss of the righting reflex and can even account for blocking some spinal components of anesthesia (Ma et al. 2002). Once again, it seems likely that anesthetic effects involve more subtle actions on ascending inputs, like this medial septal pathway – not simply a depression of signaling.

## Anesthetic Effects on Intrinsic Excitability of Hippocampal Neurons

Although some anesthetics can depress the excitability of hippocampal neurons, this occurs only at high concentrations. More importantly, not all anesthetics produce this effect (Fig. 9.3). It does not appear likely that depressed intrinsic membrane excitability can account for the major disruption produced by anesthetics on hippocampal electrical activity, especially since EEG signals are increased in amplitude at concentrations associated with loss of recall. Thus, as suggested for other brain regions, anesthetic effects on synaptic transmission are more likely to play the dominant role. To study these issues in greater detail, many investigators have turned to in

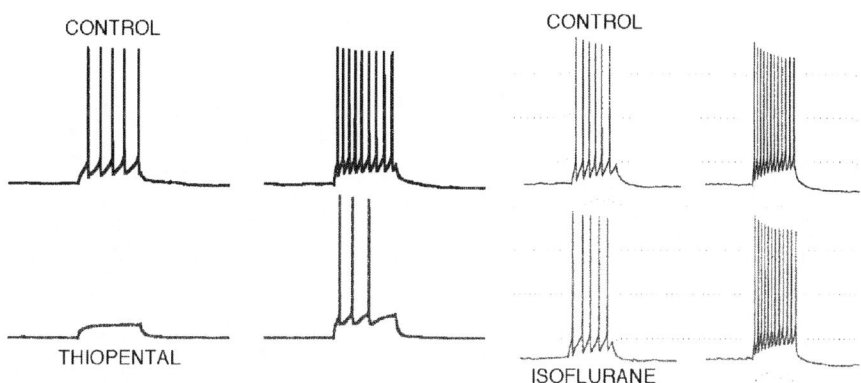

**Fig. 9.3** Thiopental and isoflurane affect the intrinsic excitability of CA1 neurons to differing degrees, although the thiopental-induced depression of action potential discharge was only evident at high concentrations (MAC equivalent for immobility). Even at MAC immobility, isoflurane had no appreciable effect on action potential discharge. For each drug, control responses at two stimulus current levels are shown on *top* and the maximum drug effect is shown at the *bottom*. Thus, it seems unlikely that anesthetic-induced depression of discharge plays a consistent role in loss of recall

vitro brain slice preparations of the hippocampus to further characterize anesthetic effects on synaptic responses.

## Anesthetics Block LTP

It has long been known that anesthetics can block LTP in the Schaffer-collateral pathway in hippocampal brain slices (MacIver, Tauck, and Kendig 1989). There appears to be a very good correlation between the concentrations of anesthetics that block recall in rats and those that block LTP (Fig. 9.4), although this has only been established for isoflurane (Turnquist and MacIver 2005). At present, it is not known how anesthetics block LTP, and it should be clear from the above discussion of LTP mechanisms that any of a number of critical stages and sites of action could be involved. For isoflurane, it is apparent that the important sites of action are upstream from the postsynaptic signaling cascade that involves intracellular $Ca^{2+}$, since it was possible to induce LTP using a nonsynaptic approach to raise intracellular $Ca^{2+}$ levels (even in the presence of high anesthetic concentrations; Cheung and MacIver 2002). So, isoflurane does not appear to alter calmodulin and/or other kinase signaling pathways, nor does it alter the ability of AMPA/kainate receptors to traffic into the subsynaptic space to support increased synaptic strength following $Ca^{2+}$ entry. Instead, the site(s) of action appear to be at the Schaffer-collateral synapse, perhaps involving depressed glutamate release, inhibited NMDA receptor responses, or enhanced GABA-mediated inhibition. Effects on any of these would prevent the initial rise in CA1 neuron calcium levels needed for LTP. Anesthetic effects on each of these possible mechanisms have been observed; the evidence in support of each is presented below.

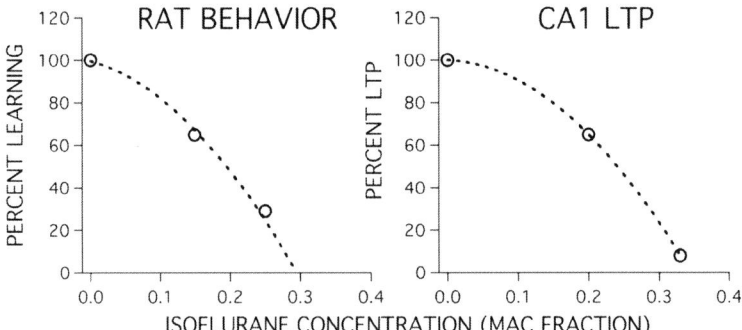

**Fig. 9.4** Graphs comparing the concentration-effect relationships for isoflurane-induced depression of learning in rats [*left*; data from Dutton et al. (2001)] and for depression of Schaffer-collateral to CA1 neuron synaptic LTP from rat brain slice experiments (unpublished data from our laboratory). Note the similarity in the effective concentrations for both, suggesting that LTP could serve as a surrogate measure for anesthetic-induced impairment of learning at the cellular level

## Anesthetic Effects on GABA Synapses
## Within the Hippocampus

The best-documented anesthetic effect seen across brain regions and all concentrations is an enhancement of GABA-mediated inhibition (Tanelian et al. 1993). By enhancing inhibition, anesthetics could depress neuronal excitability to the point that the necessary depolarization needed to allow relief of $Mg^{2+}$ block of NMDA receptors does not occur. This would block LTP induction simply by reducing calcium entry through NMDA receptors so that the postsynaptic kinase signaling pathways do not become activated and no new AMPA/kainate receptors get incorporated into synapses. Experimental evidence to support this mechanism for isoflurane-induced block of LTP has been provided using rat brain slices. Although this study used somewhat higher concentrations of isoflurane to block LTP than required for depression of learning in rats, it suggested that enhanced inhibition could play a role (Simon et al. 2001). In addition, more evidence for an involvement of enhanced GABA inhibition came from the ability to reverse the anesthetic effect by blocking GABA receptors. Other studies have shown, however, that blocking GABA receptors improves the ability to induce LTP at Schaffer-collateral synapses even in the absence of anesthetic (Barrionuevo et al. 1986). Thus, it is possible that the result simply reflects this improved ability to induce LTP. In any case, general anesthetics clearly increase GABA inhibition, and this would almost certainly contribute to a depression of LTP.

Isoflurane enhances GABA inhibition in CA1 neurons by a complex set of actions, beginning with the ability to increase GABA release from nerve terminals (Banks and Pearce 1999;Nishikawa and MacIver 2000a,b; Pittson, Himmel, and MacIver 2004). This effect occurs even when action potential invasion of nerve terminals is abolished by blocking axonal and nerve terminal sodium channels. The effect is apparent as an increase in the frequency of miniature inhibitory postsynaptic currents (mIPSCs) recorded from voltage-clamped CA1 neurons using whole-cell patch-clamp recordings. This increased mIPSC frequency would have two effects on CA1 neurons. First, having more IPSCs impinging on CA1 cells would directly inhibit any excitatory synaptic inputs needed to produce LTP, especially if they arrive at the same time as an IPSC. Second, the increased mIPSC frequency would produce higher concentrations of GABA that can diffuse onto extrasynaptic GABA receptors to increase tonic inhibitory currents (Bieda and MacIver 2004). Increased tonic GABA currents have been reported to produce depressed CA1 neuron excitability that would interfere with LTP induction, and this too has been suggested to play a role in blocking synaptic plasticity and learning (Cheng et al. 2006).

In addition to this increased GABA release, isoflurane and other anesthetics have been shown to increase the duration of GABA-mediated synaptic currents (Mody, Tanelian, and MacIver 1991; Nishikawa and MacIver 2001). This effect is thought to come about by an allosteric interaction between the anesthetic-binding site on GABA receptors that results in an increased affinity of GABA for its binding site and a slower unbinding that promotes longer duration channel open times (Li and

Pearce 2000). This results in a prolongation of IPSC decay times, such that more inhibitory current flows for a longer time (Lukatch and MacIver 1997; Banks and Pearce 1999). This would depress the induction of LTP by altering the balance of coincidence detection for NMDA receptors in favor of staying closed and preventing the $Ca^{2+}$ entry required for LTP. As suggested by Banks, White, and Pearce (2000), prolongation of $GABA_{Aslow}$ IPSCs would be particularly important for controlling NMDA-mediated synaptic transmission. A prolongation of IPSCs by anesthetics could have a major effect on LTP induction because action potentials in Schaffer-collateral fibers produce excitation of both CA1 pyramidal neurons and feedforward inhibitory interneurons. Thus, increasing the inhibitory time period could dramatically shorten the excitatory synaptic depolarization needed to open NMDA receptor-gated channels, leading to reduced $Ca^{2+}$ entry.

## Anesthetic Effects on Glutamate Synapses in the Hippocampus

Anesthetic effects on excitatory synaptic inputs to CA1 neurons could also contribute to a depression of LTP. A similar complexity of effects appears to contribute to anesthetic-induced depression of glutamate synapses, since both pre- and postsynaptic sites of action are involved (Pittson, Himmel, and MacIver 2004). Volatile anesthetics, for example, have been shown to depress glutamate release from Schaffer-collateral nerve terminals (MacIver et al. 1996; Kirson, Yaari, and Perouansky 1998). This comes about by two separate actions. First, a depression of action potential propagation into nerve terminals is evident, perhaps involving a direct depression of sodium channels (Mikulec et al. 1998; Westphalen and Hemmings 2003). This reduces glutamate release by reducing the number of active synapses and occurs with no change in paired pulse facilitation – similar to the effect seen with a local anesthetic or with TTX. Second, anesthetics produce a depression that is accompanied by an increase in synaptic facilitation, similar to the effect produced by blocking calcium entry into nerve terminals (Winegar and MacIver 2006). It is not clear at this time whether anesthetics reduce $Ca^{2+}$ entry by acting directly on calcium channels, by acting upstream of channels on any of a number of modulatory pathways that regulate nerve terminal calcium channels, and/or perhaps, indirectly by increasing nerve terminal potassium currents (Westphalen et al. 2007). There is growing evidence to implicate each of these possible sites of action, and they are not, of course, mutually exclusive (Pittson, Himmel, and MacIver 2004). A reduction in glutamate release through any of these mechanisms would depress LTP induction by reducing the postsynaptic depolarization needed to activate NMDA-gated channels.

Anesthetics have also been shown to depress NMDA-mediated synaptic transmission at Schaffer-collateral synapses, possibly via a direct action on these ligand-gated ion channels (Puil, el-Beheiry, and Baimbridge 1990; Nishikawa and MacIver 2000a,b). The depression of NMDA channels occurs at low concentrations similar to those needed to block learning in rats. This, of course, would also contribute to an anesthetic-induced block of LTP by directly reducing the amount of $Ca^{2+}$ influx into postsynaptic CA1 neurons.

# Conclusions

Anesthetics can produce a loss of recall at concentrations lower than needed to block consciousness or even to impair sensory and/or motor responses to an appreciable degree. Thus, it has been possible to establish $EC_{50}$ concentrations for this effect in humans and rodents and to study these concentrations on rodent brain slice synaptic responses. The hippocampal formation appears to be a likely brain target for the anesthetic-induced loss of recall because hippocampal electrical activity is profoundly altered by low concentrations of anesthetics, as are both excitatory and inhibitory synapses within hippocampal circuits. It remains to be determined which of a number of possible mechanisms contribute to this effect. It should be noted that it is difficult to untangle some of these effects. For example, enhanced GABA-mediated inhibition always reduces NMDA-mediated excitation by decreasing the level of depolarization that a given amount of excitatory synaptic drive will produce. Similarly, inhibiting NMDA responses always reduces the amount of GABA-mediated inhibition experienced by CA1 neurons by reducing the presynaptic excitability of inhibitory interneurons. Anesthetic effects on ascending modulatory inputs to the hippocampus appear to play a role in depressing hippocampal neuronal processing, and this will be difficult to distinguish from direct effects on hippocampal synapses. It may be impossible to ever separate all of these entangled effects, but it is already apparent that effects on glutamate and GABA transmitter systems within the hippocampus are involved in anesthetic-induced loss of recall.

# References

Banks, M. I., and R. A. Pearce. 1999. Dual actions of volatile anesthetics on GABA(A) IPSCs: dissociation of blocking and prolonging effects [see comment]. *Anesthesiology* 90(1): 120–134.

Banks, M. I., J. A. White, and R. A. Pearce. 2000. Interactions between distinct GABA(A) circuits in hippocampus. *Neuron* 25(2):449–457.

Barrionuevo, G., S. R. Kelso, D. Johnston, and T. H. Brown. 1986. Conductance mechanism responsible for long-term potentiation in monosynaptic and isolated excitatory synaptic inputs to hippocampus. *J Neurophysiol* 55(3):540–550.

Bear, M. F., and R. C. Malenka. 1994. Synaptic plasticity: LTP and LTD. *Curr Opin Neurobiol* 4(3):389–399.

Bieda, M. C., and M. B. MacIver. 2004. Major role for tonic GABAA conductances in anesthetic suppression of intrinsic neuronal excitability. *J Neurophysiol* 92(3):1658–1667.

Bland, B. H., C. E. Bland, L. V. Colom, S. H. Roth, S. DeClerk, A. Dypvik, J. Bird, and A. Deliyannides. 2003. Effect of halothane on type 2 immobility-related hippocampal theta field activity and theta-on/theta-off cell discharges. *Hippocampus* 13(1):38–47.

Bliss, T. V., and T. Lomo. 1973. Long-lasting potentiation of synaptic transmission in the dentate area of the anaesthetized rabbit following stimulation of the perforant path.[see comment]. *J Physiol* 232(2):331–356.

Cheng, V. Y., L. J. Martin, E. M. Elliott, J. H. Kim, H. T. Mount, F. A. Taverna, J. C. Roder, J. F. Macdonald, A. Bhambri, N. Collinson, K. A. Wafford, and B. A. Orser. 2006. Alpha5GABAA receptors mediate the amnestic but not sedative-hypnotic effects of the general anesthetic etomidate. *J Neurosci* 26(14):3713–3720.

Cheung, K. H.-T., and M. B. MacIver. 2002. Isoflurane-induced depression of synaptic LTP does not involve post-NMDA receptor mediated mechanisms. *Anesthesiology* 96(A815).

Dai, S., D. D. Hall, and J. W. Hell. 2009. Supramolecular assemblies and localized regulation of voltage-gated ion channels. *Physiol Rev* 89(2):411–452.

Dutton, R. C., A. J. Maurer, J. M. Sonner, M. S. Fanselow, M. J. Laster, and E. I. Eger, II. 2001. The concentration of isoflurane required to suppress learning depends on the type of learning. *Anesthesiology* 94(3):514–519.

Harris, E. W., and C. W. Cotman. 1986. Long-term potentiation of guinea pig mossy fiber responses is not blocked by N-methyl D-aspartate antagonists. *Neurosci Lett* 70(1):132–137.

Harris, E. W., A. H. Ganong, and C. W. Cotman. 1984. Long-term potentiation in the hippocampus involves activation of N-methyl-D-aspartate receptors. *Brain Res* 323(1):132–137.

Huber, R., S. Maatta, S. K. Esser, S. Sarasso, F. Ferrarelli, A. Watson, F. Ferreri, M. J. Peterson, and G. Tononi. 2008. Measures of cortical plasticity after transcranial paired associative stimulation predict changes in electroencephalogram slow-wave activity during subsequent sleep. *J Neurosci* 28(31):7911–7918.

Kirson, E. D., Y. Yaari, and M. Perouansky. 1998. Presynaptic and postsynaptic actions of halothane at glutamatergic synapses in the mouse hippocampus. *Br J Pharmacol* 124(8):1607–1614.

Li, X., and R. A. Pearce. 2000. Effects of halothane on GABA(A) receptor kinetics: evidence for slowed agonist unbinding. *J Neurosci* 20(3):899–907.

Lukatch, H. S., and M. B. MacIver. 1996. Synaptic mechanisms of thiopental-induced alterations in synchronized cortical activity. *Anesthesiology* 84(6):1425–1434.

Lukatch, H. S., and M. B. MacIver. 1997. Voltage-clamp analysis of halothane effects on GABA(A fast) and GABA(A slow) inhibitory currents. *Brain Res* 765(1):108–112.

Lynch, G., J. Larson, S. Kelso, G. Barrionuevo, and F. Schottler. 1983. Intracellular injections of EGTA block induction of hippocampal long-term potentiation. *Nature* 305(5936):719–721.

Ma, J., B. Shen, L. S. Stewart, I. A. Herrick, and L. S. Leung. 2002. The septohippocampal system participates in general anesthesia. *J Neurosci* 22(2):RC200.

MacIver, M. B., J. W. Mandema, D. R. Stanski, and B. H. Bland. 1996. Thiopental uncouples hippocampal and cortical synchronized electroencephalographic activity. *Anesthesiology* 84(6):1411–1424.

MacIver, M. B., and S. H. Roth. 1988. Inhalation anaesthetics exhibit pathway-specific and differential actions on hippocampal synaptic responses *in vitro*. *Br J Anaesthesia* 60(6):680–691.

MacIver, M. B., D. L. Tauck, and J. J. Kendig. 1989. General anaesthetic modification of synaptic facilitation and long-term potentiation in hippocampus. *Br J Anaesthesia* 62(3):301–310.

MacIver, M. B., A. A. Mikulec, S. M. Amagasu, and F. A. Monroe. 1996. Volatile anesthetics depress glutamate transmission via presynaptic actions. *Anesthesiology* 85(4):823–834.

Malenka, R. C. 2003. Synaptic plasticity and AMPA receptor trafficking. *Ann NY Acad Sci* 1003:1–11.

Malenka, R. C., D. V. Madison, and R. A. Nicoll. 1986. Potentiation of synaptic transmission in the hippocampus by phorbol esters. *Nature* 321(6066):175–177.

Malinow, R. 2003. AMPA receptor trafficking and long-term potentiation. *Philos Trans R Soc Lond Ser B: Biol Sci* 358(1432):707–714.

Malinow, R., and R. C. Malenka. 2002. AMPA receptor trafficking and synaptic plasticity. *Annu Rev Neurosci* 25:103–126.

Mikulec, A. A., S. Pittson, S. M. Amagasu, F. A. Monroe, and M. B. MacIver. 1998. Halothane depresses action potential conduction in hippocampal axons. *Brain Res* 796(1–2):231–238.

Mody, I., D. L. Tanelian, and M. B. MacIver. 1991. Halothane enhances tonic neuronal inhibition by elevating intracellular calcium. *Brain Res* 538(2):319–323.

Morris, R. G., E. Anderson, G. S. Lynch, and M. Baudry. 1986. Selective impairment of learning and blockade of long-term potentiation by an N-methyl-D-aspartate receptor antagonist, AP5. *Nature* 319(6056):774–776.

Nicoll, R. A., J. C. Eccles, T. Oshima, and F. Rubia. 1975. Prolongation of hippocampal inhibitory postsynaptic potentials by barbiturates. *Nature* 258(5536):625–627.

Nicoll, R. A., and R. C. Malenka. 1999. Expression mechanisms underlying NMDA receptor-dependent long-term potentiation. *Ann NY Acad Sci* 868:515–525.

Nishikawa, K., and M. B. MacIver. 2000a. Excitatory synaptic transmission mediated by NMDA receptors is more sensitive to isoflurane than are non-NMDA receptor-mediated responses. *Anesthesiology* 92(1):228–236.

Nishikawa, K., and M. B. MacIver. 2000b. Membrane and synaptic actions of halothane on rat hippocampal pyramidal neurons and inhibitory interneurons. *J Neurosci* 20(16): 5915–5923.

Nishikawa, K., and M. B. MacIver. 2001. Agent-selective effects of volatile anesthetics on GABAA receptor-mediated synaptic inhibition in hippocampal interneurons. *Anesthesiology* 94(2): 340–347.

Nothdurft, H. C., J. L. Gallant, and D. C. Van Essen. 1999. Response modulation by texture surround in primate area V1: correlates of "popout" under anesthesia. *Vis Neurosci* 16(1):15–34.

Pelkey, K. A., L. Topolnik, X. Q. Yuan, J. C. Lacaille, and C. J. McBain. 2008. State-dependent cAMP sensitivity of presynaptic function underlies metaplasticity in a hippocampal feedforward inhibitory circuit. *Neuron* 60(6):980–987.

Perouansky, M., H. Hentschke, M. Perkins, and R. A. Pearce. 2007. Amnesic concentrations of the nonimmobilizer 1,2-dichlorohexafluorocyclobutane (F6, 2 N) and isoflurane alter hippocampal theta oscillations *in vivo. Anesthesiology* 106(6):1168–1176.

Perouansky, M., and R. A. Pearce. 2009. Modulation of the hippocampal theta rhythm as a mechanism for anesthetic-induced amnesia. In *Suppressing the mind: anesthetic modulation of memory and consciousness*, edited by A. G. Hudetz and R. A. Pearce. New York City: Springer.

Pittson, S., A. M. Himmel, and M. B. MacIver. 2004. Multiple synaptic and membrane sites of anesthetic action in the CA1 region of rat hippocampal slices. *BMC Neurosci* 5:52.

Puil, E., H. el-Beheiry, and K. G. Baimbridge. 1990. Anesthetic effects on glutamate-stimulated increase in intraneuronal calcium. *J Pharmacol Exp Therap* 255(3):955–961.

Raymond, C. R., V. L. Thompson, W. P. Tate, and W. C. Abraham. 2000. Metabotropic glutamate receptors trigger homosynaptic protein synthesis to prolong long-term potentiation. *J Neurosci* 20(3):969–976.

Schmolck, H., E. A. Kensinger, S. Corkin, and L. R. Squire. 2002. Semantic knowledge in patient H.M. and other patients with bilateral medial and lateral temporal lobe lesions. *Hippocampus* 12(4):520–533.

Simon, W., G. Hapfelmeier, E. Kochs, W. Zieglgansberger, and G. Rammes. 2001. Isoflurane blocks synaptic plasticity in the mouse hippocampus. *Anesthesiology* 94(6):1058–1065.

Stanton, P. K., J. Winterer, X. L. Zhang, and W. Muller. 2005. Imaging LTP of presynaptic release of FM1-43 from the rapidly recycling vesicle pool of Schaffer collateral-CA1 synapses in rat hippocampal slices. *Eur J Neurosci* 22(10):2451–2461.

Tanelian, D. L., P. Kosek, I. Mody, and M. B. MacIver. 1993. The role of the GABAA receptor/chloride channel complex in anesthesia. *Anesthesiology* 78(4):757–776.

Turnquist, P. A., and M. B. MacIver. 2005. Anesthetic block of LTP involves depression of NMDA EPSPs and enhanced GABA inhibition. *Soc Neurosci* 31.

Villeneuve, M. Y., and C. Casanova. 2003. On the use of isoflurane versus halothane in the study of visual response properties of single cells in the primary visual cortex. *J Neurosci Methods* 129(1):19–31.

Vyazovskiy, V. V., C. Cirelli, M. Pfister-Genskow, U. Faraguna, and G. Tononi. 2008. Molecular and electrophysiological evidence for net synaptic potentiation in wake and depression in sleep [see comment]. *Nat Neurosci* 11(2):200–208.

Westphalen, R. I., and H. C. Hemmings, Jr. 2003. Selective depression by general anesthetics of glutamate versus GABA release from isolated cortical nerve terminals. *J Pharmacol Exp Therap* 304(3):1188–1196.

Westphalen, R. I., M. Krivitski, A. Amarosa, N. Guy, and H. C. Hemmings, Jr. 2007. Reduced inhibition of cortical glutamate and GABA release by halothane in mice lacking the K+ channel, TREK-1. *Br J Pharmacol* 152(6):939-945.

Winegar, B. D., and M. B. MacIver. 2006. Isoflurane depresses hippocampal CA1 glutamate nerve terminals without inhibiting fiber volleys. *BMC Neurosci* 7:5.

# Chapter 10
# Modulation of the Hippocampal θ-Rhythm as a Mechanism for Anesthetic-Induced Amnesia

**Misha Perouansky and Robert Pearce**

**Abstract** Understanding the mechanisms of anesthetic interference with memory and consciousness is of scientific value and clinical importance. For the clinician, the ability to precisely monitor the effect of anesthetics on the brain, the organ where anesthetic drugs exert their core effects (effects on all other organs, no matter how profound, are side effects), would be practice changing.

The last decades have witnessed a momentous increase in the interest of the (neuro) scientific community in synchronized neuronal activity, the "rhythms of the brain" (Buzsaki. 2006. *Rhythms of the Brain*. 1 ed. New York: Oxford University Press). As their links to higher cognitive function become unraveled, mechanistically grounded approaches to understanding, measuring, and predicting drug effects on the mind become fathomable. This chapter will review the evidence suggesting that the ability of the brain to form explicit memories is dependent on the precise timing of synchronized neuronal activity in the hippocampus that presents as the θ-rhythm. Based on experiments in our lab, we will then develop the argument that anesthetic-induced amnesia is, at some point, reflected by the θ-rhythm and hence can be measured and quantified.

**Keywords** Anesthesia · hippocampus · θ-rhythm · gamma oscillations · 40 Hz rhythms · amnesia · learning · memory

## Introduction

Memory and consciousness are the two manifestations of higher cognitive function that are most closely associated with our self-awareness as individuals. They are also the functions that are most dramatically affected by general anesthetics. Recent research suggests that memory and consciousness can be pharmacologically dissociated from each other in a drug concentration-dependent way by general anesthetics

M. Perouansky (✉)
Department of Anesthesiology, University of Wisconsin, SMPH B6/319 Clinical Science Center, 600 Highland Avenue, Madison, WI, 53792-3272, USA
e-mail: mperouansky@wisc.edu

A. Hudetz, R. Pearce (eds.), *Suppressing the Mind*, Contemporary Clinical Neuroscience, DOI 10.1007/978-1-60761-462-3_10,
© Humana Press, a part of Springer Science+Business Media, LLC 2010

and anesthetic-like drugs before being indiscriminately lost at deeper (so-called "surgical") levels of anesthesia. Notably, loss of consciousness requires approximately double the concentration of inhalational anesthetics as the impairment of hippocampus-dependent memory (Gonsowski et al. 1995), indicating that learning and consciousness can be dissociated using general anesthetic drugs in appropriately designed experiments.

The importance of the hippocampus for conscious memory, specifically contextual episodic or explicit memory, is well documented (for a classification of human memory systems, see Veselis and Pryor 2009). A more subtle connection between memory (though not necessarily hippocampus dependent) and consciousness exists as well: Memories provide the "mind" not only with a bridge between the past and the future but also with a powerful future-predicting engine that is an essential attribute of the state of full consciousness, aptly described by Gerald Edelman as "the remembered present" (Edelman 2001).

In this chapter, we review our work examining the effects of inhalational general anesthetics and of the experimental anesthetic-like drug 1,2-dichlorohexafluorocyclobutane (F6 or 2 N) on hippocampal network activity in adult rats. The purpose of these experiments was to investigate whether hippocampal EEG ($EEG_h$) measures reflect the behavioral effects of anesthetics on hippocampus-dependent learning and memory. We collaborated with the laboratory of Dr. E. I. Eger at the UCSF in order to quantitatively correlate electrophysiologic effects of anesthetic drugs with behavioral outcomes assessed with fear-conditioning paradigms. Based on the work discussed in this chapter, we propose that anesthetic modulation of hippocampal $\theta$-oscillations can be quantitatively correlated with, and even may play a causal role in, the impairment of hippocampus-dependent learning and memory.

## Inhaled Anesthetics

All general anesthetics in use today either enhance inhibition or block excitation – or do both. On the molecular level, the most recognized and most extensively studied anesthetic targets among the transmitter-gated receptors are the $GABA_A$ and the ionotropic glutamate receptors, which are the principal mediators of synaptic inhibition and excitation, respectively. These two groups of receptors have also been identified targets for essentially all commonly used intravenous anesthetics. However, most members of the large cys-loop superfamily of receptors, many voltage-gated channels, and some members of the large tandem-pore domain potassium and TRP channel families are also targets of inhalational general anesthetics. It is this promiscuity of the inhaled agents, paired with their generally low potency, which leads to a plethora of plausible molecular targets and renders a systematic understanding of anesthetic mechanisms – from the action at a receptor to the behavioral effect – so complicated.

## Learning the Basics from the Hippocampus

Much of the basic information that exists today regarding anesthetic modulation of the critical elements of neuronal communication came from investigations of neurons and slices obtained from the hippocampus: depression of axonal conduction (Berg-Johnsen and Langmoen 1986; Mikulec et al. 1998; Winegar and MacIver 2006); increased inhibition, primarily by postsynaptic potentiation of synaptic (Lukatch and MacIver 1997; Banks and Pearce 1999; Nishikawa and MacIver 2001) and tonic (Caraiscos et al. 2004) GABA$_A$ receptor-mediated inhibition; and block of glutamatergic excitation on both pyramidal cells and interneurons (Perouansky et al. 1995; Perouansky, Kirson, and Yaari 1996; Nishikawa and MacIver 2000a,b), primarily by presynaptic mechanisms (Berg-Johnsen and Langmoen 1992; MacIver, Mikulec et al. 1996; Kirson, Yaari, and Perouansky 1998; Hemmings 2009). Although enhanced inhibition has long been considered to be a critical component of anesthetic action, it was recognized only recently that GABA$_A$ receptor-mediated inhibition includes both tonic and phasic components (with the latter separable into GABA$_{A\text{-fast}}$ and GABA$_{A\text{-slow}}$). Again, the hippocampus served as the model structure where the differential modulation of these coexisting and complementary GABA$_A$-ergic inhibitory systems by anesthetics was discovered.

One curious and unresolved issue that may be relevant to the understanding of the mechanism of anesthetic impairment of learning and memory is an incongruence between the finding that hippocampal long-term potentiation (LTP; MacIver 2009) is not blocked by concentrations of inhalational anesthetics in vivo that caused profound unconsciousness exceeding the EC$_{50}$ for amnesia by a factor of four or more (Pearce, Stringer, and Lothman 1989). While the apparent contradiction might be explained by the limitations of the experimental paradigm (an unphysiologically intense stimulus), it is also conceivable that an important component of anesthetic action occurs either in structures upstream and downstream of the hippocampus or within the hippocampus, but affecting ongoing neuronal activity that normally forms the background for mnemonic function that is short-circuited by conventional LTP-inducing paradigms. Hence, in addition to inhibiting excitatory and enhancing inhibitory synapses within the hippocampus, anesthetics may affect exogenous or endogenous "drivers" that tune the hippocampus for processing and storing incoming information.

## Pharmacological Rationale for Our Choice of Anesthetics

While no inhaled anesthetic is specific for a receptor, some drugs do show preferential modulation of either GABA$_A$ or glutamate receptors. Among the drugs that we have chosen, nitrous oxide (N$_2$O) and halothane represent the two available extremes among established inhalational drugs. N$_2$O is not known to modulate GABA$_A$ receptors but does block the NMDA type of glutamate receptors

(Mennerick et al. 1998), while halothane potently enhances GABA$_A$ (Jones, Brooks, and Harrison 1992) but has, at clinically relevant concentrations, only minimal effects on glutamate receptor-mediated excitation (Perouansky et al. 1995). It should be remembered, however, that both agents have also common targets, e.g., the tandem-pore domain potassium channels (Franks and Lieb 1999), whose contribution to anesthetic-induced amnesia is unknown. Isoflurane has a dual effect on GABAergic inhibition. In the amnesic concentration range, it enhances GABA$_A$ receptors. At significantly higher concentrations, it blocks them (Banks and Pearce 1999). It also modestly blocks glutamate receptors at amnesic concentrations (MacIver 1997). We have chosen to also test a nonimmobilizer along with the anesthetics. Nonimmobilizers are a class of drugs (known as nonanesthetics until their amnesic properties were discovered) that disobeys the Meyer-Overton rule (linear correlation between lipid solubility and anesthetic potency), the cornerstone of anesthetic pharmacology. These agents were developed specifically for investigating mechanisms of anesthetic action (Koblin et al. 1994). We used 1,2-dichlorohexafluorocyclobutane (F6 or 2 N). We have chosen F6 because, besides being the best characterized member of its class (Perouansky 2008), F6 has useful properties: Its amnestic effect is both similar in nature (differential effect for hippocampus-dependent versus hippocampus-independent fear conditioning) and achieved at comparable concentrations (corrected for lipid solubility) as that of isoflurane (Dutton et al. 2001, 2002). In contrast to isoflurane, however, F6 has a very limited range of known molecular targets that may allow for mechanistic insights.

## The Hippocampus and Memory

The recently deceased patient known to the scientific world as "H.M." indelibly placed the hippocampus at the epicenter of the learning and memory universe. The recent surge of interest in intraoperative awareness (also termed "awareness under anesthesia") leading to postoperative recall of intraoperative events (i.e., episodic memory) illustrates how closely intertwined these manifestations of higher cognitive function are in the clinical setting (Myles et al. 2004; Avidan et al. 2008). It also underlines the importance of a mechanistic understanding of anesthetic-induced amnesia and the desirability of measurable indices of the instantaneous mnemonic potential of the brain under anesthesia.

The medial temporal lobe contains a system that is essential for declarative memory (conscious memory for facts and events; Squire, Stark, and Clark 2004). This kind of memory, which overlaps with the concepts of explicit, episodic, and associative memory (Bird and Burgess 2008; Veselis and Pryor 2009), is pertinent in the clinical context of anesthesia. Evidence from several species indicates that the hippocampal formation is a key structure within the MTL, essential for different forms of L&M (Squire, Stark, and Clark 2004), e.g., spatial learning (Morris et al. 1986; Morris and Frey 1997) and contextual learning (Kim and Fanselow

1992; Matus-Amat et al. 2004). Its elements are proposed to subserve the different computational processes required for models of declarative memory.

The hippocampus encompasses a number of subfields. Each has a distinct architecture and subserves a different function. The dentate gyrus, with its projection into the CA3, supports "pattern separation" – the ability to form distinct memories from related patterns (McHugh et al. 2007). The autoassociative network of CA3 is capable of "pattern completion" – the retrieval of full memories from partial clues (Nakazawa et al. 2002; Nakashiba et al. 2008). Inputs from dentate gyrus, CA3, and entorhinal cortex converge onto the CA1 region, which is essential for forming episodic memories (Buzsaki 2005).

## Hippocampal Rhythms

The hippocampus is a key structure for the investigation of brain rhythms. Fluctuations in extracellular field potentials arise from rhythmic changes in current flow patterns across neuronal membranes, which, when synchronized across large populations of neurons, give rise to "rhythms." The hippocampus has a characteristic rhythmicity profile that is thought to be critical to its cognitive functions. We will briefly review the most prominent rhythms observed in the hippocampal formation before discussing their modulation by pharmacologic and other means, and the consequences thereof.

## *Theta (θ) Oscillations (θ-Waves, θ-Rhythm)*

θ-Waves are rhythmic, nearly sinusoidal fluctuations of the extracellularly recorded field potentials at a frequency of approximately 7 Hz (ranging from 5 to 12 Hz, also called RSA for rhythmic slow activity). In rats, hippocampal rhythms can be correlated with the moment-to-moment behavior (Vanderwolf 1969). The behaviors during which θ-waves are most consistently observed are REM sleep (Jouvet 1969) and various types of locomotor (termed type 1) behaviors described subjectively as "voluntary," "preparatory,", "orienting," or "exploring" (Vanderwolf 1969). These findings indicate that extrahippocampal structures can tune the hippocampus for processing of information. Their presence is thought to indicate that the hippocampus is "online" (Buzsaki 2002).

While theta oscillations that are observed during active states have been termed "type 1 theta," an immobility-related θ-oscillation, termed "type 2 theta," can also be observed in its clearest form in immobile rabbits in response to sensory stimuli and in cats during "fixed staring." In rats, by contrast, type 2 theta is not easily elicited. It is recognized that type 2 theta displays a great variability between species with respect to experimental conditions under which it can be observed (Bland 1986). However, in all species tested, type 2 (immobility related) can be evoked by eserine [a cholinesterase blocker (Bland and Oddie 2001)] and largely (if not completely) blocked by atropine (Stewart and Fox 1989). Type 2 is also absent in phospholipase C knockout mice (Shin et al. 2005), pointing to an important muscarinic receptor-mediated component that is not required for type 1. Hence, type 2 theta is frequently referred to as "atropine-sensitive theta."

There are a number of interesting physiological and pharmacological differences between the two types of theta. Type 1 θ-oscillations are thought to reflect primarily the rhythmic excitation of the distal dendrites in the CA1 region and dentate gyrus produced by glutamatergic fibers that arise in the entorhinal cortex. By contrast, type 2 theta reflects primarily input from the CA3 region and the septum, which together drive oscillations in a network of local inhibitory interneurons. Type 1 theta is sensitive to urethane anesthesia; thus, θ-oscillations evoked in the urethane-anesthetized rat can be blocked by atropine. Many of the original studies of θ-oscillations were conducted under urethane, but its toxicity limits its use to experimental preparations. Halothane and other anesthetics have now also been shown to differentially impact type 1 and type 2 θ-oscillations (see below).

Field oscillations in frequency bands that overlap with theta are observed in other cortical structures (Steriade 2000). With the exception of other medial temporal lobe structures and the prefrontal cortex (Jensen 2005; Siapas, Lubenov, and Wilson 2005), these rhythms occur during different behavioral states and are not coherent with the hippocampal θ-rhythm (Kahana et al. 1999; Raghavachari et al. 2001).

## Gamma (γ) Oscillations

In the hippocampus and throughout the cortex, oscillations ranging from 40 to 100 Hz are observed primarily during sensory processing. These oscillations, which have been termed "40 Hz oscillations" or "γ-oscillations," can be synchronous over long distances. They are proposed to bind information that is distributed throughout the cortex into a unitary subjective experience referred to as a "percept" (Gray et al. 1989; Singer 1993). Although they can form large-scale ensembles, γ-oscillations themselves are generated locally by synchronous inhibitory currents originating from $GABA_A$-ergic interneurons.

Oscillations that fall into different frequency bands are not completely independent. Slower rhythms can modulate the amplitude of higher frequency oscillations, as when θ-phase modulates γ-power, a phenomenon termed "θ-γ nesting." In the hippocampus, nested θ-γ oscillations have been proposed to play a mnemonic role by organizing memories sequentially (Huerta and Lisman 1995), with θ-oscillations creating the "context" that binds information "content" (γ) together (Jensen and Lisman 2005).

## Sharp Waves and Ripples

Although they do not occur with the regularity of other "rhythms," periodic bouts of highly synchronous neuronal activity emanating from CA3 pyramidal cells are observed as brief (40–120 ms), large-amplitude potentials in the apical dendritic layer of the CA1 region. These events, known as "sharp waves," are now recognized to play an important role in learning and memory. They occur during nontheta behavioral states such as slow-wave sleep, awake immobility, and "consummatory" behaviors. Sharp waves reflect the summed postsynaptic depolarization of large numbers of pyramidal cells in the hippocampal CA1 region. Their amplitude, up to 2.5 mV in the stratum radiatum of CA1, exceeds even that of θ-waves (Buzsaki 1989).

Coincident with sharp waves in the dendritic layer, 180–220 Hz "ripples" are observed in the pyramidal cell layer. They reflect the summed activity of inhibitory interneurons that provide a barrage of high-frequency inputs to the somata of CA1 pyramidal cells (Ylinen et al. 1995). In accordance with the view that sharp wave-ripple complexes represent the transfer of stored information from the hippocampus to the neocortex, stimuli that induce LTP lead to the generation of sharp wave-ripple complexes in the in vitro hippocampal slice preparation (Behrens et al. 2005).

θ-Oscillations and sharp waves can be thought of as "companion processes," the former synchronizing the *input* pathways to the hippocampus, whereas the latter reflecting the organized *output* from the hippocampus back to cortical structures, to replay and transfer stored information into the cortex (Buzsaki 1989; Chrobak and Buzsaki 1996). Anesthetic modulation of sharp waves has, to the best of our knowledge, not been studied.

## θ-Rhythm and Memory

Evidence accumulated over the past 30 years supports the concept that θ-oscillations are a crucial substrate for hippocampal learning and memory. This body of work spans a number of species, experimental preparations, behavioral assays, and experimental interventions.

### *The Predictive Power of Theta*

The hippocampal θ-rhythm facilitates hippocampus-dependent mnemonic function in animals and humans (for reviews, see Raghavachari et al. 2001; Buzsaki 2002; Vertes 2005). Some of the earliest evidence for an important role of θ-oscillations in learning and memory came from rabbits undergoing classic eyeblink conditioning. Rabbits acquired the learning task faster if θ-rhythm dominated their $EEG_h$ just prior to the presentation of the paired stimuli (Berry and Thompson 1978; Seager et al. 2002). It should be noted that the hippocampus is not necessary for learning the most simple form of conditioning (delay), but it becomes increasingly important as the learning task becomes more complex (McEchron and Disterhoft 1999). The role of theta also changes during different phases of acquisition: High levels of theta facilitate learning especially during the early phases of conditioning (Griffin et al. 2004). Similarly, it has been found that in humans, synchronization of neuronal activity in the θ-band during the encoding phase of learning tasks predicts subsequent recall. This is the case whether oscillations occur in global, distributed networks (Sato and Yamaguchi 2007) or are more localized to the temporal lobe (Sederberg et al. 2003) and may reflect the modulatory effect of theta on the power of hippocampal γ-oscillations (Sederberg et al. 2007).

The findings from these "positive predictive" studies are complemented by a number of investigations addressing the effect of "negative" θ-modulation by various means on learning and memory. When conditioning is timed to occur during nontheta states, acquisition of the conditioned response is delayed (Seager et al. 2002; Griffin et al. 2004). Remarkably, experimental slowing of the θ-rhythm by as little as 1 Hz, either pharmacologically (injection of chlordiazepoxide into the supramammillary nucleus; Pan and McNaughton 1997) or physically (whole-body cooling; Whishaw and Vanderwolf 1971), causes comparable impairment of spatial learning and memory in rats. Selective pharmacological suppression of θ-power (without changes in θ-frequency) also impairs learning (Robbe et al. 2006), and impaired learning in the Morris water maze produced by pharmacologic block of θ-oscillations can be reversed by imposing an artificial θ-rhythm on the hippocampus (McNaughton, Ruan, and Woodnorth 2006). Together, these findings indicate that the θ-rhythm is indeed of functional importance, particularly for the early phases of learning.

## θ-*Rhythm: Cellular and Circuit Effects*

Many of the fundamental advances that led to our current understanding of the molecular basis of learning and memory came from studies of synaptic plasticity, a cellular model of memory that is amenable to study both in vitro and in vivo. The earliest observations of LTP of excitatory synapses employed high-frequency "tetanic" trains of stimuli (typically 100 Hz for 1 s). Subsequently, paradigms were developed that utilized patterned stimuli modeled on the θ-rhythm (such as the "theta burst" paradigm), with the result that plasticity can be produced more efficiently, using smaller numbers of stimuli, compared with tetanic stimulation (Larson, Wong, and Lynch 1986). This finding supported the idea of a special role for θ-oscillations in enabling synaptic plasticity. Even stronger evidence came from the observation that bursts of as few as two to four stimuli could produce LTP if they are delivered at a specific phase of an ongoing θ-oscillation; the same stimuli delivered at the opposite phase produce long-term depression (Huerta and Lisman 1995). These results, obtained during cholinergic drug-induced θ-like states in hippocampal slices in vitro, have been confirmed during naturally occurring θ-rhythm in vivo (Orr et al. 2001; McCartney et al. 2004).

How the θ-rhythm modulates and facilitates plasticity, learning, and memory has not been worked out in detail for all levels of signal integration. However, a number of working hypotheses have been proposed and corroborated to various degrees with experimental evidence. The rhythmic depolarizations of pyramidal cell dendrites occurring during θ-states provide a temporal metric of alternating increased and reduced excitability. One component of this cycle is a varying threshold for

NMDA receptor-mediated $Ca^{2+}$ influx in response to synaptically released gluta-mate. In the CA1 area, LTP is critically dependent on NMDA receptor function. It seems likely that this intermittent facilitation by depolarization contributes to the phase dependence of LTP induction (and by extrapolation of learning) during the ongoing θ-rhythm. Additionally, the finding that stimulation of Schaffer collater-als with brief bursts repeated at θ-frequency efficiently induces plasticity probably takes advantage of the intrinsic properties of CA1 pyramidal cells that favor reso-nance at θ-frequency (Leung and Yu 1998; Hu, Vervaeke, and Storm 2002) as well as the maximal suppression of dendritic inhibition at this frequency by a presynaptic mechanism (Pearce, Grunder, and Faucher 1995).

In addition to enabling plasticity, θ-rhythms can also synchronize activity in anatomically separate memory structures, e.g., between the prefrontal cortex and the hippocampus (Jones and Wilson 2005; Siapas, Lubenov, and Wilson 2005) or the hippocampus and the striatum (DeCoteau et al. 2007). In the hippocampus, θ-oscillations provide a clock for phase coding of information by objectively timing neuronal firing within the hippocampus itself (phase precession), as demonstrated by the backward shift of place-cell firing with respect to the phase of ongoing theta (spike timing) as a rat crosses the cell's place field on a linear track (O'Keefe and Recce 1993) or in two-dimensional space (Huxter et al. 2008) through place fields. Due to this tight correlation between navigation in space and memory function, it appears teleologically useful to integrate memory demands with sensorimotor sources of θ-modulation into one model. In more general terms, the interplay of θ- and γ-rhythms may serve to code and decode neuronal ensembles (Jensen and Lisman 2005).

In summary, a substantial body of literature supports an important role of this rhythm in learning and memory, in parallel to its important role in sensorimotor integration (Bland and Oddie 2001). Therefore, understanding the modulation of θ- and associated rhythms by anesthetic drugs may offer insight into network-level mechanisms of the amnestic component of anesthetic drug action.

## Theta – A Pathway to Anesthetic Amnesia?

### Background: $EEG_h$ and Anesthetics

A limited number of studies investigated the effects of anesthetic drugs on the $EEG_h$. Early reports on the effects of urethane, ether, and pentobarbital came from the lab-oratories in which θ-rhythms were investigated: Many experiments are done under anesthesia and most anesthetics affect the EEG, so these observations were con-ducted not necessarily from an "anesthetic mechanism" perspective. Nevertheless, the observation that runs of θ-waves could be completely blocked by muscarinic antagonists under anesthesia, but not in the awake, exploring animal, contributed to the separation of type 1 from type 2 theta. Hence, anesthetics supplied a pharma-cological criterion for discriminating between two types of θ-oscillations (Kramis, Vanderwolf, and Bland 1975; Leung 1984).

A more detailed investigation into the effect of the injectable anesthetic thiopental on EEG, which set out to correlate EEG changes with behavioral endpoints, was published by MacIver, Mandema et al. (1996). They noted that the cortical EEG activity gradually slowed after a 5-min infusion of an anesthetic dose of thiopental. The dominant frequency of the $EEG_h$, which was in the $\theta$-frequency range in the alert animal, slowed during the rapid transition from awake to comatose state (MacIver, Mandema et al. 1996). However, the experiments were not designed to analyze the EEG at intermediate levels, such as at concentrations that cause amnesia without loss of consciousness, but rather to compare fully awake versus deeply anesthetized. The authors noted that, during profound anesthetic-induced unconsciousness, thiopental caused a generalized suppression across all resolved frequency bands (30 Hz and below were measured) followed by burst suppression. This pattern was observed in both cortex and hippocampus (MacIver, Mandema et al. 1996). Later, the slowing of ongoing activity followed by burst-suppression pattern of thiopental's effects on the EEG was reproduced in neocortical slices in which $\theta$-like activity was pharmacologically generated (Lukatch, Doze, and MacIver 1996).

Recently, Bland et al. (2003) published an investigation into halothane's effect on hippocampal network activity (halothane is a halogenated alkane with strong $GABA_A$-enhancing properties). The "minimum alveolar concentration" (MAC) that prevents movement in response to a noxious stimulus (in rats, approximately 1% halothane, 1.45% isoflurane, and 160–240% $N_2O$) delineates the lower margin of the so-called "surgical" concentration range and is the accepted index for anesthetic potency. Bland and others found that 0.5% halothane induced $\theta$-activity in the absence of exogenous stimulation. Increasing its concentration from 0.5 to 2% slowed $\theta$-peak frequency from 6.5 to 4.0 Hz. Compared to movement-related $\theta$- (i.e., type 1) recorded in the absence of halothane, $\theta$-amplitude was increased. Atropine completely abolished $\theta$-oscillations under halothane, identifying them as type 2. The authors concluded that, as halothane is known to enhance $GABA_A$ receptor-mediated currents, type 2 theta was generated by $GABA_A$-ergic mechanisms in the presence of a cholinergic tone.

Anesthetics also affect oscillations that occur at higher frequencies than theta. The discovery that synchronized activity in the $\gamma$-range could be pharmacologically induced in the hippocampal slice led to experiments on the effects of general anesthetics on these oscillations in vitro. The finding that general anesthetics disrupted these oscillations was attributed to the drug-induced slowing of $GABA_A$ receptor-mediated synaptic currents that synchronize oscillating interneuronal networks (Whittington, Jefferys, and Traub 1996; Faulkner, Traub, and Whittington 1998, 1999). The nature of the experiments precluded a direct correlation of electrophysiological observations with behavioral endpoints, but they did offer a molecular explanation for observations of anesthetic effects on $\gamma$-oscillations that have been observed in vivo.

In summary, many studies have demonstrated that synchronized neuronal activity that gives rise to brain rhythms is modulated by anesthetics. It is not known, however, whether the observed changes are causally related to the behavioral effects of general anesthetics. Specifically, with respect to θ-rhythm, the existing literature does not inform us on the degree to which anesthetics modulate the θ-rhythm at concentrations that selectively impair hippocampal learning and memory. Moreover, no attempts have been made to systematically compare anesthetics with different receptor-level activity profiles, at behaviorally comparable concentrations, in order to examine whether they share overlapping mechanisms at the network level.

## Hypothesis: Anesthetic Modulation of the Hippocampal θ-Rhythm Contributes to Anesthetic-Induced Amnesia

Based on the evidence outlined above indicating that θ-oscillations are important for hippocampus-dependent learning and memory, we hypothesized that anesthetic-induced changes of the θ-rhythm contribute to their impairment of learning and memory. Thus, we have begun to test whether these two variables (theta and amnesia) can be quantitatively interrelated (are correlated).

The most complete set of studies that we have completed to date, and which we summarize below, was conducted using young adult Sprague-Dawley rats. We studied the effects of a range of inhaled anesthetics that have different receptor-level activity profiles, targeting the anesthetic concentrations to bracket the $EC_{50}$ for amnesia, as determined in behavioral experiments conducted by our collaborators at UCSF (Drs. Rao and Eger) or taken from the literature (Dutton et al. 2001, 2002). These included the "promiscuous" agent isoflurane; halothane, which preferentially enhances $GABA_A$ receptors; $N_2O$, which has no effect on $GABA_A$ receptors; and the nonimmobilizer F6 (Perouansky, Perkins, and Pearce 2007; Perouansky, Zurn, et al. 2007). As the occurrence and characteristics of theta are strongly influenced by behavior, we analyzed the $EEG_h$ separately for behaviors classified as "exploring" (i.e., type 1 theta) and "immobile"(i.e., predominantly type 2 theta).

## Type 1 Theta: Anesthetics, but not F6, Slow θ-Rhythm

As described above, θ-oscillations during exploration are likely to be important for the mnemonic function of the hippocampus, as the animal "maps" the space available to it. Therefore, if (anesthetic) dugs converge on theta, measurable effects on some characteristic of θ-rhythms would be expected.

We tested isoflurane, halothane, $N_2O$, and F6 at concentrations that are comparable with respect to their amnestic effect, as measured by the ability to inhibit fear conditioning to context, a hippocampus-dependent learning task. The $EC_{50}$ of isoflurane for this purpose, determined using exactly the same paradigm as for F6, is 0.37% (Dutton et al. 2001). For halothane and $N_2O$, data obtained under directly comparable conditions have not been published, but preliminary results were made available by Drs. Eger and Rau. Hence, we used 0.25% for halothane (the equivalent, in terms of MAC fraction, of 0.37% isoflurane) and 60% for N₂O. The MAC of N₂O in rats has been reported to range from 150 to 235%, depending on strain and experimental design (Gonsowski and Eger 1994). Similarly to the previously reported results for isoflurane, halothane slowed θ-oscillations (Fig. 10.1A)

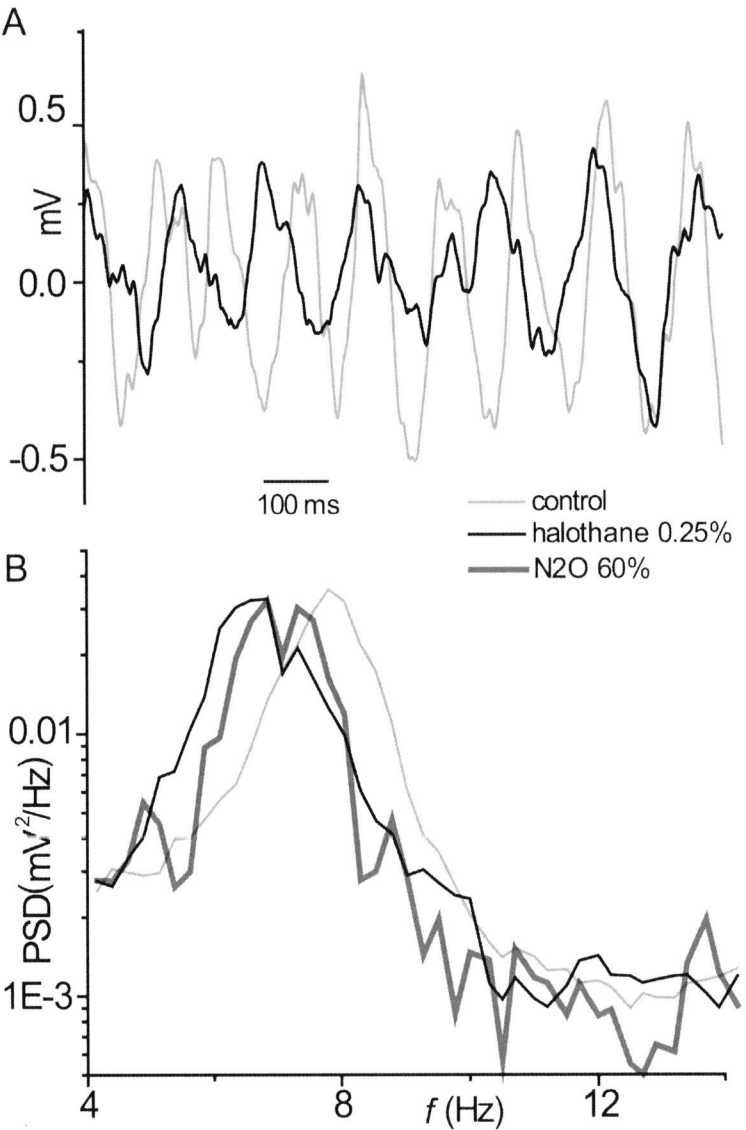

**Fig. 10.1** Halothane and N₂O slow θ-peak frequency during exploration. **A** θ-Oscillations are slowed by 0.25% halothane. Note that the trace in halothane (*black*) accrues an increasing delay compared to the drug-free condition (*gray*). **B** The spectrograms of EEG$_h$ activity under control conditions, 0.25% halothane and 60% N₂O (obtained from the same animal), show a comparable slowing of the θ-rhythm by the agents at concentrations that are amnestic, but not hypnotic

as did $N_2O$. Figure 10.1B illustrates the effects of both drugs obtained from the same animal. The spectra illustrate that both halothane and $N_2O$ slowed the peak of θ-oscillations without significantly affecting the peak amplitude. On average, both 0.25% halothane and 60% $N_2O$ significantly slowed θ-peak frequency by approximately 11% or 0.7 Hz ($p < 0.05$ compared to control). As previously reported (Perouansky, Perkins, and Pearce 2007; Perouansky, Zurn, et al. 2007), a comparable concentration of isoflurane (0.32%) slowed θ-peak by 14%. These findings indicate that the amnestic effect of the anesthetics cannot be attributed to decreased power of θ-oscillations, but to a slowing of the rhythm. Notably, at roughly comparable concentrations, the agents had a quantitatively comparable effect on θ-frequency.

F6 also had a marked and distinctive effect on hippocampal field potentials during exploration. As illustrated in the spectra and unlike the anesthetics we tested, F6 suppressed $EEG_h$ power in the θ-band but without slowing the θ-rhythm (Fig. 10.2A, dotted lines). At a concentration close to the $EC_{50}$ for inhibiting fear conditioning to context ($2 \pm 0.1\%$, a hippocampus-dependent learning task; Dutton et al. 2002), the principal effect of F6 on the $EEG_h$ was a suppression of the θ-peak by 40%. Power in the whole θ-band (4–12 Hz) was also reduced. Remarkably, the frequency of θ-peak was unchanged ($7.6 \pm 0.2$ Hz versus $7.4 \pm 0.4$ Hz for F6 and placebo, respectively). These findings are consistent with the hypothesis that F6 exerts its amnestic effect also by modulating the θ-rhythm but, unlike the anesthetics, by suppressing θ-power.

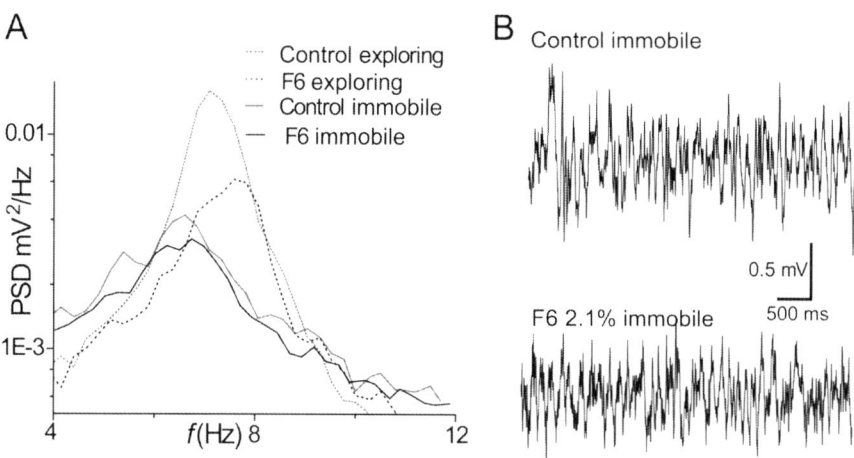

**Fig. 10.2** F6 suppresses θ-power during exploration, but not during immobility. **A** The $EEG_h$ during awake immobility looks indistinguishable under control conditions and in the presence of amnestic concentrations of F6. **B** The spectrograms show that F6 caused no significant difference either in power or at the peak frequency of θ-oscillations during immobility. By contrast, F6 suppressed the power of the θ-rhythm without slowing its frequency during exploration. Note the similarity and difference with the effect of $N_2O$ during immobility and exploring, respectively, illustrated in Fig. 10.3

## Type 2 Theta

Activity in the EEG$_h$ during immobility includes both large-amplitude irregular activity (LIA; Leung 1980) and the type 2 theta of "alert immobility" that has been linked to sensory processing either preceding movement (Bland et al. 2006) or in response to patterned stimuli (Shin et al. 2005). Sharp waves can be occasionally observed as well, but were not targeted by our analysis. While we were primarily interested in type 1 theta, we also analyzed the effects of these four agents on type 2 theta, with the caveat that under our experimental conditions and mode of time-averaged analysis we obtain a mix of EEG$_h$ patterns present during awake immobility that we termed θ-immobile.

Neither F6 (straight lines in 2A) nor N$_2$O (Fig. 10.3) had any discernible effect on either θ-power or θ-frequency of θ-immobile, i.e., on those parameters of type 1 theta that these agents altered. Hence, F6 and N$_2$O differed in this regard from the potent halogenated anesthetics. Figure 10.3 illustrates the effects of the three anesthetics at roughly comparable amnesic concentrations. Halothane induced the most characteristic effect on EEG$_h$ (Fig. 10.3A). The EEG under 0.25% halothane (red trace, top right) displayed a regular rhythmic activity that is absent from the traces recorded under control conditions (black trace, top left) and also differed from the activity under isoflurane (purple, bottom right) as well as N$_2$O (light blue, bottom left). The visual impression from the raw EEG$_h$ traces is confirmed by the averaged data shown in the power spectra (Fig. 10.3B).

We confirmed Bland's results with respect to halothane's effect on theta during immobility at high (unconsciousness inducing) concentrations (Bland et al. 2003), and we extended his findings into the lower, amnestic range: At 0.25% (a concentration that has amnestic effects but does not cause unconsciousness), halothane induced a highly regular, continuous θ-rhythm with a peak at 6 Hz, i.e., significantly slower than in drug-free conditions. In addition, we found that the halogenated anesthetic isoflurane differed from halothane in its effect on EEG$_h$ (Fig. 10.3: under amnestic isoflurane, we observed occasional large-amplitude slow complexes coupled with a generalized suppression of power in the EEG$_h$ above 5 Hz, but no ongoing rhythmicity). N$_2$O and F6, agents with receptor-level activities that differ from the halogenated anesthetics, had no effect on θ-immobile.

These results are compatible with Bland's suggestion that halothane's distinctive effect on EEG$_h$ during immobility is due to its pronounced GABA$_A$-ergic activity, as isoflurane has a more distributed receptor-level activity and N$_2$O does not enhance GABA$_A$ but blocks NMDA receptors (Mennerick et al. 1998). F6 has no known effect on either receptor type, but it blocks both nicotinic and muscarinic receptors, the latter effect being compatible with selective suppression of θ-power only in the presence of a cholinergic "tone" in the hippocampus, as during exploration. In summary, the effects on type 1 theta of all four drugs are compatible with the hypothesis that they interfere with learning and memory via modulation of the hippocampal θ-rhythm. The effects on both θ-frequency and θ-power observed under amnestic but not hypnotic concentrations are qualitatively and quantitatively sufficient (Whishaw and Vanderwolf 1971; Pan and McNaughton 1997; Robbe et al. 2006)

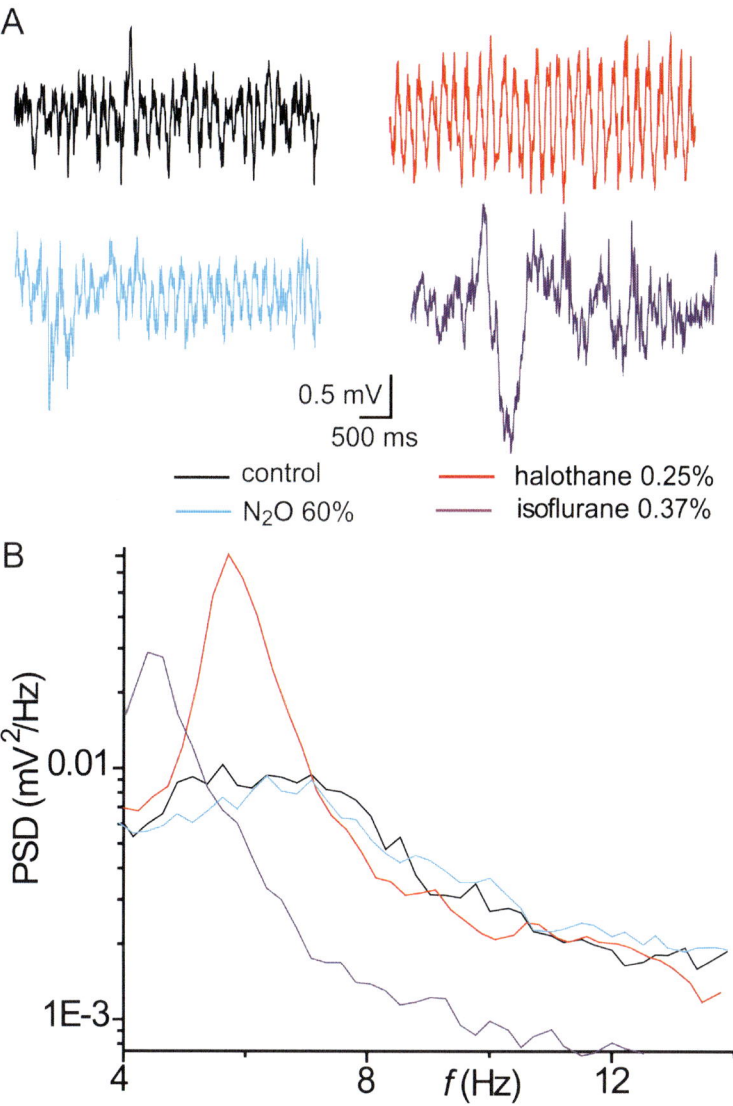

**Fig. 10.3** Anesthetic effects on θ-immobile. During awake immobility, the dominant pattern recorded from the hippocampus is LIA (*black trace*). This pattern was sculpted into regular θ-oscillations by 0.25% halothane (*red trace*) but remained largely unchanged under 60% $N_2O$ or 0.37% isoflurane (*light blue* and *purple traces*, respectively). The spectrograms, averaged over the time behaviorally rated as "immobile" during 15 min of anesthetic drug exposure quantify the effects illustrated in the traces. Note that $N_2O$ exerted no visible change in the power spectrum while isoflurane increased power in the low frequency band, possibly due to the cumulative effect of large slow events (sharp waves?) as captured in the trace shown in **A**

to cause the degree of learning impairment that has been observed in behavioral experiments.

We cannot assign a behavioral relevance to the effect or lack thereof on theta type 2 ($\theta$-immobile) that we observed with the different agents. However, the differences between the modulation of the two types of theta by individual drugs, especially $N_2O$ and F6, support and extend the notion of pharmacological differences in the generation and expression of type 1 and type 2 $\theta$-oscillations.

## Hippocampal $\gamma$-Oscillations: A Network Correlate of Sedation?

Although the primary purpose of the experiments described above was to test whether there is a correlation between modulation of memory and $\theta$-oscillations by anesthetics, we also examined the effect of all the agents on $\gamma$-oscillations (30–90 Hz; Perouansky, Perkins, and Pearce 2007; Perouansky, Zurn, et al. 2007). In contrast to $\theta$-oscillations, F6 did not suppress $\gamma$-power, even at the highest concentrations tested (3.85% or 0.85 MAC predicted). Behaviorally, this corresponded to "vigilance" and active exploration (above baseline) at concentrations of F6 that are reliably amnestic. By contrast, all anesthetics caused some reduction in exploratory activity (sedation) that was accompanied by suppression of the higher frequency component of $\gamma$-power. Only 60% $N_2O$ reliably suppressed $\gamma$-oscillation across the whole $\gamma$-band (from 30 to 90 Hz, Fig. 10.4). Halothane and isoflurane suppressed only the higher frequency component of $\gamma$-oscillations (50–90 Hz) at concentrations of 0.25 MAC and higher, but had inconsistent effects on the lower frequency

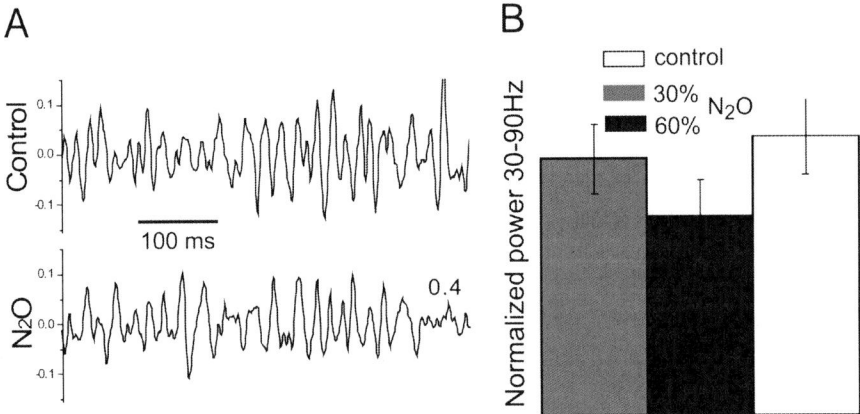

**Fig. 10.4** $N_2O$ suppressed power in the $\gamma$-band. **A** $\gamma$-Waves appear degraded in the presence of 60% $N_2O$ (original traces bandpass filtered at 20–100 Hz). **B** $N_2O$ (60%) significantly reduced power in the $\gamma$-frequency band (by 26–38% for 30–50 and 50–90 Hz, respectively, $p < 0.01$ paired $t$-test). F6 had no effect on $\gamma$-power at any concentration tested. At 0.5 MAC, isoflurane and halothane depressed only the higher frequency component of the $\gamma$-spectrum (not shown)

component (30–50 Hz). These findings indicate a possible correlation between the sedative/amnestic properties and suppression of γ-oscillations in the hippocampus. F6, with its sedation-free amnestic action and no effect on γ-power, seems to differ substantially. For details on anesthetic effects on sensory-evoked γ-oscillations, see Hudetz (2009).

## Summary

We have presented evidence that three anesthetics with different molecular activity profiles slow hippocampal θ-oscillations to comparable degrees at concentrations that impair hippocampus-dependent learning and memory. This correlation is consistent with the hypothesis that anesthetic modulation of θ-oscillations contributes to anesthetic-induced amnesia. Even if there is not a causal relationship (as the hypothesis implies), these results indicate that slowing of θ-oscillations might serve as a signature effect for general anesthetic-induced prevention of declarative memory formation.

The clinical importance of preventing the recollection of distinct episodes (the what, when, and where of personal experience) when a patient is assumed to be under general anesthesia is apparent. The importance of the hippocampus and associated structures of the medial temporal lobe for the processing (generation or consolidation) of such memory traces also appears to be well established (Eichenbaum 2004). Evidence obtained with more specific interventions supports the conjecture that a change of θ-oscillations can affect the ability of the brain to learn and/or remember. In the future, it will be useful to assess more quantitatively whether the observed degree of θ-slowing produced by anesthetics of different classes and compared to other interventions produce quantitatively comparable degrees of disruption of learning and memory.

Our observations with respect to γ-activity are admittedly limited, as the analytic approach we used disregarded the localized, highly dynamic, and short-lived nature of fluctuations in γ-activity. Nevertheless, the observation that F6 was devoid of any effect on γ-power, with the animal maintained in a highly alert but profoundly amnestic state, is remarkable if viewed through the prism of anesthetic mechanisms: What molecular targets and neuronal circuits account for the differential behavioral effects of the nonimmobilizing cyclobutane F6 (1,2-dichlorohexafluorocyclobutane, amnesia, and seizures) versus anesthetics? How can a highly alert but profoundly amnesic state be generated when the majority of amnesia-inducing drugs cause varying degrees of sedation? Is it possible that F6 targets "memory-selective" receptors/pathways and allows a dissociation of learning and memory versus the neural correlates of consciousness?

**Acknowledgements**  Supported by National Institutes of Health (Bethesda, MD, USA) GM47818 (to M. Perouansky and R. Pearce) and NS056411 (to R. Pearce) and the Department of Anesthesiology, University of Wisconsin, Madison, WI, USA

# References

Avidan, M. S., L. Zhang, B. A. Burnside, K. J. Finkel, A. C. Searleman, J. A. Selvidge, L. Saager, M. S. Turner, S. Rao, M. Bottros, C. Hantler, E. Jacobsohn, and A. S. Evers. 2008. Anesthesia awareness and the bispectral index. *N Engl J Med* 358(11):1097–1108.

Banks, M. I., and R. A. Pearce. 1999. Dual actions of volatile anesthetics on GABA(A) IPSCs: dissociation of blocking and prolonging effects. *Anesthesiology* 90(1):120–134.

Behrens, C. J., L. P. van den Boom, L. de Hoz, A. Friedman, and U. Heinemann. 2005. Induction of sharp wave-ripple complexes in vitro and reorganization of hippocampal networks. *Nat Neurosci* 8(11):1560–1567.

Berg-Johnsen, J., and I. A. Langmoen. 1986. The effect of isoflurane on unmyelinated and myelinated fibres in the rat brain. *Acta Physiol Scand* 127(1):87–93.

Berg-Johnsen, J., and I. A. Langmoen. 1992. The effect of isoflurane on excitatory synaptic transmission in the rat hippocampus. *Acta Anaesthesiol Scand* 36(4):350–355.

Berry, S. D., and R. F. Thompson. 1978. Prediction of learning rate from the hippocampal electroencephalogram. *Science* 200(4347):1298–1300.

Bird, C. M., and N. Burgess. 2008. The hippocampus and memory: insights from spatial processing. *Nat Rev Neurosci* 9(3):182–194.

Bland, B. H. 1986. The physiology and pharmacology of hippocampal formation theta rhythms. *Prog Neurobiol* 26(1):1–54.

Bland, B. H., C. E. Bland, L. V. Colom, S. H. Roth, S. DeClerk, A. Dypvik, J. Bird, and A. Deliyannides. 2003. Effect of halothane on type 2 immobility-related hippocampal theta field activity and theta-on/theta-off cell discharges. *Hippocampus* 13(1):38–47.

Bland, B. H., J. Jackson, D. Derrie-Gillespie, T. Azad, A. Rickhi, and J. Abriam. 2006. Amplitude, frequency, and phase analysis of hippocampal theta during sensorimotor processing in a jump avoidance task. *Hippocampus* 16(8):673–681.

Bland, B. H., and S. D. Oddie. 2001. Theta band oscillation and synchrony in the hippocampal formation and associated structures: the case for its role in sensorimotor integration. *Behav Brain Res* 127(1–2):119–136.

Buzsaki, G. 1989. Two-stage model of memory trace formation: a role for "noisy" brain states. *Neuroscience* 31(3):551–570.

Buzsaki, G. 2002. Theta oscillations in the hippocampus. *Neuron* 33(3):325–340.

Buzsaki, G. 2005. Theta rhythm of navigation: link between path integration and landmark navigation, episodic and semantic memory. *Hippocampus* 15(7):827–840.

Buzsaki, G. 2006. *Rhythms of the Brain*. 1 ed. New York: Oxford University Press.

Caraiscos, V. B., J. G. Newell, K. E. You-Ten, E. M. Elliott, T. W. Rosahl, K. A. Wafford, J. F. MacDonald, and B. A. Orser. 2004. Selective enhancement of tonic GABAergic inhibition in murine hippocampal neurons by low concentrations of the volatile anesthetic isoflurane. *J Neurosci* 24(39):8454–8458.

Chrobak, J. J., and G. Buzsaki. 1996. High-frequency oscillations in the output networks of the hippocampal-entorhinal axis of the freely behaving rat. *J Neurosci* 16(9):3056–3066.

DeCoteau, W. E., C. Thorn, D. J. Gibson, R. Courtemanche, P. Mitra, Y. Kubota, and A. M. Graybiel. 2007. Learning-related coordination of striatal and hippocampal theta rhythms during acquisition of a procedural maze task. *Proc Natl Acad Sci USA* 104(13): 5644–5649.

Dutton, R. C., A. J. Maurer, J. M. Sonner, M. S. Fanselow, M. J. Laster, and E. I. Eger, II. 2001. The concentration of isoflurane required to suppress learning depends on the type of learning. *Anesthesiology* 94(3):514–519.

Dutton, R. C., A. J. Maurer, J. M. Sonner, M. S. Fanselow, M. J. Laster, and E. I. Eger, II. 2002. Short-term memory resists the depressant effect of the nonimmobilizer 1-2-dichlorohexafluorocyclobutane (2 N) more than long-term memory. *Anesth Analg* 94(3): 631–639; table of contents.

Edelman, G. 2001. Consciousness: the remembered present. *Ann N Y Acad Sci* 929:111–122.

Eichenbaum, H. 2004. Hippocampus: cognitive processes and neural representations that underlie declarative memory. *Neuron* 44(1):109–120.

Faulkner, H. J., R. D. Traub, and M. A. Whittington. 1998. Disruption of synchronous gamma oscillations in the rat hippocampal slice: a common mechanism of anaesthetic drug action. *Br J Pharmacol* 125(3):483–492.

Faulkner, H. J., R. D. Traub, and M. A. Whittington. 1999. Anaesthetic/amnesic agents disrupt beta frequency oscillations associated with potentiation of excitatory synaptic potentials in the rat hippocampal slice. *Br J Pharmacol* 128(8):1813–1825.

Franks, N. P., and W. R. Lieb. 1999. Background K+ channels: an important target for volatile anesthetics? *Nat Neurosci* 2(5):395–396.

Gonsowski, C. T., B. S. Chortkoff, E. I. Eger, II, H. L. Bennett, and R. B. Weiskopf. 1995. Subanesthetic concentrations of desflurane and isoflurane suppress explicit and implicit learning. *Anesth Analg* 80(3):568–572.

Gonsowski, C. T., and E. I. Eger, II. 1994. Nitrous oxide minimum alveolar anesthetic concentration in rats is greater than previously reported. *Anesth Analg* 79(4):710–712.

Gray, C. M., P. Konig, A. K. Engel, and W. Singer. 1989. Oscillatory responses in cat visual cortex exhibit inter-columnar synchronization which reflects global stimulus properties. *Nature* 338(6213):334–337.

Griffin, A. L., Y. Asaka, R. D. Darling, and S. D. Berry. 2004. Theta-contingent trial presentation accelerates learning rate and enhances hippocampal plasticity during trace eyeblink conditioning. *Behav Neurosci* 118(2):403–411.

Hemmings, H. C. 2009. Molecular targets of general anesthetics in the nervous system. In *Suppressing the mind: anesthetic modulation of memory and consciousness*, edited by A. G. Hudetz and R. A. Pearce. New York City: Springer.

Hu, H., K. Vervaeke, and J. F. Storm. 2002. Two forms of electrical resonance at theta frequencies, generated by M-current, h-current and persistent Na+ current in rat hippocampal pyramidal cells. *J Physiol* 545(Pt 3):783–805.

Hudetz, A. G. 2009. Cortical disintegration mechanism of anesthetic-induced unconsciousness. In *Suppressing the mind: anesthetic modulation of memory and consciousness*, edited by A. G. Hudetz and R. A. Pearce. New York City: Springer.

Huerta, P. T., and J. E. Lisman. 1995. Bidirectional synaptic plasticity induced by a single burst during cholinergic theta oscillation in CA1 *in vitro*. *Neuron* 15 (5):1053-63.

Huxter, J. R., T. J. Senior, K. Allen, and J. Csicsvari. 2008. Theta phase-specific codes for two-dimensional position, trajectory and heading in the hippocampus. *Nat Neurosci* 11(5): 587–594.

Jensen, O. 2005. Reading the hippocampal code by theta phase-locking. *Trends Cogn Sci* 9(12):551–553.

Jensen, O., and J. E. Lisman. 2005. Hippocampal sequence-encoding driven by a cortical multi-item working memory buffer. *Trends Neurosci* 28(2):67–72.

Jones, M. V., P. A. Brooks, and N. L. Harrison. 1992. Enhancement of gamma-aminobutyric acid-activated Cl-currents in cultured rat hippocampal neurones by three volatile anaesthetics. *J Physiol* 449:279–293.

Jones, M. W., and M. A. Wilson. 2005. Theta rhythms coordinate hippocampal-prefrontal interactions in a spatial memory task. *PLoS Biol* 3(12):e402.

Jouvet, M. 1969. Biogenic amines and the states of sleep. *Science* 163(862):32–41.

Kahana, M. J., R. Sekuler, J. B. Caplan, M. Kirschen, and J. R. Madsen. 1999. Human theta oscillations exhibit task dependence during virtual maze navigation. *Nature* 399(6738): 781–784.

Kim, J. J., and M. S. Fanselow. 1992. Modality-specific retrograde amnesia of fear. *Science* 256(5057):675–677.

Kirson, E. D., Y. Yaari, and M. Perouansky. 1998. Presynaptic and postsynaptic actions of halothane at glutamatergic synapses in the mouse hippocampus. *Br J Pharmacol* 124(8): 1607–1614.

Koblin, D. D., B. S. Chortkoff, M. J. Laster, E. I. Eger, II, M. J. Halsey, and P. Ionescu. 1994. Polyhalogenated and perfluorinated compounds that disobey the Meyer-Overton hypothesis. *Anesth Analg* 79(6):1043–1048.

Kramis, R., C. H. Vanderwolf, and B. H. Bland. 1975. Two types of hippocampal rhythmical slow activity in both the rabbit and the rat: relations to behavior and effects of atropine, diethyl ether, urethane, and pentobarbital. *Exp Neurol* 49(1 Pt 1):58–85.

Larson, J., D. Wong, and G. Lynch. 1986. Patterned stimulation at the theta frequency is optimal for the induction of hippocampal long-term potentiation. *Brain Res* 368(2):347–350.

Leung, L. S. 1980. Behavior-dependent evoked potentials in the hippocampal CA1 region of the rat. I. Correlation with behavior and EEG. *Brain Res* 198(1):95–117.

Leung, L. S. 1984. Theta rhythm during REM sleep and waking: correlations between power, phase and frequency. *Electroencephalogr Clin Neurophysiol* 58(6):553–564.

Leung, L. S., and H. W. Yu. 1998. Theta-frequency resonance in hippocampal CA1 neurons *in vitro* demonstrated by sinusoidal current injection. *J Neurophysiol* 79(3):1592–1596.

Lukatch, H. S., V. A. Doze, and M. B. MacIver. 1996. Halothane prolongs GABA(A) fast and slow inhibitory currents. *Anesthesiology* 85:A673.

Lukatch, H. S., and M. B. MacIver. 1997. Voltage-clamp analysis of halothane effects on GABA(A fast) and GABA(A slow) inhibitory currents. *Brain Res* 765(1):108–112.

MacIver, M. B. 1997. General anesthetic actions on transmission at glutamate and GABA synapses. In *Anesthesia: biologic foundations*, edited by T. L. e. a. Yaksh. Philadelphia: Lippincott-Raven.

MacIver, M. B. 2009. Loss of recall and the hippocampal circuit effects produced by anesthetics. In *Suppressing the mind: anesthetic modulation of memory and consciousness*, edited by A. G. Hudetz and R. A. Pearce. New York City: Springer.

MacIver, M. B., J. W. Mandema, D. R. Stanski, and B. H. Bland. 1996. Thiopental uncouples hippocampal and cortical synchronized electroencephalographic activity. *Anesthesiology* 84(6):1411–1424.

MacIver, M. B., A. A. Mikulec, S. M. Amagasu, and F. A. Monroe. 1996. Volatile anesthetics depress glutamate transmission via presynaptic actions. *Anesthesiology* 85(4):823–834.

Matus-Amat, P., E. A. Higgins, R. M. Barrientos, and J. W. Rudy. 2004. The role of the dorsal hippocampus in the acquisition and retrieval of context memory representations. *J Neurosci* 24(10):2431–2439.

McCartney, H., A. D. Johnson, Z. M. Weil, and B. Givens. 2004. Theta reset produces optimal conditions for long-term potentiation. *Hippocampus* 14(6):684–687.

McEchron, M. D., and J. F. Disterhoft. 1999. Hippocampal encoding of non-spatial trace conditioning. *Hippocampus* 9(4):385–396.

McHugh, T. J., M. W. Jones, J. J. Quinn, N. Balthasar, R. Coppari, J. K. Elmquist, B. B. Lowell, M. S. Fanselow, M. A. Wilson, and S. Tonegawa. 2007. Dentate gyrus NMDA receptors mediate rapid pattern separation in the hippocampal network. *Science* 317(5834):94–99.

McNaughton, N., M. Ruan, and M. A. Woodnorth. 2006. Restoring theta-like rhythmicity in rats restores initial learning in the Morris water maze. *Hippocampus* 16(12):1102–1110.

Mennerick, S., V. Jevtovic-Todorovic, S. M. Todorovic, W. Shen, J. W. Olney, and C. F. Zorumski. 1998. Effect of nitrous oxide on excitatory and inhibitory synaptic transmission in hippocampal cultures. *J Neurosci* 18(23):9716–9726.

Mikulec, A. A., S. Pittson, S. M. Amagasu, F. A. Monroe, and M. B. MacIver. 1998. Halothane depresses action potential conduction in hippocampal axons. *Brain Res* 796(1–2): 231–238.

Morris, R. G., E. Anderson, G. S. Lynch, and M. Baudry. 1986. Selective impairment of learning and blockade of long-term potentiation by an N-methyl-D-aspartate receptor antagonist, AP5. *Nature* 319(6056):774–776.

Morris, R. G., and U. Frey. 1997. Hippocampal synaptic plasticity: role in spatial learning or the automatic recording of attended experience? *Philos Trans R Soc Lond B Biol Sci* 352(1360):1489–1503.

Myles, P. S., K. Leslie, J. McNeil, A. Forbes, and M. T. Chan. 2004. Bispectral index monitoring to prevent awareness during anaesthesia: the B-Aware randomised controlled trial. *Lancet* 363(9423):1757–1763.

Nakashiba, T., J. Z. Young, T. J. McHugh, D. L. Buhl, and S. Tonegawa. 2008. Transgenic inhibition of synaptic transmission reveals role of CA3 output in hippocampal learning. *Science* 319(5867):1260–1264.

Nakazawa, K., M. C. Quirk, R. A. Chitwood, M. Watanabe, M. F. Yeckel, L. D. Sun, A. Kato, C. A. Carr, D. Johnston, M. A. Wilson, and S. Tonegawa. 2002. Requirement for hippocampal CA3 NMDA receptors in associative memory recall. *Science* 297(5579):211–218.

Nishikawa, K., and M. B. MacIver. 2000a. Excitatory synaptic transmission mediated by NMDA receptors is more sensitive to isoflurane than are non-NMDA receptor-mediated responses. *Anesthesiology* 92(1):228–236.

Nishikawa, K., and M. B. MacIver. 2000b. Membrane and synaptic actions of halothane on rat hippocampal pyramidal neurons and inhibitory interneurons. *J Neurosci* 20(16):5915–5923.

Nishikawa, K., and M. B. MacIver. 2001. Agent-selective effects of volatile anesthetics on GABAA receptor-mediated synaptic inhibition in hippocampal interneurons. *Anesthesiology* 94(2): 340–347.

O'Keefe, J., and M. L. Recce. 1993. Phase relationship between hippocampal place units and the EEG theta rhythm. *Hippocampus* 3(3):317–330.

Orr, G., G. Rao, F. P. Houston, B. L. McNaughton, and C. A. Barnes. 2001. Hippocampal synaptic plasticity is modulated by theta rhythm in the fascia dentata of adult and aged freely behaving rats. *Hippocampus* 11(6):647–654.

Pan, W. X., and N. McNaughton. 1997. The medial supramammillary nucleus, spatial learning and the frequency of hippocampal theta activity. *Brain Res* 764(1–2):101–108.

Pearce, R. A., S. D. Grunder, and L. D. Faucher. 1995. Different mechanisms for use-dependent depression of two GABAA-mediated IPSCs in rat hippocampus. *J Physiol* 484(Pt 2): 425–435.

Pearce, R. A., J. L. Stringer, and E. W. Lothman. 1989. Effect of volatile anesthetics on synaptic transmission in the rat hippocampus. *Anesthesiology* 71(4):591–598.

Perouansky, M. 2008. Modern Anesthetics. *Handb Exp Pharmacol* (182):209–223.

Perouansky, M., D. Baranov, M. Salman, and Y. Yaari. 1995. Effects of halothane on glutamate receptor-mediated excitatory postsynaptic currents. A patch-clamp study in adult mouse hippocampal slices. *Anesthesiology* 83(1):109–119.

Perouansky, M., H. Hentschke, M. Perkins, and R. A. Pearce. 2007. Amnesic concentrations of the nonimmobilizer 1,2-dichlorohexafluorocyclobutane (F6, 2 N) and isoflurane alter hippocampal theta oscillations *in vivo*. *Anesthesiology* 106(6):1168–1176.

Perouansky, M., E. D. Kirson, and Y. Yaari. 1996. Halothane blocks synaptic excitation of inhibitory interneurons. *Anesthesiology* 85(6):1431–1438; discussion 29A.

Perouansky, M., Perkins, M. G., Ford, T. M., Pearce, R. A. 2007. A comparison of the effects of halothane and nitrous oxide on hippocampal network activity. *Soc Neurosci Annual Meeting Abstract* # 937.1/AAA8.

Perouansky, M., Perkins, M. G., Pearce, R. A. 2007. Isoflurane and F6 differentially affect hippocampal γ-oscillations. *Am Soc Anesth Annual Meeting Abstract* # A1217.

Perouansky, M., Zurn, J., Perkins, M. G., Pearce, R. A. 2007. Nitrous oxide alters hippocampal theta and gamma oscillations in rats. *Am Soc Anesth Annual Meeting Abstract* # A1217.

Raghavachari, S., M. J. Kahana, D. S. Rizzuto, J. B. Caplan, M. P. Kirschen, B. Bourgeois, J. R. Madsen, and J. E. Lisman. 2001. Gating of human theta oscillations by a working memory task. *J Neurosci* 21(9):3175–3183.

Robbe, D., S. M. Montgomery, A. Thome, P. E. Rueda-Orozco, B. L. McNaughton, and G. Buzsaki. 2006. Cannabinoids reveal importance of spike timing coordination in hippocampal function. *Nat Neurosci* 9(12):1526–1533.

Sato, N., and Y. Yamaguchi. 2007. Theta synchronization networks emerge during human object-place memory encoding. *Neuroreport* 18(5):419–424.

Seager, M. A., L. D. Johnson, E. S. Chabot, Y. Asaka, and S. D. Berry. 2002. Oscillatory brain states and learning: impact of hippocampal theta-contingent training. *Proc Natl Acad Sci USA* 99(3):1616–1620.

Sederberg, P. B., M. J. Kahana, M. W. Howard, E. J. Donner, and J. R. Madsen. 2003. Theta and gamma oscillations during encoding predict subsequent recall. *J Neurosci* 23(34): 10809–10814.

Sederberg, P. B., A. Schulze-Bonhage, J. R. Madsen, E. B. Bromfield, D. C. McCarthy, A. Brandt, M. S. Tully, and M. J. Kahana. 2007. Hippocampal and neocortical gamma oscillations predict memory formation in humans. *Cereb Cortex* 17(5):1190–1196.

Shin, J., D. Kim, R. Bianchi, R. K. Wong, and H. S. Shin. 2005. Genetic dissection of theta rhythm heterogeneity in mice. *Proc Natl Acad Sci USA* 102(50):18165–18170.

Siapas, A. G., E. V. Lubenov, and M. A. Wilson. 2005. Prefrontal phase locking to hippocampal theta oscillations. *Neuron* 46(1):141–151.

Singer, W. 1993. Synchronization of cortical activity and its putative role in information processing and learning. *Annu Rev Physiol* 55:349–374.

Squire, L. R., C. E. Stark, and R. E. Clark. 2004. The medial temporal lobe. *Annu Rev Neurosci* 27:279–306.

Steriade, M. 2000. Corticothalamic resonance, states of vigilance and mentation. *Neuroscience* 101(2):243–276.

Stewart, M., and S. E. Fox. 1989. Detection of an atropine-resistant component of the hippocampal theta rhythm in urethane-anesthetized rats. *Brain Res* 500(1–2):55–60.

Vanderwolf, C. H. 1969. Hippocampal electrical activity and voluntary movement in the rat. *Electroencephalogr Clin Neurophysiol* 26(4):407–418.

Vertes, R. P. 2005. Hippocampal theta rhythm: a tag for short-term memory. *Hippocampus* 15(7):923–935.

Veselis, R. A., and K. O. Pryor. 2009. Propofol amnesia – what is going on in the brain? In *Suppressing the mind: anesthetic modulation of memory and consciousness*, edited by A. G. Hudetz and R. A. Pearce. New York City: Springer.

Whishaw, I. Q., and C. H. Vanderwolf. 1971. Hippocampal EEG and behavior: effects of variation in body temperature and relation of EEG to vibrissae movement, swimming and shivering. *Physiol Behav* 6(4):391–397.

Whittington, M. A., J. G. Jefferys, and R. D. Traub. 1996. Effects of intravenous anaesthetic agents on fast inhibitory oscillations in the rat hippocampus *in vitro*. *Br J Pharmacol* 118(8): 1977–1986.

Winegar, B. D., and M. B. MacIver. 2006. Isoflurane depresses hippocampal CA1 glutamate nerve terminals without inhibiting fiber volleys. *BMC Neurosci* 7:5.

Ylinen, A., A. Bragin, Z. Nadasdy, G. Jando, I. Szabo, A. Sik, and G. Buzsaki. 1995. Sharp wave-associated high-frequency oscillation (200 Hz) in the intact hippocampus: network and intracellular mechanisms. *J Neurosci* 15(1 Pt 1):30–46.

# Chapter 11
# Propofol Amnesia – What is Going on in the Brain?

**Robert A. Veselis and Kane O. Pryor**

**Abstract** By necessity, all anesthetic drugs produce amnesia (lack of memory) at high enough concentrations. However, certain amnesic drugs, exemplified by propofol or midazolam, produce dense amnesia at low drug concentrations while a person is still awake. This chapter reviews human memory systems, with a focus on episodic memory. This memory system is the one most affected by amnesic drugs. It is also the one most difficult to define and study in animal models. Potential mechanisms by which amnesia is produced in humans are defined in this chapter using propofol as an exemplar drug. Episodic memory is considered in terms of information flow from the outside world into long-term memory. Propofol has little effect on encoding, thus permitting memories to be formed in its presence. However, these memories are quickly forgotten, and electrophysiologic measures (event-related potentials, ERPs) of recognition memory reveal effects of propofol on memory within seconds of encoding. Potential targets for propofol's amnesic actions on the chain of physiologic events activated by learning, termed consolidation, are discussed.

**Keywords** Memory · episodic · human · propofol · midazolam · amnesia · EEG/ERP

## Memory and Consciousness: Two and the Same

There is a close relationship between consciousness and memory. Some consciousness must be present to form a memory, with the most complex memories requiring the highest functioning of consciousness. These most complex memories define us as being uniquely human – we can plan, we can project ourselves into the future, and we can consciously modify our recollections (Tulving 2001). Controversy surrounds whether the family pet possesses the same abilities (Aggleton and Pearce 2001). As

R.A. Veselis (✉)
Department of Anesthesiology and Critical Care Medicine, Memorial Sloan-Kettering, Cancer Center, New York, USA
e-mail: veselisr@mskcc.org

A. Hudetz, R. Pearce (eds.), *Suppressing the Mind*, Contemporary Clinical Neuroscience, DOI 10.1007/978-1-60761-462-3_11,

memory and consciousness intersect in numerous ways, insights into understanding consciousness can illuminate memory function and vice versa. From an evolutionary standpoint, the brain is likely to solve similar problems in similar ways. Both consciousness and memory are subserved by distributed networks of overlapping brain regions in constant, almost instantaneous communications with each other (Ungerleider 1995; Vincent et al. 2006; Greicius et al. 2008). This form of communication is descriptively termed the "brainweb," with communications between different brain regions dependent on the properties that emerge from the electrophysiologic interactions of large groups of neurons (John 2001; Varela et al. 2001). Communication between brain regions is more complex than simple transmission along the axons of neurons. The electroencephalogram (EEG), either surface or intracranially recorded, can be analyzed using well-described, though complex, methods that can index communications between brain regions using coherence, synchrony, or phase parameters of multiple EEG signals. One such index is phase synchrony between different brain regions, locking these regions for a period of time during which transfer of information can occur (Lachaux et al. 1999). Because of these complex interactions in the functioning brain, simple knowledge of anatomical connections is not enough to understand memory function. Equally important, anatomical localization of anesthetic drug effects on the brain will only provide partial insight into how these drugs affect memory.

A caveat needs to be inserted here. Great and, importantly, critical insights into drug actions on memory formation and function can be obtained by careful study of defined memory paradigms in animals or analogs such as long-term potentiation (LTP) [1] in simpler neuronal systems. However, translation to understanding human memory is multilayered in complexity. A good example is translation of LTP, induced in the laboratory by exogenous electrical stimulation to modify synaptic connections in experimental preparations of neurons, to memory function in humans (Malenka and Nicoll 1999); see MacIver, Chap. 9. Despite the suggestion that synchronous stimulation of neurons could modify synaptic connections in 1948, only recently has it been generally accepted that such changes are the basis of learning in animals and presumably man, with learning being defined as the acquisition of new information from the environment into memory (Konorski 1948; Hebb 1949; Leuner, Falduto, and Shors 2003).

Importantly, the effects of amnesic drugs in humans principally relate to episodic memory, which is poorly defined in animal models. Thus, a review of taxonomy, or classification of memory systems in humans will be presented in this chapter, in order to better understand how drugs, principally propofol, affect human memory. Historically, the actions of anesthetic drugs on human memory were understood in a cognitive psychological framework, as described for many years for benzodiazepines (Ghoneim and Mewaldt 1990). In order to understand how propofol produces amnesia, any framework of understanding needs to incorporate recent advances in memory research. Thus, specific cognitive processes, interactions of

---

[1] The permanent modification of synaptic function on the basis of complex interactions between electrical signals, genes, proteins, and membranes.

brain regions, and their relationships to the underlying complex molecular machinery of learning are crucial to this effort. Importantly, the glue that holds these pieces together resides in the oscillatory rhythms present in the brain (Buzsaki and Draguhn 2004); see Perouansky, Chap. 10. Critical insights can be obtained by measuring these properties of the brain using the EEG, as described shortly.

The complexities of human memory are embedded in multiple, separable, but interacting systems (Tulving and Schacter 1990). Though the term "systems" does not imply anatomical structures in the brain per se, much of what we understand in terms of human memory, and drug actions on these, depends on anatomical localization of memory processes and drug-induced changes in memory function. Differing topographies[2] of memory processes are important in identification of inherently distinct processes, such as the familiarity and recollection processes supporting recognition in episodic memory (Friedman and Johnson 2000; Curran et al. 2006; Bowles et al. 2007). Frequently, a given brain region is active in any number of distinct memory functions (Cabeza and Nyberg 2000).

One way of measuring memory-related brain activity in humans is by recording EEG or event-related potentials (ERPs) during performance of defined memory tasks. A commonly used paradigm is recognition of previously presented items, which elicits the "old/new" effect – the enhanced brain response when an item is consciously recognized (see Fig. 11.1).

The ERP measures responses of neuronal populations to an outside stimulus, such as a word or picture. ERPs are averaged EEG recordings time-locked to the onset of the stimulus in question. As more and more ERP responses are averaged, the common activity related to stimulus processing becomes larger, while other unrelated EEG activity is canceled out. What remains is the ERP waveform, which represents activity of neuronal populations related to stimulus processing[3] (Johnson 1995). ERP waveforms that vary over different EEG electrodes define the topographical characteristics of memory processes and their underlying neuronal populations. In a similar fashion, hemodynamic markers of neuronal activity are used. One such marker is the BOLD signal measured in functional magnetic resonance imaging (fMRI), derived from transient changes in hemoglobin saturation localized closely to neuronal activity (Logothetis 2003, 2008). The high resolution of fMRI imaging provides more accurate spatial localization of effects measured with ERPs (Kahn, Davachi, and Wagner 2004; Iidaka et al. 2006). The sensitivity of BOLD to very small changes in neuronal, or, more correctly physiologic surrogates of this activity allows very high spatial and, increasingly, temporal resolution of memory processes. The use of this methodology to examine the effects of

---

[2] Separable locations over or in the brain.

[3] The electrophysiology of brain response to item presentation is complex. It can consist of resetting of ongoing oscillatory activity without amplitude response, amplitude response time locked to the stimulus (evoked response), amplitude response not time locked to the stimulus (induced response), or a combination of all these. These concepts are further discussed in the section titled "Consolidation versus forgetting – the importance of learning-related electrophysiologic processes."

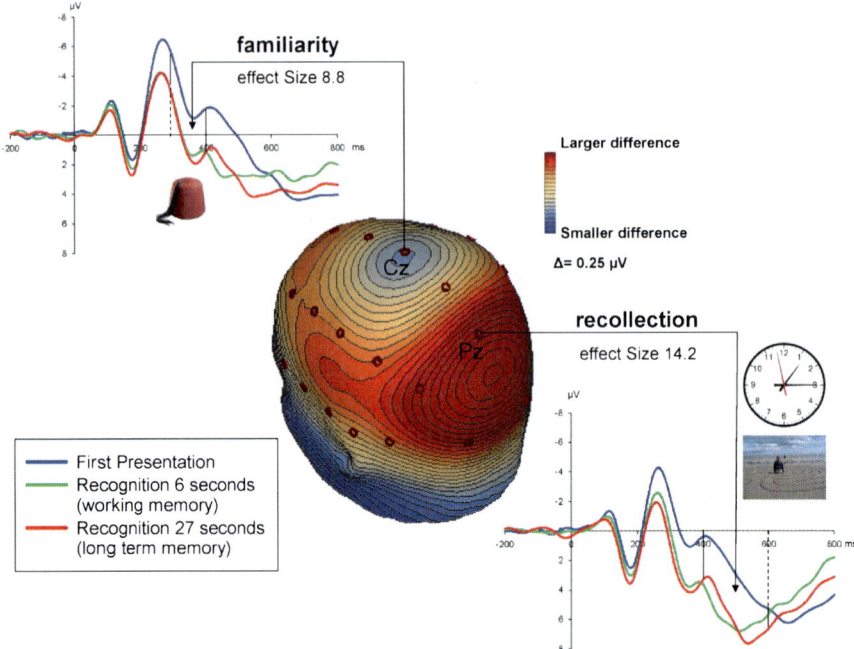

**Fig. 11.1** The old/new effect is a very robust measure of recognition of conscious memories. The ERP produced by conscious, successful recognition of a stimulus is more positive than a novel, first presentation (*green* and *red lines* vs. *blue*, positive voltage is down). The largest difference between waveforms occurs from 400 to 600 ms in the parietal region, close to the Pz electrode (diagrammed as the red "blob" on the back of the head, nose is in front). This represents recollection of a memory in a context of time and place, i.e., episodic memory. A more undifferentiated sense of familiarity recognition occurs a few hundred milliseconds before recollection (the "fez is familiar...," without recall of specific details). Familiarity and recollection are separate recognition processes in conscious memory, being localized to different regions in the medial temporal lobe (familiarity in rhinal cortex, opposed to recollection in the hippocampus). In this figure, arrows point to areas between old and new waveforms. The strength of memory is related to the size of these areas

anesthetic drugs on memory processes is relatively new. The underlying assumption has been that if BOLD activation is changed in the presence of an anesthetic drug, then the drug has some effect on neuronal populations involved in these processes. However, it is now becoming apparent that even at low concentrations of drugs, anesthetic effects on the relationship of the BOLD signal to underlying physiology may be disturbed. This, in turn, would bring into question the interpretation of anesthetic effects on BOLD activation as reflection of changes in neural activity versus other physiologic effects (Qiu, Ramani, Swetye, and Constable 2008; Qiu, Ramani, Swetye, Rajeevan et al. 2008) (see Fig. 11.2). [4]

---

[4] Veselis et al. (2009) The amnesic drug propofol decreases BOLD response: change in cognition or physiology? A761, Human Brain Mapping, June 18–22, San Francisco.

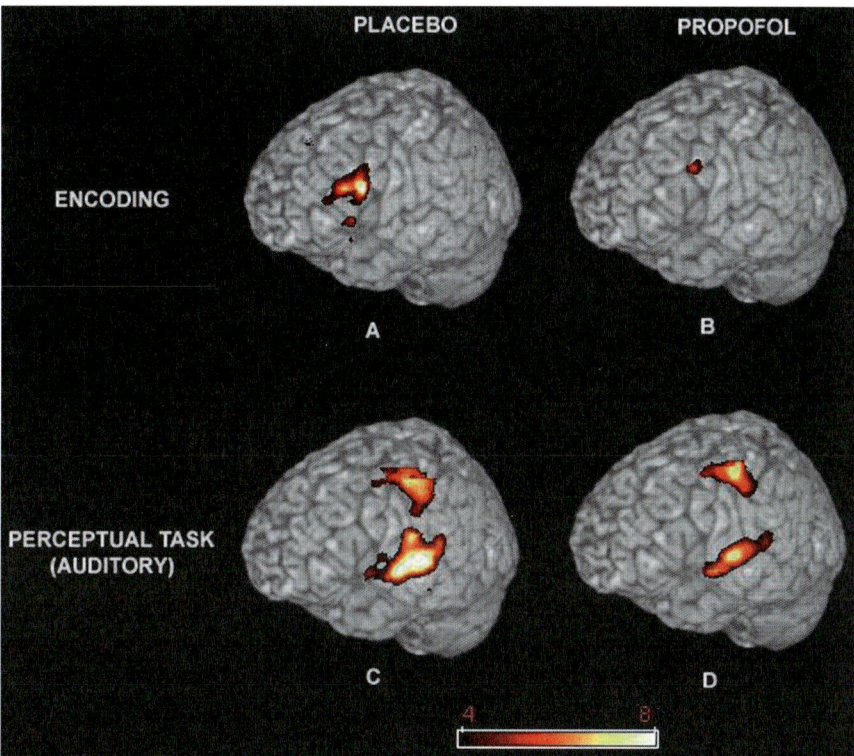

**Fig. 11.2** fMRI measures of the BOLD signal are very sensitive to small changes in neuronal and associated physiologic activity (this study was done with a 3 T scanner). At this level of sensitivity, the effect of even small doses of anesthetics, in this case propofol, on the relationship between neuronal and BOLD responses is evident. **A,B** Represent encoding activity, which in a previous study using PET imaging of rCBF was found not to be affected by propofol. **C,D** Show perceptual processing of auditory words (is the word spoken in a male or female voice?). Despite volunteers hearing the words clearly and making correct categorizations, the group receiving propofol demonstrated reduced activation of the auditory cortex ($T = 8.07$, 994 voxels activated) compared with placebo ($T = 10.07$, 1123). A previous study using PET measures of rCBF demonstrated, if anything, an increase in auditory activation to word stimuli with propofol

## Taxonomy (Classification) of Human Memory Systems

Terminologies used to describe specific memory systems and their components are not rigorously defined. For instance, short-term memory means somewhat different things to a neuropsychologist testing a postoperative patient, a researcher using mice in a learning paradigm, and a cognitive psychologist studying ERPs in volunteers. Thus, the descriptors of memory used in this chapter are to a greater or lesser degree nebulous. For the purposes of describing the effects of anesthetics on memory, a definition of memory as being short or long, unconscious or conscious is a good starting point.

## Conscious Memories – Closely Related Terms: Explicit, Declarative, Episodic, Semantic, and Autobiographical

Conscious memories are those that we know we have and that we can manipulate. For example, I remember seeing a friend last night, and I can imagine them wearing a particular item of clothing. I can enhance that memory, or I can actively suppress it (Miller 2004). This form of memory is referred to as declarative or explicit memory. The term declarative arises from the fact that these memories can be "declared" to be true or false or to have occurred at a particular time or place. Declarative memories can be divided into two broad categories – contextual memories or general knowledge memories. Contextual memories occur at a particular time and place (i.e., contextual information is a part of the memory) and are called episodic memories. It is this contextually rich memory remembered in a given time and place that is particularly affected by amnesic drugs (Ghoneim and Mewaldt 1990). Then there are memories that can be imagined (e.g., Paris in the spring), but where or when those particular memories of Paris or the meaning of spring were obtained are not known. Such "knowledge of the world" memories are semantic memories. These memories are not particularly affected by amnesic drugs. The temporal lobe is to a large extent the necessary, but not sufficient seat of those memories (Levy, Bayley, and Squire 2004). Distant, long-term memories (e.g., first day of kindergarten) are termed autobiographical, as these are usually personally relevant memories (otherwise they are likely to be semantic memories) (Burianova and Grady 2007; Cabeza and St Jacques 2007). These are the conscious memories least affected by almost any intervention or disease, indicating that a widely distributed, robust system must support these memories (Smith and Squire 2009).

### "Short or Long": Perception and Working Memory

Information from the environment must be perceived and processed before conscious memories can be formed. Many streams of sensory information are combined into a conscious image of the world called the percept (Mashour 2006). Newly perceived information is held in a temporary store called working memory (though some refer to this as short-term memory). Information in working memory can either be transferred (encoded) into longer term memory, or not. Memories in working memory disappear on the order of seconds, being replaced by newer information (Lisman and Idiart 1995). Working memory is thus of limited capacity, as demonstrated by the ability to remember and then write down a new seven-digit phone number, but not a new credit card number.

## Unconscious Memories; Closely Related Terms: Implicit, Procedural, Subliminal, Priming, and Conditioning

Unconscious memories are, as follows from above, memories we possess but have no conscious knowledge of. Unconscious memories are termed implicit or non-declarative memories (they can't be "declared" to be true or false, or whether in fact they are present). These most fascinating memories are learned unconsciously and serve as the "workhorses" of surviving in a particular environment. Unconscious memories are formed without the involvement of working memory processes. Examples of such unconscious learning and influence include conditioning, therapeutic suggestion, emotional aversion, i.e., phobias, and subliminal advertising.

## Procedural Memory and Conditioning

Similar to conscious semantic memories of world knowledge, there are unconscious memories of how to do things in the world – i.e., procedural memories such as playing a musical instrument. Though initially learned as conscious memories, these actions become semi-automatic and stereotypical. This explains the difficulty in, for example, playing a familiar piece of music starting at the third bar. The reader may wonder why procedural memories are "unconscious," especially as many conscious memories may surround learning a certain skill, such as driving. Indeed, conscious actions can influence unconscious memories and vice versa.

Procedural memory is mentioned here, as it embodies Pavlovian conditioning, a very commonly used model of memory in animal studies. Aversive conditioning is inherently dependent on the amygdala, the modulator of fear memory. Thus, conditioning is a "visceral" type of memory. For example, a rat quickly learns (even in one trial) that a foot shock awaits it in an otherwise attractive environment such as a dark corridor. This type of learning, mediated by the emotional memory system, demonstrates the strongest link between memory, anatomy, and animal–human behavior that exists. Emotional memories are modulated via the amygdala, and in turn are influenced by multiple inputs such as stress hormone levels, both in animals and humans (Cahill et al. 1995). Memory systems can affect each other, and emotional influences on episodic memory are large and well known (Cahill et al. 1996; Morris et al. 1996; Cahill and McGaugh 1998; Morris, Ohman, and Dolan 1998; Kim et al. 2001; McGaugh, McIntyre, and Power 2002). Emotionally laden episodic memories are more resistant to the amnesic effects of intravenous anesthetics (Pryor et al. 2004).

## *Memory as a Neuro-mechanistic Process*

In the past, memory was understood in terms of models of the mind and not necessarily related to brain structure. For a long time, it was known that memory somehow resided in the brain, but no clear link to a specific portion of the brain was evident. That all changed in 1954 when an operation for the control of epilepsy in the patient "H.M." excised a small part of the brain called the hippocampus (Scoville and Milner 1957). The bilateral excision of H.M.'s temporal poles resulted in the loss of the ability to form any new episodic memories, despite otherwise normal cognitive abilities. The dramatic changes in memory following the operation in H.M. transformed the study of memory in humans to a more neuro-mechanistic-based approach. This approach has exploded in the past decades largely due to advanced research techniques ranging from careful study of molecular basis of hippocampal function using genetic knockouts (animals missing a specific gene) to fMRI and ERP imaging of networks used in support human memory function (Sonner et al. 2007). In many ways, the hippocampus and surrounding brain can be considered as the seat of conscious memory. Quite naturally, cellular and molecular investigations have focused on this particular brain region, as depicted in Fig. 11.3.

## Drug Actions on Memory – A System of Understanding

Recent insight is available into how anesthetic agents such as propofol, midazolam, and diazepam produce their effects on memory at low doses. As opposed to drugs that impact memory from purely sedative effects, such as alcohol and thiopental, propofol can impair memory with little or no sedation being present (Veselis et al. 1997, 2004, 2009; Veselis, Reinsel, and Feshchenko 2001). Under these conditions, fairly normal behavior is possible, such as continuing a conversation or performing working memory tasks. Though performance is quite good, even normal, there is no conscious memory of these events at a later time. As these examples illustrate, the memory effect typified by propofol and benzodiazepines is really one of forgetting. Apparently normal cognition under the influence of drugs implies that memories must be present for a short period of time in order to function in a changing environment. In order to better understand how anesthetic drugs affect memory, a conducive taxonomy of memory is needed. The taxonomy of memory systems described by Squire is the one most commonly used and, as summarized above, is useful to understand episodic memory in relation to other systems (Squire and Zola-Morgan 1991). However, an alternative system described by Tulving and Schacter may be more relevant in terms of drug actions on human memory. The serial parallel independent (SPI) system is useful in the sense that it allows unconscious memory formation to occur without consciousness being present, as happens in learning during anesthesia (Schacter and Tulving 1994; Ghoneim and Block 1997).

### *The Serial Parallel Independent Taxonomy of Memory*

The SPI model classifies memory systems in a hierarchical fashion. Grossly, these can be related to underlying anatomical structures. "S" depicts information flow as being hierarchical, i.e., the systems are serial, while the ability of one system to be operational while another is not represented by "PI" – parallel, independent. Three modules comprise the SPI model – perceptual, semantic, and episodic, in ascending order of complexity (see Fig. 11.4).

Learning can be thought of as the flow of information from the outside world to the final resting place of archival memory. At the lowest level of the SPI model, a stimulus is perceived and initially processed in the sensory and perisensory cortices.

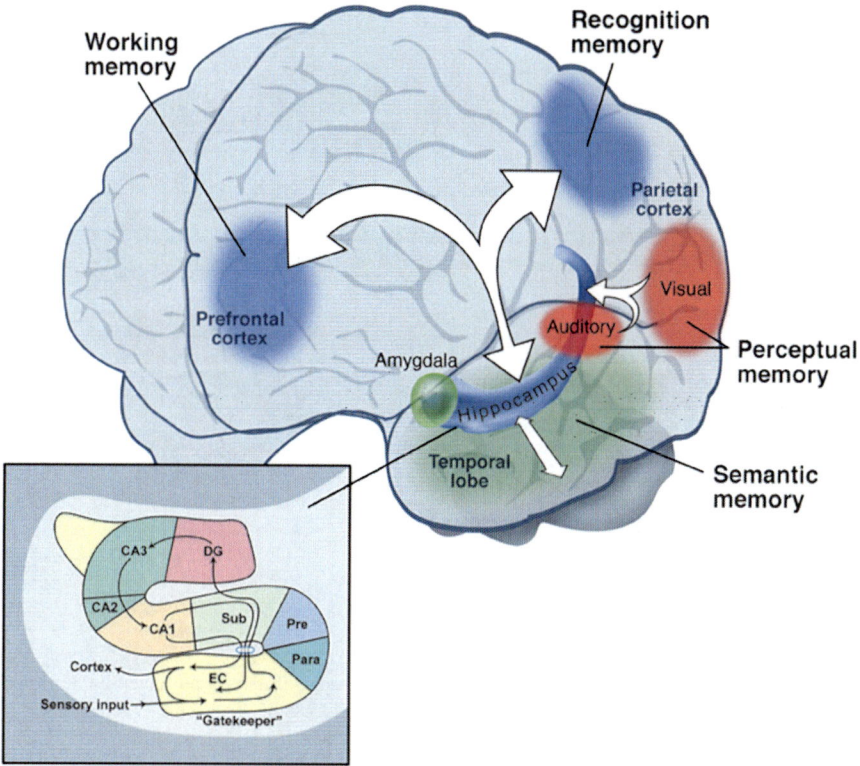

**Fig. 11.3** A neuro-mechanistic description of conscious memory in humans: conscious memory includes semantic and episodic memory (and can also be classified according to working and long-term memory processes, see Fig. 11.4). Perceptual memories can be formed in regions close to primary sensory cortices, but are unconscious memories (the perceptual "rung" of the SPI system, see Fig. 11.4). Information flows from auditory and visual streams to the hippocampus, which is the seat of conscious memory. The hippocampus is located close to the amygdala, which modulates the influence of emotion and fear on episodic memory. The hippocampus communicates with widespread brain regions in support of episodic memory function. Working memory function is located to a large extent in the prefrontal cortex. Recognition from episodic memory involves the parietal cortex, where maximal surface-recorded ERP effects occur during recognition. A large amount of research has focused on the internal workings of the hippocampus (exploded view). Input arises from close-by regions in the medial temporal lobe [entorhinal cortex (EC) the "gatekeeper" of information flow to and from the hippocampus]. Sensory input via the EC projects to the dentate gyrus (DG), the CA3 and CA1 fields of the hippocampus and the subiculum (Sub) via the perforant pathway. The dentate gyrus projects to the CA3 field of the hippocampus via Mossy fibers. CA3 neurons project to the CA1 field of the hippocampus, which in turn projects back to the subiculum. The subiculum feeds back to the EC. In the EC, superficial and deep layers are arranged to produce a recurrent loop for incoming sensory information. Some of this processed information is transferred to the hippocampus via the perforant pathways. In turn, after processing in the hippocampus, output influences information in the entorhinal reverberating circuit, which in turn repetitively activates the hippocampal formation or is transmitted to other regions of the cerebral cortex (pre, presubiculum; para, parasubiculum; based on Andersen P. The hippocampus book. Oxford ; New York: Oxford University Press 2007, Fig. 3-1 p 38 and Iijima T, Witter MP, Ichikawa M, Tominaga T, Kajiwara R, Matsumoto G. Entorhinal–hippocampal interactions revealed by real-time imaging. Science. 1996 May 24;272(5265):1176–1179)

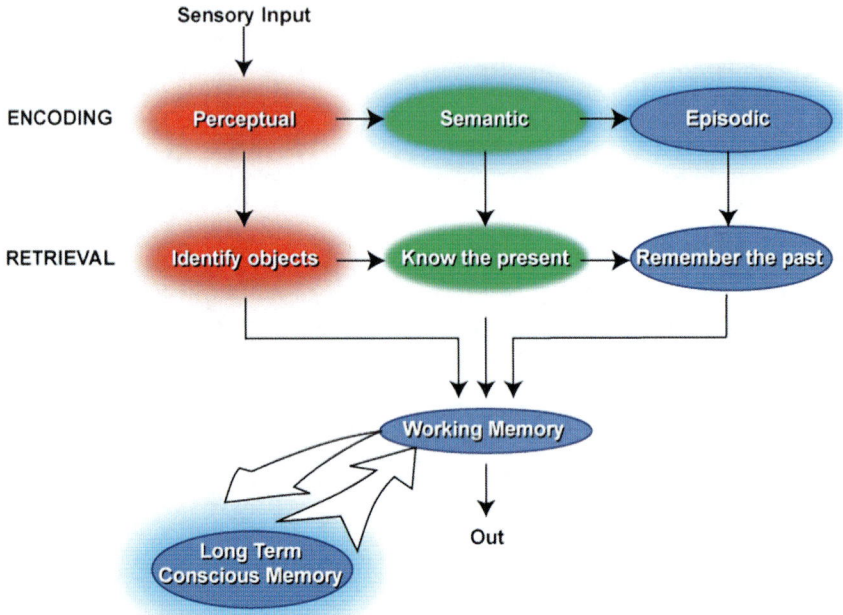

**Fig. 11.4** This is one representation of the taxonomy of human memory presented by Schachter and Tulving (the SPI classification, Schacter DL, Tulving E. Memory systems 1994. Cambridge, MA: MIT Press 1994). It is particularly useful to understand how learning during general anesthesia can occur without conscious awareness. The three main systems (perceptual, semantic, and episodic) are hierarchical, parallel, and independent and can roughly be related to the anatomical structures in Fig. 11.3. Thus, perceptual learning during anesthesia can occur without the involvement of episodic memory. The semantic system refers to the matching of information from the environment to stored memory templates, which largely occurs in the anterior temporal lobes. "Semantic" in this taxonomy is used in a different sense than "semantic" in language processing. Long-term conscious memory (*blue blob*) refers to the episodic module at the *top right* of the diagram. There is a close relationship between active manipulation of information in working memory and successful encoding into long-term memory. After retrieval from long-term memory, information is held in working memory before action occurs. The prefrontal cortex is involved particularly with the evaluation of novel information, evaluation of different choices, and other high-order "executive" functions, whereas more automated processing can occur in different brain regions (e.g., medial temporal lobe)

Following this, attempts to match the newly perceived information with previously stored "templates" (a memory, either episodic or semantic) allow the stimulus to be sorted out as novel or already known information (John, Easton, and Isenhart 1997). Note that "semantic" does not imply language processing per se but rather matching of information to templates in memory before higher levels of decoding and comprehension. Before acquisition into conscious long-term memory, information is held, decoded, and manipulated in working memory, mainly in regions in the prefrontal cortex (Ranganath, Cohen, and Brozinsky 2005; Blumenfeld and Ranganath 2006). Higher cognitive processing can then proceed in the upper rungs of memory (episodic memory), with flow of information between the hippocampus and widely distributed regions throughout cerebral cortex (Eichenbaum 2000, 2004).

# Drug Actions on Conscious, Episodic Memory – Accelerated Memory Decay

## *Procedural Memory in Animals Versus Episodic Memory in Humans*

As alluded to previously with Pavlovian conditioning and LTP, the translation of drug actions in animal and simpler models of memory systems to human episodic memory is not linear. Even though amygdalar-based fear-mediated memory systems are very similar between animals and man, care needs to be taken in extrapolation. It is known that propofol and benzodiazepines impair memory of aversive learning in animals (Alkire et al. 2001). It is also known that this effect disappears with lesions in the basolateral nucleus of the amygdala (Dickinson-Anson and McGaugh 1997; Alkire et al. 2001). But does this mean that the amnesic effects of propofol and benzodiazepines on the human require amygdalar function or that the locus of action of these drugs resides in the amygdala? As there is a distinction between episodic and procedural memory in humans, exemplified by the difference between learning mirror reading and having a memory of performing this task, there may well be a dissociation between the amnesic effects of drugs on aversive learning in animals and on episodic memory in humans (Cohen and Squire 1980; Heindel et al. 1989). Though critical knowledge stands to be gained by studying the effects of amnesic drugs on animals and more basic neuronal systems, the final answers will evolve from studies in humans (Alkire et al. 2008).

## *Drug Actions on Episodic Memory in Humans; Focus on Propofol*

A simplified model of episodic memory is a sufficient starting point from which to examine drug effects on human memory. Information is processed in working memory, from which it either disappears or is subsequently encoded into long-term episodic memory. Note that long-term memory is defined here as conscious memory lasting longer than the time frame of working memory, namely about 20–30 s after stimulus perception. Clearly, there are different phases of long-term memory. After processing in working memory, various processes that help retain memories come into play and continue for hours, if not days and years. The cascade of processes put into motion after acquisition of information into memory is collectively termed consolidation. Two characteristics differentiate the vast number of consolidation processes: their location in the brain and the time period after learning during which they are active (McGaugh 2000; Abel and Lattal 2001; Izquierdo et al. 2006).

The nebulous term "short-term memory" can refer to working memory and/or the initial stages of long-term memory. This chapter focuses on the effects of propofol and midazolam. However, evidence points to similar effects of low concentrations of inhalational agents, even though these studies are not as detailed as those presented here (Gonsowski et al. 1995; Galinkin et al. 1997; Ghoneim, Block, and Dhanaraj 1998; Ramani, Qiu, and Constable 2007). Careful behavioral testing reveals that the memory effects of propofol and midazolam are dissociable from sedation effects (Veselis et al. 1997). This chapter will not focus on how this dissociation has been established, but new behavioral and ERP data will be presented, which lend further support (see below) (Veselis et al. 2009). Note that sedation itself will impair memory, thus separating sedative from amnesic actions is not straightforward. The sedative, or sleep-like, properties of many anesthetics may be mediated through the natural, endogenous sleep pathway in the hypothalamus (Nelson et al. 2002). This region is closely related to the medial temporal lobe and hippocampus and linked to these via the fornix, which may explain the close relation between sedation and nonsedative amnesic effects of drugs on memory (Ghoneim and Hinrichs 1997).

Conceptually, the range of memory effects associated with the sedative and nonsedative effects of drugs can be understood in terms of the SPI system. At the highest doses of drug, where unresponsiveness is present, no episodic memory is formed, as the upper rungs (semantic, episodic) of the SPI model are nonoperational. This is not to say that the perceptual system is not functional at anesthetic drug concentrations when unresponsiveness is present. Sensory stimuli are perceived by the brain, and some perceptual memories may be formed (Plourde et al. 2006; Hudetz and Imas 2007). These are revealed in priming tasks, where behavior is affected by memories formed during general anesthesia (i.e., faster response or stimulus preference; Deeprose et al. 2004, 2005). At lower concentrations (responsive, but "drunken"), information from the outside world is perceived through a haze, but then is insufficiently processed to be remembered, and thus, again, episodic memories are not formed. Information that is not attended to is less likely to be remembered (Fernandes and Moscovitch 2000). The sedative effect typified by drugs such as thiopental, dexmedetomidine, or alcohol impairs memory by inhibiting the smooth flow of information from perceptual through semantic processing (template match ing) to episodic memory. Another conceptualization of sedation is that information may reside in working memory, but is not transferred to long-term memory (Veselis et al. 2004). At still lower concentrations of certain drugs, notably propofol or benzodiazepines, consciousness is present with fairly normal cognition, minimal impairment on task performance, and normal working memory function. Memory is encoded successfully into episodic memory (Veselis et al. 2008). Amnesia in this situation is characterized not by problems in encoding, but by the loss of memories, as shown in Fig. 11.5.

The accelerated memory decay of learned information fits in nicely with the clinical observation of being awake and aware at low concentrations of propofol, but without subsequent recall (Nordstrom and Sandin 1996).

It is helpful to understand the rationale in using the continuous recognition task (CRT) as the key memory paradigm to obtain data presented in Figs. 11.1, 11.5, and

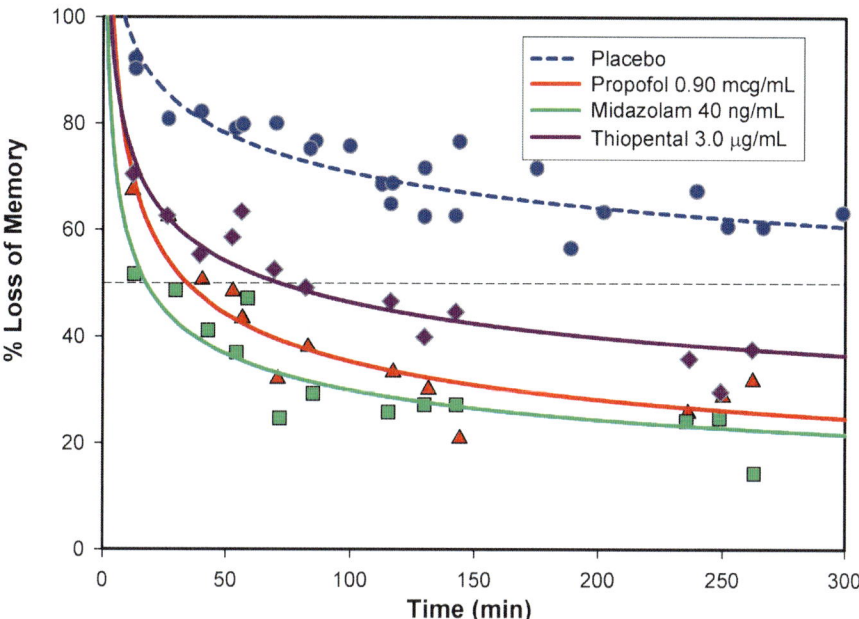

**Fig. 11.5** Accelerated memory decay occurs when information is learned in the presence of anesthetic drugs, in this case propofol, midazolam, or thiopental. The *top line* represents the normal rate of forgetting in the placebo group. This is a normal process, as it is important that irrelevant information is forgotten over time. These curves represent the loss of memories encoded into long-term memory during a CRT (described in Fig. 11.6) just before time = 0. When active drug was present at low concentrations during this task, items encoded into long-term memory were forgotten at an accelerated rate compared to placebo. Most of the differences in forgetting happened in the first 45 min. Careful analysis of the rate of decay reveals important differences among drugs, with arousal and amnesic effects affecting indices of the forgetting model differently (see Fig. 11.7)

11.7–11.9. The effect of drug on memory performance is traditionally assessed in a 2 by 2 design, testing recognition of items presented previously for study (learning phase) in the presence of either a placebo or a drug, i.e., a study-test paradigm. Testing is delayed in relation to study, and as our data demonstrate in Fig. 11.5, significant memory loss can occur by the time first recognition testing occurs. The continuous recognition task solves this dilemma, as initial recognition of items occurs during the study (learning) task, albeit in the presence of the drug. Items are presented twice for study recognition in a randomized, continuous stream of presentations. Manipulation of the interval from short to longer periods between study-test presentations determines whether working or long-term memory processes are used in recognition of the item (see Fig. 11.6, where details of the CRT are explained).

Recognition of an item during the CRT documents that the item was perceived and is present either in working or in long-term memory. The assumption that the

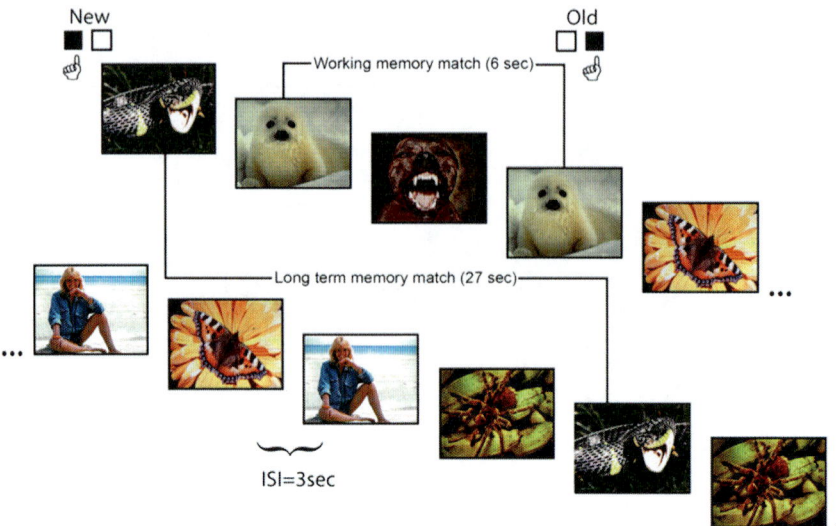

**Fig 11.6** The CRT tests encoding (study) and recognition (test) in the same paradigm. Recognition of pictures was tested either at a 6-s (recognition occurring from working memory; WM) or at a 27-s interval after initial presentation. Recognition at 27 s must be from "long-term memory" (LTM), as the contents of working memory are continually replaced by new material that must be attended to in performance of the CRT. The CRT consisted of a standardized series of photographic pictures from the International Affective Picture System, with one presented every 3 s. Most pictures were repeated once (to test either WM at 6 s or LTM at 27 s) to maximize the efficiency of the task. The volunteer categorized each picture as either old (i.e., a repeat) or new (first presentation) using a button press. Sedation from drug on task performance was evident as increases in reaction times (all drugs increased RT compared with placebo, $p < 0.002$, with an increase of 50–200 ms depending on the drug) and errors in categorization of pictures (increased 6%, $p < 0.02$)

presence of drugs does not affect recognition of previously encoded memories holds over short periods of time, as demonstrated by high performance of participants for recognition of items, even from long-term memory tested after 27 s. Imaging of encoding processes supports the normality of these in the presence of propofol (Veselis et al. 2008).

After the end of drug infusion, recognition of successfully encoded items can be tested at increasing time intervals after completion of the CRT. Successive recognition testing defines the decay of memories learned on the CRT in the presence of the drug. Normally, results from one recognition task (e.g., at 30 min after end of CRT) are averaged to produce a single value, which is compared with an average performance value on the CRT. However, this negates any variations in timing between encoding and recognition of individual items. These differences are critical, particularly at short intervals after encoding where memory loss is greatest. Figures 11.5 and 11.7 were derived by curve fitting of recognition of items against exact retention intervals from when the item was presented on the CRT to when it was presented for subsequent recognition testing after the CRT was finished. Rapid decay of memories, particularly in the case of propofol and midazolam, is best captured when this method of data analysis is used.

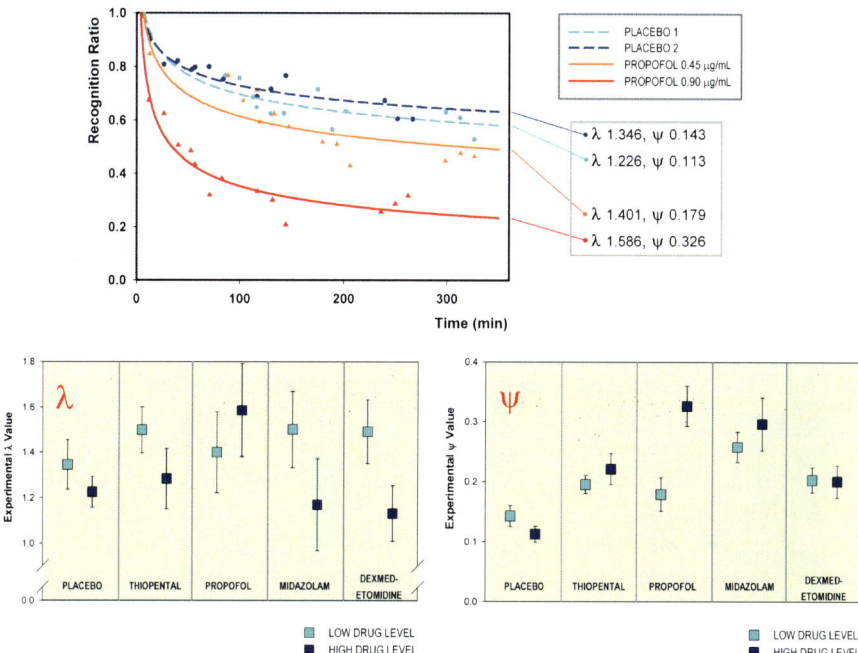

**Fig 11.7** In this study, participants performed two CRTs at sequentially increasing dose concentrations, at which point drug infusion was stopped. The loss of memories formed during CRT was defined as a function of time for each dose, fitted to a decay function $m_t = \lambda t^{-\psi}$ as derived by nonlinear regression of averaged group data, with corresponding values for $\lambda$ and $\psi$ shown for placebo and propofol. The lower *two panels* show the derived values for the coefficients $\lambda$ and $\psi$ in each drug and dose condition. *Error bars* represent SEM for the coefficient estimate. Note that the amnesic drugs, propofol, and midazolam are associated with the largest effects on $\psi$, whereas all drugs produce some change in $\lambda$, which relates to arousal (larger values indicating more arousal, i.e., less sedation). $\lambda$ represents the initial strength of encoding and is dose dependent (propofol produced a seemingly paradoxical response on $\lambda$, or the relative dose effect of propofol versus midazolam was less). Thus, $\psi$ seems to index processes early in memory consolidation, and this is the quality that is most inhibited by the amnesic actions of propofol and midazolam

Behavioral results from the CRT lead to the initial impression that accelerated memory decay may be a defining characteristic of amnesic drugs, i.e., propofol and midazolam. However, when memory for items presented in the presence of drugs is tested at some relatively long time point after learning (e.g., 4 h), participants receiving thiopental and dexmedetomidine also exhibit significant memory decrements in relation to the placebo. Depending on specific conditions of testing, differences among drugs at the end of the study could be significant (e.g., see Veselis et al. 2004) or not (e.g., Fig. 11.5). In general, it could be said that anesthetics at low doses produce accelerated memory decay of conscious long-term memories. In reality, careful analysis is required to reveal differences in memory decay among drugs.

The loss of memories encoded during the CRT over time can be described more rigorously. Successful encoding of information into long-term memory, i.e., correct recognition of items 27 s after encoding on the CRT, serves as a starting point for subsequent recognition testing after drug infusion is stopped at the completion of the CRT. A number of decay functions were tested, but the best fit was provided by a simple two-parameter function of the form

$$m_t = \lambda t^{-\Psi},$$

where $m$ is the memory strength at time $t$. Specifically, $\lambda$ indexes the signal strength of the initial memory trace and $\psi$ indexes the subsequent rate of decay of that memory trace. Differences between drugs emerge in terms of their effects on $\lambda$ and $\psi$ indices, where $\psi$ is most affected by propofol and midazolam and $\lambda$ is affected by all drugs. In other words, $\lambda$ seems to relate to the attentional effects of the drug, affecting the initial signal strength of encoding, whereas $\psi$ relates to memory loss after encoding. Differences of drug effects on these indices support the separation of memory impairment into sedative and nonsedative mechanisms (Veselis et al. 1997, 2004; see Fig. 11.7).

Thus, at low, minimally sedative concentrations of propofol and midazolam, episodic memories do form and are registered in long-term memory (i.e., not working memory), but then these are subsequently quickly forgotten, with the majority of forgetting occurring in the first 45 min (Veselis et al. 2004, 2008). Though careful analysis of behavioral data is most revealing, additional important insights are provided by ERPs (Veselis et al. 2009).

## ERP Measures of Working and Long-Term Memory Processes in the Presence of Drug

ERPs obtained from recognition of items on the CRT index brain processes associated with recognition memory. Short intervals on the CRT, i.e., less than 10 s, represent recognition of items from working memory. Longer intervals index retrieval processes from long-term memory as at that point working memory contains newer information (Friedman 1990; Veselis et al. 2004). Retrieval processes are indexed by the fact that ERP potentials are more positive when items previously presented on the CRT are successfully recognized as being experienced before, i.e., "old," as compared with first presentations of "new" items [note that in the figures down represents more positive voltage, with maximal differences occurring in the parietal region (Pz electrode) in Figs. 11.8 and 11.9, with similar, but smaller differences being present near the Cz electrode]. This is the well-described old/new recognition effect (Rugg and Curran 2007). The enhanced brain response to old items is dependent on temporal lobe function and is also revealed in BOLD measures in fMRI studies (Rugg et al. 1991; Guillem et al. 1995).

Working memory processes are preserved in the presence of sedative concentrations of propofol, midazolam, thiopental, and dexmedetomidine. This was revealed

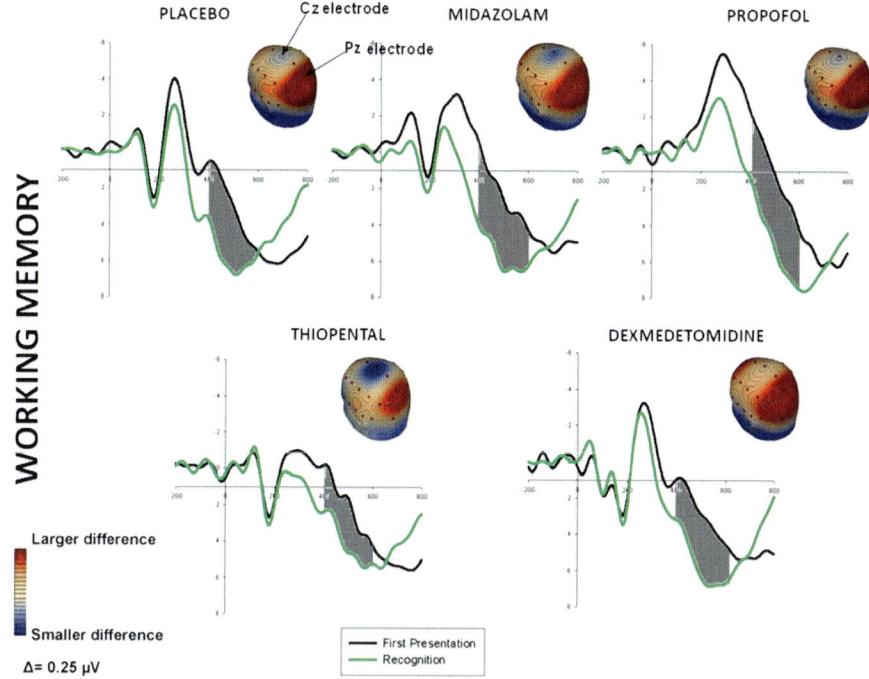

**Fig. 11.8** ERP waveforms in response to "new" (first presentation on the CRT, *black line*) and "old" pictures (second presentation, successful recognition, *green line*) were recorded from the Pz electrode and diagrammed in this figure. ERP "old" waveforms in the time window 400–600 ms represent recollection of episodic memories, and the area between old and new waveforms indicates the strength of those memories. The topographical distribution of the old–new differences measured at 484 ms (the middle of the area time window) is shown over the head (nose in front). The large *red blob* in the back of the head represents recognition activity in the parietal region. Recognition occurred 6 safter item presentation, and ERPs reveal essentially unchanged working memory processes in the presence of any of the study drugs ($p = 0.29$)

by no change in the ERP response to old stimuli (Veselis et al. 2009). A quantitative measure of the function of recognition memory is the area between old and new ERP waveforms in the time window representative of recollection (i.e., 400–600 ms, see Fig. 11.8). Inhibition of memory processes is revealed by decreases in this area (Allan, Robb, and Rugg 2000; Iidaka et al. 2006).

ERPs revealed unequivocally that working memory processes were well preserved in the presence of these drugs, as there was no change in these areas. Thus, cognitive slowing as revealed by small increases in reaction times and error rates did not impair working memory. These findings support previous behavioral data that indicate normal working memory in the presence of benzodiazepines (e.g., preserved performance on the digit symbol substation task; Ghoneim and Mewaldt 1990).

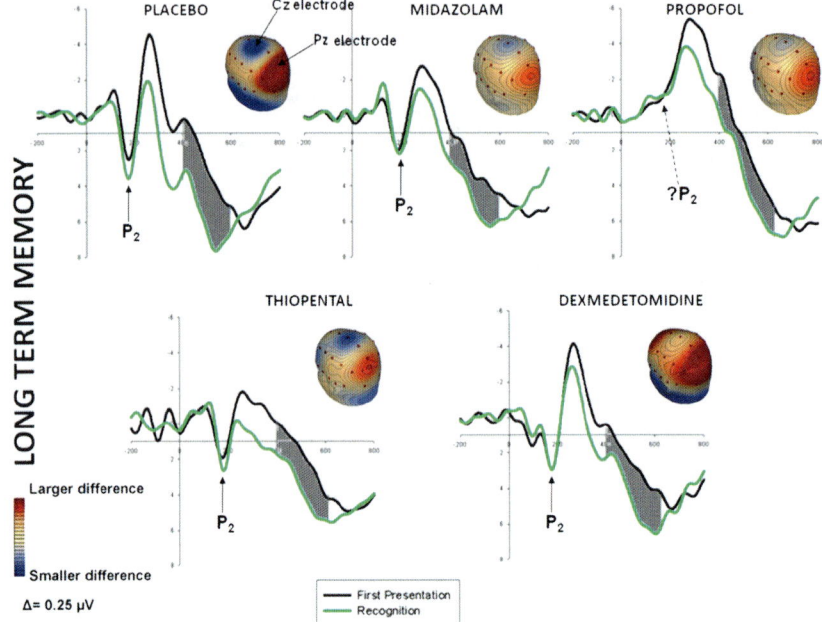

**Fig. 11.9** As opposed to working memory, recognition of items at a slightly longer time interval on the CRT (27 s, from "LTM") reveals that the memory processes supporting recognition are already impaired in the case of midazolam and propofol. Compared with placebo, the areas were smaller for propofol ($p < 0.003$) and possibly midazolam ($p < 0.09$). Note that ERPs represent only correct responses (in other words recognition was successful), but the brain response supporting recognition was already abnormal for propofol. Besides the obvious differences in areas for propofol and midazolam, there are additional effects present for propofol, strongly suggesting different mechanisms of action despite virtually identical behavioral effects. The P2 is remarkably diminished with propofol in comparison with midazolam and other study groups. Additionally, propofol has a strong effect on shifting overall voltage to more negative potentials (up in the graph) than other drugs. ERP measures of the old/new effect may index electrophysiologic memory processes in the hippocampus (see the section titled "Consolidation versus forgetting – the importance of learning-related electrophysiologic processes")

ERPs paint a different picture for retrieval from long-term memory on the CRT. As opposed to working memory, recognition from long-term memory is impaired in the presence of propofol or midazolam. ERP analyses were performed only for items that were successfully categorized, not for those associated with behavioral errors (however, most items were correctly classified in the presence of drug). Despite correct recognition, the amplitude of the ERP response to "old" items was diminished in the presence of propofol ($p = 0.003$) or midazolam ($p = 0.09$) before 30 s after item learning (see Fig. 11.9). Thus, the old/new effect measured in ERPs revealed that memory processes are affected by propofol or midazolam very soon after encoding.

Though both drugs resulted in decreased amplitude of "old" ERP waveforms during recognition compared with placebo, propofol had additional effects on ERPs.

Propofol inhibited the P2 component of the ERP more than did midazolam. These early components of the ERP may be representative of theta oscillatory activity, which is reset and synchronized with stimulus presentation (described below). Propofol had a much stronger effect on ERP voltage, shifting N3 voltage to more negative potentials in comparison to midazolam.

Thus, the effects of propofol and midazolam on memory processes are evident in ERP morphology earlier in time than behavioral measures. These require detailed analyses in order to quantify drug actions in terms of effects on $\lambda$ and $\psi$ indices. How can the effects of propofol and midazolam on these measures be interpreted in terms of underlying physiology?

## Consolidation Versus Forgetting – The Importance of Learning-Related Electrophysiologic Processes

Most memories are lost over time, a fact which seems to preserve sanity (Anderson et al. 2004; Wixted 2005; Parker, Cahill, and McGaugh 2006). Whether forgetting is a lack of consolidation or an active process of pruning memories is yet to be determined. Likely, both processes are at play.

On the other hand, memories to be remembered over time need to undergo consolidation. Consolidation refers to a large variety of physiologic processes, including gene expression, protein synthesis, and synaptic and morphologic changes in dendrites that produce a brain that is different from what it was before (McGaugh 2000; Shimizu et al. 2000; Abel and Lattal 2001; Igaz et al. 2002; Leuner, Falduto, and Shors 2003; Lamprecht and LeDoux 2004; Lynch 2004; Crochet et al. 2006; Draganski et al. 2006; Izquierdo et al. 2006; May et al. 2006). Before these "hard" changes in physiology occur, information is represented in the brain by changes in the properties of oscillatory rhythms present in the brain at all times (Ward 2003; Buzsaki and Draguhn 2004; Duzel, Neufang, and Heinze 2005; Caplan and Glaholt 2007; Meltzer et al. 2008; Mormann et al. 2008). Learning-related changes have their start in the medial temporal lobe, with similar effects active during retrieval of previously stored memories (Lachaux et al. 1999; Guderian and Duzel 2005; Mainy et al. 2008;). An example of information being coded in the electrophysiologic properties of neuronal ensembles is described in detail for pyramidal cells in the entorhinal cortex in support of working memory (Koene and Hasselmo 2006). These processes can be modeled as gamma oscillations (greater than 40 Hz) nested within theta oscillations (about 7 Hz) and provide for first in, first out item replacement of limited capacity, congruent with behavioral observations (the classic $7 \pm 2$ items) (Miller 1956; Lisman 2005). Unless information held in working memory is transferred to other brain regions for processing (e.g., hippocampal CA1 neurons), this information is lost forever (Montgomery and Buzsaki 2007). Notably, the rhinal cortex can be considered to be the gatekeeper for information transfer into and out of the hippocampus (Fernandez and Tendolkar 2006). Such transfer of information can be visualized in real time between rhinal and hippocampal regions (Iijima et al. 1996; Fernandez et al. 1999; Fell et al. 2001; Montgomery and Buzsaki 2007). Subsequent memories are identified at the time of encoding as increased coherence or synchronization between these regions when the item is presented. Normally, these regions demonstrate separate delta and theta rhythms, with a "gap" present at the entorhinal-hippocampal border (Mormann et al. 2008). Theta rhythms in particular are associated with mnemonic processes (Fell et al. 2003; Kahana 2006). Theta rhythms are closely associated with and modulate LTP, the prototypical process underlying the learning of new information (Klimesch et al. 2000; Vertes 2005; Axmacher et al. 2006; Montgomery, Betancur, and Buzsaki 2009). In humans, induced gamma and theta activities are stronger for subsequently remembered stimuli (encoding) and successfully recognized stimuli (old/new effect) (Duzel, Neufang, and Heinze 2005; Axmacher et al. 2006; Osipova et al. 2006).

Mnemonic-related oscillatory rhythms are most clearly measured by direct recording of cortical potentials using implanted electrodes in humans undergoing treatments for epilepsy (Mormann et al. 2005; Sederberg et al. 2006; Ludowig et al. 2008; Mainy et al. 2008; Meltzer et al. 2008; Mormann et al. 2008). However, it is much easier to record surface EEG and the goal is to relate findings from intracranial recordings to surface recorded EEG. Importantly, memory processes based on temporal lobe function (i.e., conscious memories) depend on interaction of the hippocampal/temporal with other brain regions, notably frontal and parietal cortices (Rugg et al. 1991; Guillem et al. 1995; Guderian and Duzel 2005; Maccotta et al. 2006; Mainy et al. 2008). Because of these relationships, surface-recorded ERPs can index hippocampal-based memory processes. This is demonstrated in the old/new effect, maximal over parietal regions (Pz labeled electrode in Figs. 11.8 and 11.9) (Klimesch et al. 2006).

The old/new ERP effect is mirrored in the hippocampal and rhinal regions in intracranially recorded ERPs (icERPs) (Mormann et al. 2005; Fernandez and Tendolkar 2006; Ludowig et al. 2008). Items successfully recognized as being old result in different icERP responses than new items. These differences occur in a similar time window after item presentation to recordings obtained with surface electrodes, more than 300 ms after stimulus presentation. Further support that hippocampal processes are involved in the generation of the old/new ERP effect is provided by patients with hippocampal lesions, who demonstrate diminished recognition ERPs for contextual recollection (as opposed to less confident familiarity recognitions) (Smith and Halgren 1989; Duzel et al. 2001). The relationship between hippocampal and cortical ERPs is not straightforward, as medial temporal brain and cortical regions influence each other (Klimesch et al. 2000; Fell et al. 2004). Learning-related gamma oscillations are modulated by theta activity, which phase locks gamma activity throughout different cortical regions (Kirk and Mackay 2003; Canolty et al. 2006; Osipova et al. 2006). Top-down influence on hippocampal theta rhythms from cortical regions may provide temporally synchronous input which initiates early LTP-like processes, which are critically affected by the specific nature of theta rhythms (Vertes 2005; Osipova et al. 2006). Cognitive-related icERPs in the medial temporal lobe likely arise from multiple discrete sources, indicated by polarity reversals over small distances, whereas widely distributed, interacting sources are typically involved in surface-recorded ERPs (Fell et al. 2004; Mormann et al. 2008; Montgomery, Betancur, and Buzsaki 2009). Thus, the polarity of icERP responses is frequently different from the analogous ERP.

High fidelity recordings from intracranial electrodes reveal that cognitive-related ERPs can be influenced by two fundamental electrophysiologic events in response to a stimulus, namely phase reset of oscillatory rhythms and evoked responses (Fell et al. 2004; Duzel, Neufang, and Heinze 2005). Induced power changes are related but independent effects, as described at the end of this paragraph. icERPs reveal that word recognition in a continuous recognition or working memory task resets phase locking of theta rhythms in the medial temporal lobe within a few hundred milliseconds (Rizzuto et al. 2003; Fell et al. 2004). This phase resetting can result in changes in ERP waveforms associated with mnemonic processing of stimuli independent of the traditionally described increase in activity of neuronal assemblies evoked by the stimulus item (the evoked power response, described in Sect. 1) (Duzel, Neufang, and Heinze 2005).[5] In addition to these time-locked, stimulus-related changes, power changes can be observed that are not time locked to the stimulus (Duzel, Neufang, and Heinze 2005). Such induced activity is measured best when related to individual EEG patterns (i.e., theta activity is measured in the band 4–6 Hz below an individual's alpha peak). Induced band power (IBP) is calculated as averaged electrical activity in a given time window (e.g., 375–750 ms after item presentation) remaining after evoked responses to stimuli are removed (Klimesch et al. 2000). Theta and delta IBP displays a similar old/new effect both in timing and topography as evoked responses (Klimesch et al. 2000, 2006). ERPs and IBP provide separate measures of functionally related processes dependent, in this case, on hippocampal function.

## Summary and Hypothesis: Propofol Affects Learning-Related Electrophysiologic Processes, Upstream from Consolidation Processes

The specific meaning of the word "memory" is nebulous and encompasses a series of closely related systems and processes, which are still being delineated. The largest differences in memory between humans and other animals occur in the systems supporting episodic memory, the most "human" of memory function. These memory processes, however, are those most affected by low concentrations of anesthetic drugs. The electrophysiologic processes that support learning and memory are not as well understood as those of gene expression, protein synthesis, and

---

[5]Resetting oscillatory activity may or may not be accompanied with changes in EEG power. Phase reset without change in power will be visible in the ERP, but measures of neuronal activity (e.g., BOLD signal in fMRI) may be unchanged. Thus, ERP measures may provide additional information not available with other neuroimaging techniques.

morphologic change in consolidation of memories. In support of conscious memory function, the hippocampus communicates with both nearby and more distant brain regions. A window into hippocampal processes may be provided by measuring electrical activity in cortical regions. Such long-distance communications are supported by specific properties of oscillatory rhythms in the brain (Pulvermuller 2001; Varela et al. 2001). Information can be transferred in the electronic soup of the brain faster than traditional axonal conduction can happen (John 2001). The importance of understanding the nature of these communications is that anesthetics produce their effect on consciousness, in part, by interference with anterior–posterior coherence in the brain (John et al. 2001). Likely, similar effects on neuronal polarization, their oscillatory properties and participation in network dynamics could underlie some of the effects of anesthetics on memory function at lower concentrations (Perouansky et al. 2007).

Early evidence, presented in this chapter, indicates that the electrophysiologic processes supporting memory are likely targets in anesthetic-induced memory impairment. Changes in these early memory processes could quite conceivably result in downstream effects on the "hard" physiology of memory consolidation mentioned above. Any one or any number of points in these interrelated cascades that support learning and consolidation could be targets of anesthetic actions on memory function (e.g., see O'Gorman et al. 1998; Cheng et al. 2006; Kozinn et al. 2006). Likely, different anesthetic agents, or the same agent at different concentrations, target different processes. The final result is that conscious memory is never encoded at higher anesthetic concentrations or is encoded but not consolidated at lower anesthetic concentrations. Absence of memory is the final stage of most anesthetic drugs when they are given at the appropriate concentrations. However, pathways to this point are different depending on the agent and the dose used. Amnesic drugs, such as propofol or midazolam, do not affect learning of information, but they do affect normal memory processes very early after that, as revealed seconds later by changes in brain electrophysiology measured in the ERP.

A key question is which of the multitude of memory processes are critically affected by a given drug to produce eventual memory impairment. The dissociation of sedation versus accelerated memory decay as differing mechanisms for memory impairment is now well established. The specific details of drug actions underlying these effects now need to be elucidated. Accelerated memory decay seems to be presaged by changes in the electrophysiology of memory processes (i.e., changes in and/or inhibition of ERP waveforms). These are present before behavioral changes become apparent over increasing time. The effects of anesthetic drugs on electrophysiologic processes need to be thoroughly understood in order to critically assess anesthetic effects on other processes involved with memory consolidation (e.g., gene expression and protein synthesis). This is a particularly relevant issue when the effects of anesthetics on human memory are measured in surrogate systems, such as isolated neuronal systems or even animal models.

# References

Abel, T., and K. M. Lattal. 2001. Molecular mechanisms of memory acquisition, consolidation and retrieval. *Curr Opin Neurobiol* 11(2):180–187.

Aggleton, J. P., and J. M. Pearce. 2001. Neural systems underlying episodic memory: insights from animal research. *Philos Trans R Soc Lond B Biol Sci* 356(1413):1467–1482.

Alkire, M. T., A. Vazdarjanova, H. Dickinson-Anson, N. S. White, and L. Cahill. 2001. Lesions of the basolateral amygdala complex block propofol-induced amnesia for inhibitory avoidance learning in rats. *Anesthesiology* 95(3):708–715.

Alkire, M. T., R. Gruver, J. Miller, J. R. McReynolds, E. L. Hahn, and L. Cahill. 2008. Neuroimaging analysis of an anesthetic gas that blocks human emotional memory. *Proc Natl Acad Sci USA* 105(5):1722–1727.

Allan, K., W. G. Robb, and M. D. Rugg. 2000. The effect of encoding manipulations on neural correlates of episodic retrieval. *Neuropsychologia* 38(8):1188–1205.

Anderson, M. C., K. N. Ochsner, B. Kuhl, J. Cooper, E. Robertson, S. W. Gabrieli, G. H. Glover, and J. D. Gabrieli. 2004. Neural systems underlying the suppression of unwanted memories. *Science* 303(5655):232 235.

Axmacher, N., F. Mormann, G. Fernandez, C. E. Elger, and J. Fell. 2006. Memory formation by neuronal synchronization. *Brain Res Rev* 52(1):170–182.

Blumenfeld, R. S., and C. Ranganath. 2006. Dorsolateral prefrontal cortex promotes long-term memory formation through its role in working memory organization. *J Neurosci* 26(3): 916–925.

Bowles, B., C. Crupi, S. M. Mirsattari, S. E. Pigott, A. G. Parrent, J. C. Pruessner, A. P. Yonelinas, and S. Kohler. 2007. Impaired familiarity with preserved recollection after anterior temporal-lobe resection that spares the hippocampus. *PNAS* 104(41):16382–16387.

Burianova, H., and C. L. Grady. 2007. Common and unique neural activations in autobiographical, episodic, and semantic retrieval. *J Cogn Neurosci* 19(9):1520–1534.

Buzsaki, G., and A. Draguhn. 2004. Neuronal oscillations in cortical networks. *Science* 304(5679):1926–1929.

Cabeza, R., and L. Nyberg. 2000. Imaging cognition II: an empirical review of 275 PET and fMRI studies. *J Cogn Neurosci* 12(1):1–47.

Cabeza, R., and P. St Jacques. 2007. Functional neuroimaging of autobiographical memory. *Trends Cogn Sci* 11(5):219–227.

Cahill, L., and J. L. McGaugh. 1998. Mechanisms of emotional arousal and lasting declarative memory. *Trends Neurosci* 21(7):294–299.

Cahill, L., R. Babinsky, H. J. Markowitsch, and J. L. McGaugh. 1995. The amygdala and emotional memory. *Nature* 377(6547):295–296.

Cahill, L., R. J. Haier, J. Fallon, M. T. Alkire, C. Tang, D. Keator, J. Wu, and J. L. McGaugh. 1996. Amygdala activity at encoding correlated with long-term, free recall of emotional information. *Proc Natl Acad Sci USA* 93(15):8016–8021.

Canolty, R. T., E. Edwards, S. S. Dalal, M. Soltani, S. S. Nagarajan, H. E. Kirsch, M. S. Berger, N. M. Barbaro, and R. T. Knight. 2006. High gamma power is phase-locked to theta oscillations in human neocortex. *Science* 313(5793):1626–1628.

Caplan, J. B., and M. G. Glaholt. 2007. The roles of EEG oscillations in learning relational information. *Neuroimage* 38(3):604–616.

Cheng, V. Y., L. J. Martin, E. M. Elliott, J. H. Kim, H. T. J. Mount, F. A. Taverna, J. C. Roder, J. F. MacDonald, A. Bhambri, N. Collinson, K. A. Wafford, and B. A. Orser. 2006. {alpha}5GABAA receptors mediate the amnestic but not sedative-hypnotic effects of the general anesthetic etomidate. *J Neurosci* 26(14):3713–3720.

Cohen, N. J., and L. R. Squire. 1980. Preserved learning and retention of pattern-analyzing skill in amnesia: dissociation of knowing how and knowing that. *Science* 210(4466): 207–210.

Crochet, S., P. Fuentealba, Y. Cisse, I. Timofeev, and M. Steriade. 2006. Synaptic plasticity in local cortical network in vivo and its modulation by the level of neuronal activity. *Cereb Cortex* 16(5):618–631.

Curran, T., C. DeBuse, B. Woroch, and E. Hirshman. 2006. Combined pharmacological and electrophysiological dissociation of familiarity and recollection. *J Neurosci* 26(7):1979–1985.

Deeprose, C., J. Andrade, S. Varma, and N. Edwards. 2004. Unconscious learning during surgery with propofol anaesthesia. *Br J Anaesth* 92(2):171–177.

Deeprose, C., J. Andrade, D. Harrison, and N. Edwards. 2005. Unconscious auditory priming during surgery with propofol and nitrous oxide anaesthesia: a replication. *Br J Anaesth* 94(1):57–62.

Dickinson-Anson, H., and J. L. McGaugh. 1997. Bicuculline administered into the amygdala after training blocks benzodiazepine-induced amnesia. *Brain Res* 752(1–2):197–202.

Draganski, B., C. Gaser, G. Kempermann, H. Georg Kuhn, J. Winkler, C. Buchel, and A. May. 2006. Temporal and spatial dynamics of brain structure changes during extensive learning. *J Neurosci* 26(23):6314–6317.

Duzel, E., M. Neufang, and H.-J. Heinze. 2005. The oscillatory dynamics of recognition memory and its relationship to event-related responses. *Cereb Cortex* 15(12):1992–2002.

Duzel, E., F. Vargha-Khadem, H. J. Heinze, and M. Mishkin. 2001. Brain activity evidence for recognition without recollection after early hippocampal damage. *Proc Natl Acad Sci USA* 98(14):8101–8106.

Eichenbaum, H. 2000. A cortical-hippocampal system for declarative memory. *Nat Rev Neurosci* 1:41–50.

Eichenbaum, H. 2004. Hippocampus: cognitive processes and neural representations that underlie declarative memory. *Neuron* 44(1):109–120.

Fell, J., P. Klaver, K. Lehnertz, T. Grunwald, C. Schaller, C. E. Elger, and G. Fernandez. 2001. Human memory formation is accompanied by rhinal-hippocampal coupling and decoupling. *Nat Neurosci* 4(12):1259–1264.

Fell, J., P. Klaver, H. Elfadil, C. Schaller, C. E. Elger, and G. Fernandez. 2003. Rhinal-hippocampal theta coherence during declarative memory formation: interaction with gamma synchronization? *Eur J Neurosci* 17(5):1082–1088.

Fell, J., T. Dietl, T. Grunwald, M. Kurthen, P. Klaver, P. Trautner, C. Schaller, C. E. Elger, and G. Fernandez. 2004. Neural bases of cognitive ERPs: more than phase reset. *J Cogn Neurosci* 16(9):1595–1604.

Fernandes, M. A., and M. Moscovitch. 2000. Divided attention and memory: evidence of substantial interference effects at retrieval and encoding. *J Exp Psychol Gen* 129(2): 155–176.

Fernandez, G., and I. Tendolkar. 2006. The rhinal cortex: 'gatekeeper' of the declarative memory system. *Trends Cogn Sci* 10(8):358–362.

Fernandez, G., A. Effern, T. Grunwald, N. Pezer, K. Lehnert, M. Duempelmann, D. Van Roost, and C. E. Elger. 1999. Real-time tracking of memory formation in the human rhinal cortex and hippocampus. *Science* 285(3 September 1999):1582–1585.

Friedman, D. 1990. ERPs during continuous recognition memory for words. *Biol Psychol* 30: 61–87.

Friedman, D., and R. Johnson, Jr. 2000. Event-related potential (ERP) studies of memory encoding and retrieval: a selective review. *Microsc Res Technique* 51(1):6–28.

Galinkin, J. L., D. Janiszewski, C. J. Young, J. M. Klafta, P. A. Klock, D. W. Coalson, J. L. Apfelbaum, and J. P. Zacny. 1997. Subjective, psychomotor, cognitive, and analgesic effects of subanesthetic concentrations of sevoflurane and nitrous oxide. *Anesthesiology* 87(5):1082–1088.

Ghoneim, M. M., and R. I. Block. 1997. Learning and memory during general anesthesia. *Anesthesiology* 87(2):387–410.

Ghoneim, M. M., and J. V. Hinrichs. 1997. Drugs, memory and sedation: specificity of efects. *Anesthesiology* 87(Oct):734–736.

Ghoneim, M. M., and S. P. Mewaldt. 1990. Benzodiazepines and human memory: a review. *Anesthesiology* 72(5):926–938.

Ghoneim, M. M., R. I. Block, and V. J. Dhanaraj. 1998. Interaction of a subanaesthetic concentration of isoflurane with midazolam: effects on responsiveness, learning and memory [see comments]. *Br J Anaesthesia* 80(5):581–587.

Gonsowski, C. T., B. S. Chortkoff, E. I. Eger, II, H. L. Bennett, and R. B. Weiskopf. 1995. Subanesthetic concentrations of desflurane and isoflurane suppress explicit and implicit learning. *Anesthesia Analgesia* 80(3):568–572.

Greicius, M. D., K. Supekar, V. Menon, and R. F. Dougherty. 2008. Resting-state functional connectivity reflects structural connectivity in the default mode network. *Cereb Cortex Adv Access* Apr 9, bhn059.

Guderian, S., and E. Duzel. 2005. Induced theta oscillations mediate large-scale synchrony with mediotemporal areas during recollection in humans. *Hippocampus* 15(7):901–912.

Guillem, F., B. N'Kaoua, A. Rougier, and B. Claverie. 1995. Effects of temporal versus temporal plus extra-temporal lobe epilepsies on hippocampal ERPs: physiopathological implications for recognition memory studies in humans. *Brain Res Cogn Brain Res* 2(3):147–153.

Hebb, D. O. 1949. *The organization of behavior; a neuropsychological theory.* New York: Wiley.

Heindel, W. C., D. P. Salmon, C. W. Shults, P. A. Walicke, and N. Butters. 1989. Neuropsychological evidence for multiple implicit memory systems: a comparison of Alzheimer's, Huntington's, and Parkinson's disease patients. *J Neurosci* 9(2):582–587.

Hudetz, A. G., and O. A. Imas. 2007. Burst activation of the cerebral cortex by flash stimuli during isoflurane anesthesia in rats. *Anesthesiology* 107(6):983–991.

Igaz, L. M., M. R. Vianna, J. H. Medina, and I. Izquierdo. 2002. Two time periods of hippocampal mRNA synthesis are required for memory consolidation of fear-motivated learning. *J Neurosci* 22(15):6781–6789.

Iidaka, T., A. Matsumoto, J. Nogawa, Y. Yamamoto, and N. Sadato. 2006. Frontoparietal network involved in successful retrieval from episodic memory. Spatial and temporal analyses using fMRI and ERP. *Cereb Cortex* 16(9):1349–1360.

Iijima, T., M. P. Witter, M. Ichikawa, T. Tominaga, R. Kajiwara, and G. Matsumoto. 1996. Entorhinal-hippocampal interactions revealed by real-time imaging. *Science* 272(5265): 1176–1179.

Izquierdo, I., L. R. Bevilaqua, J. I. Rossato, J. S. Bonini, J. H. Medina, and M. Cammarota. 2006. Different molecular cascades in different sites of the brain control memory consolidation. *Trends Neurosci* 29(9):496–505.

John, E. R. 2001. A field theory of consciousness. *Conscious Cogn* 10(2):184–213.

John, E. R., P. Easton, and R. Isenhart. 1997. Consciousness and cognition may be mediated by multiple independent coherent ensembles [published erratum appears in Conscious Cogn 1997 Dec; 6(4):598–599]. *Consci Cogn* 6(1):3–39; discussion 40-1, 50-5, 65-6.

John, E. R., L. S. Prichep, W. Kox, P. Valdes-Sosa, J. Bosch-Bayard, E. Aubert, M. Tom, F. diMichele, and L. D. Gugino. 2001. Invariant reversible qEEG effects of anesthetics. *Consci Cogn* 10(2):165–183.

Johnson, R., Jr. 1995. Event-related potential insights into the neurobiology of memory systems. In *Handbook of Neuropsychology, vol. 10,* edited by R. J. Johnson and J. C. Baron. Amsterdam and New York: Elsevier Science B.V.

Kahana, M. J. 2006. The cognitive correlates of human brain oscillations. *J Neurosci* 26(6): 1669–1672.

Kahn, I., L. Davachi, and A. D. Wagner. 2004. Functional-neuroanatomic correlates of recollection: implications for models of recognition memory. *J Neurosci* 24(17):4172–4180.

Kim, J. J., H. J. Lee, J. S. Han, and M. G. Packard. 2001. Amygdala is critical for stress-induced modulation of hippocampal long-term potentiation and learning. *J Neurosci* 21(14): 5222–5228.

Kirk, I. J., and J. C. Mackay. 2003. The role of theta-range oscillations in synchronising and integrating activity in distributed mnemonic networks. *Cortex* 39(4–5):993–1008.

Klimesch, W., M. Doppelmayr, J. Schwaiger, T. Winkler, and W. Gruber. 2000. Theta oscillations and the ERP old/new effect: independent phenomena? *Clin Neurophysiol* 111(5):781–793.

Klimesch, W., S. Hanslmayr, P. Sauseng, W. Gruber, C. J. Brozinsky, N. E. Kroll, A. P. Yonelinas, and M. Doppelmayr. 2006. Oscillatory EEG correlates of episodic trace decay. *Cereb Cortex* 16(2):280–290.

Koene, R. A., and M. E. Hasselmo. 2006. First-in-first-out item replacement in a model of short-term memory based on persistent spiking. *Cereb Cortex* eAP:Oct 9 bhl088.

Konorski, J. 1948. *Conditioned reflexes and neuron organization.* Cambridge [Eng.]: Cambridge University Press.

Kozinn, J., L. Mao, A. Arora, L. Yang, E. E. Fibuch, and J. Q. Wang. 2006. Inhibition of gluta-matergic activation of extracellular signal-regulated protein kinases in hippocampal neurons by the intravenous anesthetic propofol. *Anesthesiology* 105(6):1182–1191.

Lachaux, J. P., E. Rodriguez, J. Martinerie, and F. J. Varela. 1999. Measuring phase synchrony in brain signals. *Hum Brain Mapp* 8(4):194–208.

Lamprecht, R., and J. LeDoux. 2004. Structural plasticity and memory. *Nat Rev Neurosci* 5(1): 45–54.

Leuner, B., J. Falduto, and T. J. Shors. 2003. Associative memory formation increases the observation of dendritic spines in the hippocampus. *J Neurosci* 23(2):659–665.

Levy, D. A., P. J. Bayley, and L. R. Squire. 2004. The anatomy of semantic knowledge: medial vs. lateral temporal lobe. *PNAS* 101(17):6710–6715.

Lisman, J. E. 2005. Hippocampus, II: memory connections. *Am J Psychiatry* 162(2):239.

Lisman, J. E., and M. A. Idiart. 1995. Storage of 7 +/− 2 short-term memories in oscillatory subcycles. *Science* 267(5203):1512–1515.

Logothetis, N. K. 2003. The underpinnings of the BOLD functional magnetic resonance imaging signal. *J Neurosci* 23(10):3963–3971.

Logothetis, N. K. 2008. What we can do and what we cannot do with fMRI. *Nature* 453(7197): 869–878.

Ludowig, E., P. Trautner, M. Kurthen, C. Schaller, C. G. Bien, C. E. Elger, and T. Rosburg. 2008. Intracranially recorded memory-related potentials reveal higher posterior than anterior hippocampal involvement in verbal encoding and retrieval. *J Cogn Neurosci* 20(5):841–851.

Lynch, M. A. 2004. Long-term potentiation and memory. *Physiol Rev* 84(1):87–136.

Maccotta, L., R. L. Buckner, F. G. Gilliam, and J. G. Ojemann. 2006. Changing frontal contribu-tions to memory before and after medial temporal lobectomy. *Cereb Cortex* eAP: March 17 bhj161.

Mainy, N., J. Jung, M. Baciu, P. Kahane, B. Schoendorff, L. Minotti, D. Hoffmann, O. Bertrand, and J. P. Lachaux. 2008. Cortical dynamics of word recognition. *Hum Brain Mapp* 29(11):1215–1230.

Malenka, R. C., and R. A. Nicoll. 1999. Long-term potentiation – a decade of progress? *Science* 285(5435):1870–1874.

Mashour, G. A. 2006. Integrating the science of consciousness and anesthesia. *Anesth Analg* 103(4):975–982.

May, A., G. Hajak, S. Ganssbauer, T. Steffens, B. Langguth, T. Kleinjung, and P. Eichhammer. 2006. Structural brain alterations following 5 days of intervention: dynamic aspects of neuroplasticity. *Cereb Cortex* eAP:Feb 15.

McGaugh, J. L. 2000. Memory – a century of consolidation. *Science* 287(5451):248–251.

McGaugh, J. L., C. K. McIntyre, and A. E. Power. 2002. Amygdala modulation of memory consolidation: interaction with other brain systems. *Neurobiol Learn Mem* 78(3):539–552.

Meltzer, J. A., H. P. Zaveri, Goncharova, II, M. M. Distasio, X. Papademetris, S. S. Spencer, D. D. Spencer, and R. T. Constable. 2008. Effects of working memory load on oscillatory power in human intracranial EEG. *Cereb Cortex* 18(8):1843–1855.

Miller, G. A. 1956. The magical number seven, plus or minus two: some limits on our capacity for processing information. *Psychol Rev* 63:81–97.

Miller, G. 2004. Forgetting and remembering. Learning to forget. *Science* 304(5667):34–36.

Montgomery, S. M., and G. Buzsaki. 2007. Gamma oscillations dynamically couple hippocampal CA3 and CA1 regions during memory task performance. *Proc Natl Acad Sci* 104(36): 14495–14500.

Montgomery, S. M., M. I. Betancur, and G. Buzsaki. 2009. Behavior-dependent coordination of multiple theta dipoles in the hippocampus. *J Neurosci* 29(5):1381–1394.

Mormann, F., J. Fell, N. Axmacher, B. Weber, K. Lehnertz, C. E. Elger, and G. Fernandez. 2005. Phase/amplitude reset and theta-gamma interaction in the human medial temporal lobe during a continuous word recognition memory task. *Hippocampus* 15(7):890–900.

Mormann, F., H. Osterhage, R. G. Andrzejak, B. Weber, G. Fernandez, J. Fell, C. E. Elger, and K. Lehnertz. 2008. Independent delta/theta rhythms in the human hippocampus and entorhinal cortex. *Front Hum Neurosci* 2:3.

Morris, J. S., A. Ohman, and R. J. Dolan. 1998. Conscious and unconscious emotional learning in the human amygdala [see comments]. *Nature* 393(6684):467–470.

Morris, J. S., C. D. Frith, D. I. Perrett, D. Rowland, A. W. Young, A. J. Calder, and R. J. Dolan. 1996. A differential neural response in the human amygdala to fearful and happy facial expressions. *Nature* 383(6603):812–815.

Nelson, L. E., T. Z. Guo, J. Lu, C. B. Saper, N. P. Franks, and M. Maze. 2002. The sedative component of anesthesia is mediated by GABA(A) receptors in an endogenous sleep pathway. *Nat Neurosci* 5(10):979–984.

Nordstrom, O., and R. Sandin. 1996. Recall during intermittent propofol anaesthesia. *Br J Anaesthesia* 76(5):699–701.

O'Gorman, D. A., A. W. O'Connell, K. J. Murphy, D. C. Moriarty, T. Shiotani, and C. M. Regan. 1998. Nefiracetam prevents propofol-induced anterograde and retrograde amnesia in the rodent without compromising quality of anesthesia. *Anesthesiology* 89(3):699–706.

Osipova, D., A. Takashima, R. Oostenveld, G. Fernandez, E. Maris, and O. Jensen. 2006. Theta and gamma oscillations predict encoding and retrieval of declarative memory. *J Neurosci* 26(28):7523–7531.

Parker, E. S., L. Cahill, and J. L. McGaugh. 2006. A case of unusual autobiographical remembering. *Neurocase* 12(1):35–49.

Perouansky, M., H. Hentschke, M. Perkins, and R. A. Pearce. 2007. Amnesic concentrations of the nonimmobilizer 1,2-dichlorohexafluorocyclobutane (F6, 2 N) and isoflurane alter hippocampal theta oscillations *in vivo*. *Anesthesiology* 106(6):1168–1176.

Plourde, G., P. Belin, D. Chartrand, P. Fiset, S. B. Backman, G. Xie, and R. J. Zatorre. 2006. Cortical processing of complex auditory stimuli during alterations of consciousness with the general anesthetic propofol. *Anesthesiology* 104(3):448–457.

Pryor, K. O., R. A. Veselis, R. A. Reinsel, and V. A. Feshchenko. 2004. Enhanced visual memory effect for negative versus positive emotional content is potentiated at sub-anaesthetic concentrations of thiopental. *Br J Anaesth* 93(3):348–355.

Pulvermuller, F. 2001. Brain reflections of words and their meaning. *Trends Cogn Sci* 5(12): 517–524.

Qiu, M., R. Ramani, M. Swetye, and R. T. Constable. 2008. Spatial nonuniformity of the resting CBF and BOLD responses to sevoflurane: in vivo study of normal human subjects with magnetic resonance imaging. *Hum Brain Mapp* 29(12):1390–1399.

Qiu, M., R. Ramani, M. Swetye, N. Rajeevan, and R. T. Constable. 2008. Anesthetic effects on regional CBF, BOLD, and the coupling between task-induced changes in CBF and BOLD: an fMRI study in normal human subjects. *Magn Reson Med* 60(4):987–996.

Ramani, R., M. Qiu, and R. T. Constable. 2007. Sevoflurane 0.25 MAC preferentially affects higher order association areas: a functional magnetic resonance imaging study in volunteers. *Anesth Analg* 105(3):648–655.

Ranganath, C., M. X. Cohen, and C. J. Brozinsky. 2005. Working memory maintenance contributes to long-term memory formation: neural and behavioral evidence. *J Cogn Neurosci* 17(7): 994–1010.

Rizzuto, D. S., J. R. Madsen, E. B. Bromfield, A. Schulze-Bonhage, D. Seelig, R. Aschenbrenner-Scheibe, and M. J. Kahana. 2003. Reset of human neocortical oscillations during a working memory task. *Proc Natl Acad Sci USA* 100(13):7931–7936.

Rugg, M. D., and T. Curran. 2007. Event-related potentials and recognition memory. *Trends Cogn Sci* 11(6):251–257.

Rugg, M. D., R. C. Roberts, D. D. Potter, C. D. Pickles, and M. E. Nagy. 1991. Event-related potentials related to recognition memory. Effects of unilateral temporal lobectomy and temporal lobe epilepsy. *Brain* 114(Pt 5):2313–2332.

Schacter, D. L., and E. Tulving. 1994. *Memory systems 1994*. Cambridge, MA: MIT Press.

Scoville, W.B., and B Milner. 1957. Loss of recent memory after bilateral hippocampal lesions. *J Neurol Neurosurg Psychiatry* 20:11–21.

Sederberg, P. B., A. Schulze-Bonhage, J. R. Madsen, E. B. Bromfield, D. C. McCarthy, A. Brandt, M. S. Tully, and M. J. Kahana. 2006. Hippocampal and neocortical gamma oscillations predict memory formation in humans. *Cereb Cortex* eAP:jul10 bhl030.

Shimizu, E., Y. P. Tang, C. Rampon, and J. Z. Tsien. 2000. NMDA receptor-dependent synaptic reinforcement as a crucial process for memory consolidation. *Science* 290(5494):1170–1174.

Smith, M. E., and E. Halgren. 1989. Dissociation of recognition memory components following temporal lobe lesions. *J Exp Psychol Learn Mem Cogn* 15(1):50–60.

Smith, C. N., and L. R. Squire. 2009. Medial temporal lobe activity during retrieval of semantic memory is related to the age of the memory. *J Neurosci* 29(4):930–938.

Sonner, J. M., D. F. Werner, F. P. Elsen, Y. Xing, M. Liao, R. A. Harris, N. L. Harrison, M. S. Fanselow, E. I. Eger, II, and G. E. Homanics. 2007. Effect of isoflurane and other potent inhaled anesthetics on MAC, learning, and the righting reflex in mice engineered to express alpha1 GABA-A receptors unresponsive to isoflurane. *Anesthesiology* 106(1):107–113.

Squire, L. R., and S. Zola-Morgan. 1991. The medial temporal lobe memory system. *Science* 253(5026):1380–1386.

Tulving, E. 2001. Episodic memory and common sense: how far apart? *Philos Trans R Soc Lond B Biol Sci* 356(1413):1505–1515.

Tulving, E., and D. L. Schacter. 1990. Priming and human memory systems. *Science* 247(4940):301–306.

Ungerleider, L. G. 1995. Functional brain imaging studies of cortical mechanisms for memory. *Science* 270(5237):769–775.

Varela, F., J. P. Lachaux, E. Rodriguez, and J. Martinerie. 2001. The brainweb: phase synchronization and large-scale integration. *Nat Rev Neurosci* 2(4):229–239.

Vertes, R. P. 2005. Hippocampal theta rhythm: a tag for short-term memory. *Hippocampus* 15(7):923–935.

Veselis, R. A., R. A. Reinsel, and V. A. Feshchenko. 2001. Drug-induced amnesia is a separate phenomenon from sedation: electrophysiologic evidence. *Anesthesiology* 95(4):896–907.

Veselis, R. A., R. A. Reinsel, V. A. Feshchenko, and M. Wronski. 1997. The comparative amnestic effects of midazolam, propofol, thiopental, and fentanyl at equisedative concentrations. *Anesthesiology* 87(4):749–764.

Veselis, R. A., R. A. Reinsel, V. A. Feshchenko, and R. Johnson, Jr. 2004. Information loss over time defines the memory defect of propofol: A comparative response with thiopental and dexmedetomidine. *Anesthesiology* 101(4):831–841.

Veselis, R. A., K. O. Pryor, R. A. Reinsel, M. Mehta, H. Pan, and R. Johnson, Jr. 2008. Low-dose propofol-induced amnesia is not due to a failure of encoding: left inferior prefrontal cortex is still active. *Anesthesiology* 109(2):213–224.

Veselis, R. A., K. O. Pryor, R. A. Reinsel, Y. Li, M. Mehta, and R. Johnson, Jr. 2009. Propofol and midazolam inhibit conscious memory processes very soon after encoding: an event-related potential study of familiarity and recollection in volunteers. *Anesthesiology* 110(2):295–312.

Vincent, J. L., A. Z. Snyder, M. D. Fox, B. J. Shannon, J. R. Andrews, M. E. Raichle, and R. L. Buckner. 2006. Coherent spontaneous activity identifies a hippocampal-parietal memory network. *J Neurophysiol* 96(6):3517–3531.

Ward, L. M. 2003. Synchronous neural oscillations and cognitive processes. *Trends Cogn Sci* 7(12):553–559.

Wixted, J. T. 2005. A theory about why we forget what we once knew. *Curr Directions Psychol Sci* 14(1):6–9.

# Subject Index

A. Hudetz, R. Pearce (eds.), *Suppressing the Mind,* Contemporary Clinical
Neuroscience, DOI 10.1007/978-1-60761-462-3,
© Humana Press, a part of Springer Science+Business Media, LLC 2010